D1765442

This book is due for return on or before the last date shown below.

Don Gresswell Ltd., London, N.21   Cat. No. 1208

DG 02242/71

SOUTH BIRMINGHAM MENTAL HEALTH TRUST

Q02978

Prion Diseases

METHODS IN MOLECULAR MEDICINE™

John M. Walker, SERIES EDITOR

METHODS IN MOLECULAR MEDICINE™

Prion Diseases

Edited by

Harry F. Baker
and Rosalind M. Ridley

Cambridge University, Cambridge, UK

With a Foreword by

Stanley B. Prusiner

University of California, San Francisco, CA

Humana Press ✴ Totowa, New Jersey

© 1996 Humana Press Inc.
999 Riverview Drive, Suite 208
Totowa, New Jersey 07512

All rights reserved.

No part of this book may be reproduced, stored in a retrieval system, or transmitted in any form or by any means, electronic, mechanical, photocopying, microfilming, recording, or otherwise without written permission from the Publisher. Methods in Molecular Medicine™ is a trademark of The Humana Press Inc.

All authored papers, comments, opinions, conclusions, or recommendations are those of the author(s), and do not necessarily reflect the views of the publisher.

This publication is printed on acid-free paper. ∞
ANSI Z39.48-1984 (American Standards Institute) Permanence of Paper for Printed Library Materials.

Cover illustration: Fig. 2 from Chapter 18, "Immunohistochemistry of Resinated Tissues for Light and Electron Microscopy," by Martin Jeffrey and Caroline M. Goodsir.

Photocopy Authorization Policy:
Authorization to photocopy items for internal or personal use, or the internal or personal use of specific clients, is granted by Humana Press Inc., provided that the base fee of US $5.00 per copy, plus US $00.25 per page, is paid directly to the Copyright Clearance Center at 222 Rosewood Drive, Danvers, MA 01923. For those organizations that have been granted a photocopy license from the CCC, a separate system of payment has been arranged and is acceptable to Humana Press Inc. The fee code for users of the Transactional Reporting Service is: [0-89603-342-2/96 $5.00 + $00.25].

Printed in the United States of America. 10 9 8 7 6 5 4 3 2 1

Library of Congress Cataloging in Publication Data

Main entry under title:

Methods in molecular medicine™.

Prion diseases/edited by Harry F. Baker and Rosalind M. Ridley.
    p.   cm.—(Methods in molecular medicine™)
  Includes index.
  ISBN 0-89603-342-2 (alk. paper)
  1. Prion diseases.   I. Baker, Harry F.   II. Ridley, Rosalind M.   III. Series.
  [DNLM: 1. Prion Diseases. WL 300 P958 1996]
  QR201.P73P74   1996
  616.8—dc20
  DNLM/DLC
  for Library of Congress   96-3408
                                              CIP

Foreword

Harry Baker and Rosalind Ridley have done an admirable job in assembling this collection of articles that describe the methodology frequently used to study a group of CNS illnesses often referred to as the "prion diseases." Research on prions and the disorders that they cause has progressed relatively rapidly over the last decade since the discovery of the prion protein (PrP) that allowed the application of modern molecular biological and genetic tools. The power of these techniques is awesome and their use in deciphering the once mysterious prion diseases has brought a wealth of new information.

Although prions are unprecedented pathogens, appearing to consist only of PrPSc molecules, the diseases that they cause are no less remarkable. The prion diseases in animals include scrapie of sheep and goats as well as "mad cow" disease or bovine spongiform encephalopathy (BSE). In the United Kingdom, the epidemic of BSE has heightened public awareness of this previously obscure group of diseases such that any work in the field is likely to stir up interest in the media and become a subject of public debate. It has been difficult for British investigators to work on prion diseases without being involved in these controversies. As such, several chapters have been included that deal with political and social issues surrounding prion diseases.

The human prion diseases present an equally fascinating saga in which these CNS degenerations present as genetic, sporadic, and infectious illnesses. Because the first human prion disease to be discovered was kuru, which apppears to have been transmitted among the Foré people of New Guinea by ritualistic cannibalism, the focus for many years was on the infectious forms of prion disease. Shortly after the initial descriptions of Creutzfeldt-Jakob disease (CJD) in the early 1920s, familial forms of the disease were reported, but their significance remained obscure until the discovery of the mutations in the PrP genes of patients dying of these inherited illnesses. The report of transmission of CJD to apes by Carleton Gajdusek and Joe Gibbs two decades before the discovery of PrP gene mutations certainly must be considered a turning point in the rich history of the prion diseases. Several chapters describe these disorders in detail and the methodologies that have been developed for their investigation.

Studies of the biochemistry of prions have been equally as fascinating as the molecular genetics. Much of this work has focused on the structures of the cellular (PrPC) and scrapie isoform (PrPSc) of the prion protein. These studies have shown that PrPC and PrPSc have very different conformations despite possessing the same primary structure. Such findings run counter to the dictum that the amino acid sequence of a protein dictates a single tertiary structure. These unexpected results have given rise to the notion that prion diseases are diseases of protein conformation. It seems likely that this concept will have broader implications and that some of the more common neurodegenerative diseases, such as Alzheimer's disease, Parkinson's disease, or amyotrophic lateral sclerosis, may share some features with the prion diseases in their pathogenesis.

As investigations of the prion diseases expand, it seems reasonable to expect that the unprecedented principles elucidated through studies of prions will be found to operate more widely throughout biology and medicine. It is hoped that the wide-ranging survey presented here will facilitate studies of prion biology and diseases in the future.

Stanley B. Prusiner

Preface

When we were asked to contribute a volume on *Prion Diseases* for the new series, *Methods in Molecular Medicine*, our hope (and that of the Series Editor) was to put together a selection of chapters that would cover the various research areas in the study of the spongiform encephalopathies in such experimental detail that a competent scientist, unfamiliar with these conditions, could carry out experiments. This is the philosophy underlying the companion *Methods in Molecular Biology* series, in which detailed protocols for each experimental procedure or assay are given. However, it soon became clear to us that, with one or two exceptions, such a format was unlikely to be successful in the areas of research we wished to cover. It is true, of course, that a great deal of the recent progress in our understanding of this group of diseases has come about by applying the techniques of molecular biology, molecular genetics, and transgenetics, and details of these techniques can be found in any number of "recipe" books. What we have tried to produce here is a book that covers a number of different aspects of research into prion diseases. Rather than detailing precisely how experimental procedures are carried out, each chapter gives the reader a view of the approaches taken by researchers when addressing specific problems.

For those who have worked on prion diseases it is clear that research interests and approaches are not compartmentalized. So it is that those who are engaged in molecular biological studies are very familiar with the work of the geneticists and epidemiologists, and those who concentrate on animal diseases like BSE and scrapie have an interest in the human spongiform encephalopathies. Although the prion disease research community is quite small, it continues to generate controversy, and although there are a number of views concerning the nature of the causative agent in these diseases, two have come to dominate the field. The first argues that the agent consists of, or contains, an informational molecule (independent of the affected animal) that is capable of mutation. The other view proposes that the agent probably contains only protein coded by the affected animal and derives its "information" from the secondary and tertiary structure of that protein. In the first case, developed by Alan Dickinson and colleagues at the MRC/BBSRC Neuropathogenesis Unit in Edinburgh, the agent has been given the name "virino." This they conceive of as a small, informational nucleic acid molecule that is susceptible to mutation, which then combines with a host protein, prion protein (PrPC), after the infecting event. The incubation time and neuropathological profile is determined by this combination. In the protein-only case, first proposed by Griffiths almost 30 years ago and elaborated by Stan Prusiner at the University of California (San Francisco), it is argued that there is no informational nucleic acid. Disease progression is accompanied by the conversion of normal host-coded PrPC to an abnormal form (PrPSc). According to the protein-only hypothesis, when this abnormal protein is transferred to a new host, it catalyzes the modification of the new host's PrPC so that it in turn becomes abnormal. Clear arguments have been made for each view and we hope they are reflected in this volume. Our introductory chapter attempts to lay the groundwork for the rest of the book.

Chapters 2–5 deal with different aspects of diagnosis of prion diseases in both humans and animals. There is a discussion of the clinical diagnosis of human disease by Rajith de Silva (Chapter 2) and detailed accounts of the neuropathological examination of human brains by James Ironside and Jeanne Bell (Chapters 3 and 4). Mick Stack and colleagues (Chapter 5) detail the laboratory techniques currently used to confirm the presence of PrPSc either by immunoblotting or by electron microscopy examination of the scrapie-associated fibrils prepared from the brains of affected animals.

David Taylor (Chapter 6) has carried out detailed investigations of the methods for inactivating the agents involved in prion diseases and the implications for laboratory safety and working practice of their remarkable resistance to chemical treatments.

A number of chapters cover epidemiology. In Chapter 7, Bob Will describes the epidemiology of Creutzfeldt-Jakob disease, and in Chapter 8, Paul Brown gives an account of the cases of human spongiform encephalopathy resulting from iatrogenic contamination. Our understanding of the BSE epidemic in the United Kingdom owes much to the work of John Wilesmith and his group, which is described in Chapter 9. This growing understanding was matched by appropriate action by the UK government, which is presented chronologically by David Tyrrell and Kevin Taylor in Chapter 10. These authors also describe their interactions with the media.

Chapters 11 and 12 deal with different aspects of genetics. In the first, Jonathon Gray and colleagues detail the protocols they have developed for counseling those at risk of developing Huntington's disease or of carrying a fetus at risk of Huntington's disease. The same considerations apply to any adult-onset, dominantly inherited, fatal disorder, such as Gerstmann-Sträussler-Scheinker disease. In Chapter 12, Nora Hunter details her work in PrP genotyping sheep and relating gene variants to scrapie susceptibility. This work is relevant to the development of breeding strategies that might then lead to scrapie-resistant flocks. The possible pitfalls of such a strategy are also pointed out.

Possibly the most uncomfortable aspect of the spongiform encephalopathies for the "protein-only" hypothesis is strain variation. There is little doubt that there are a number of strains of the scapie agent and it is difficult to see how these strains come about without the involvement of an informational molecule capable of undergoing mutational events. Those who adhere to the protein-only view argue that strain differences could depend on variations in protein conformation (which is now known to differ between certain strains) although they recognize that this explanation will only suffice for a small number of strains. In Chapter 13, Moira Bruce presents her analysis of the strain variation in mouse-adapted scrapie and the arguments for multiple strains. She also describes her recent work on strain-typing of BSE agent and agents from other affected species.

The final group of chapters (Chapters 14–18) cover different approaches toward an understanding of the pathogenesis of prion diseases. Jean Manson (Chapter 14) describes the production of mice whose PrP genes have been mutated so that they produce little or no protein and the use of these PrP-deficient mice in studies of the role of PrP in agent replication. In Chapter 15, David Westaway shows how transgenic mice carrying PrP genes from other species have been of immense value in elucidating "species barrier" effects. All researchers now accept an important role for prion protein in the pathogenesis of these diseases and that the conversion of PrPC to PrPSc is central to the disease process. In kuru, Gerstmann-Sträussler-Scheinker syndrome, some cases of Creutzfeldt-Jakob disease, and some mouse-scrapie models, PrPSc is found deposited in amyloid form in congophilic plaques within brain parenchyma. This propensity of PrPSc to form amyloid is of great interest and its importance is discussed by Tagliavini and colleagues in Chapter 16. Much excitement has accompanied the recent publications by Byron Caughey and his colleagues who have demonstrated that PrPSc can induce the conversion of PrPC to PrPSc in cell-free systems and that "strain characteristics" of the catalyzing PrPSc species (as determined by the pattern of cleavage by proteinase K) are preserved in the newly converted PrPSc. These experiments are described in some detail in Chapter 17. In Chapter 18 Martin Jeffrey and Caroline Goodsir detail the methods of immunohistochemistry at both the light and electron microscope levels that they employ in their elegant studies of the localization of PrPSc during the course of disease in murine scrapie.

This selection of topics is by no means exhaustive, but will give the reader an introduction to the breadth of the effort currently spent on trying to understand prion diseases. Many in the field are convinced that what we learn of the pathogenesis of these diseases will be of relevance to an understanding of more common neurodegenerative diseases. For example, it is already clear that there are a number of similarities between the abnormal accumulation of PrPSc amyloids in some cases of prion disease and the accumulation of β-amyloid in cases of Alzheimer's disease. Furthermore, there are families with dominantly inherited prion disease in which PrPSc amyloid and β-amyloid are colocalized within plaques.

Finally, we would like to dedicate this volume to the memory of Nigel Denton, who died at the age of 39. Nigel was a member of the family in which a PrP gene mutation (PrP102 Pro → Leu) was first shown to be linked to Gerstmann-Sträussler-Scheinker disease. He was very committed to research directed toward an understanding of a disease that has afflicted many members of his extended family. Although not a scientist, he was very knowledgeable about prion diseases and read widely on the subject. He acted as a "liaison officer" between the family and us and made possible the collection of blood samples that were used in the DNA analyses. He helped us to understand what these diseases are really about. We are privileged to have known him and his family.

Harry F. Baker
Rosalind M. Ridley

Contents

Contributors

HARRY F. BAKER • *MRC Comparative Cognition Team, Department of Experimental Psychology, School of Clinical Veterinary Medicine, Cambridge, UK*

JEANNE E. BELL • *Neuropathology Laboratory, National CJD Surveillance Unit, Department of Pathology, University of Edinburgh, Western General Hospital, Edinburgh, UK*

RICHARD A. BESSEN • *Laboratory of Persistent Viral Diseases, Rocky Mountain Laboratories, National Institutes of Allergies and Infectious Diseases, Hamilton, MT*

PAUL BROWN • *Laboratory of Central Nervous System Studies, National Institute of Neurological Disorders and Stroke, National Institutes of Health, Bethesda, MD*

MOIRA E. BRUCE • *BBSRC/MRC Neuropathogenesis Unit, Institute for Animal Health, Edinburgh, UK*

ORSO BUGIANI • *Istituto Nazionale Neurologico Carlo Besta, Milano, Italy*

BYRON CAUGHEY • *Laboratory of Persistent Viral Diseases, Rocky Mountain Laboratories, National Institutes of Allergies and Infectious Diseases, Hamilton, MT*

BRUCE CHESEBRO • *Laboratory of Persistent Viral Diseases, Rocky Mountain Laboratories, National Institutes of Allergies and Infectious Diseases, Hamilton, MT*

RAJITH DE SILVA • *CJD Surveillance Unit, Western General Hospital, Edinburgh, UK*

GIANLUIGI FORLIONI • *Istituto di Ricerche Farmacologiche Mario Negri, Milano, Italy*

BERNARDINO GHETTI • *Department of Pathology, Indiana University School of Medicine, Indianapolis, IN*

GIORGIO GIACCONE • *Istituto Nazionale Neurologico Carlo Besta, Milano, Italy*

CAROLINE GOODSIR • *Lasswade Veterinary Laboratory, Midlothian, UK*

JONATHON R. GRAY • *Institute of Medical Genetics, University of Wales College of Medicine, Cardiff, UK*

PETER S. HARPER • *Institute of Medical Genetics, University of Wales College of Medicine, Cardiff, UK*

NORA HUNTER • *BBSRC/MRC Neuropathogenesis Unit, Institute for Animal Health, Edinburgh, UK*

JAMES W. IRONSIDE • *Neuropathology Laboratory, National CJD Surveillance Unit, Department of Pathology, University of Edinburgh, Western General Hospital, Edinburgh, UK*

MARTIN JEFFERY • *Lasswade Veterinary Laboratory, Midlothian, UK*

PAULA KEYES • *Central Veterinary Laboratory, Surrey, UK*

DAVID A. KOCISKO • *Laboratory of Persistent Viral Diseases, Rocky Mountain Laboratories, National Institutes of Allergies and Infectious Diseases, Hamilton, MT; Department of Chemistry, Massachusetts Institute of Technology, Cambridge, MA*

PETER T. LANSBURY, JR. • *Department of Chemistry, Massachusetts Institute of Technology, Cambridge, MA*

JEAN C. MANSON • *BBSRC/MRC Neuropathogenesis Unit, Institute for Animal Health, Edinburgh, UK*

PEDRO PICCARDO • *Department of Pathology, Indiana University School of Medicine, Indianapolis, IN*

FRANCES PRELLI • *Department of Pathology, New York University Medical Center, New York, NY*

SUZETTE A. PRIOLA • *Laboratory of Persistent Viral Diseases, Rocky Mountain Laboratories, National Institutes of Allergies and Infectious Diseases, Hamilton, MT*

RICHARD E. RACE • *Laboratory of Persistent Viral Diseases, Rocky Mountain Laboratories, National Institutes of Allergies and Infectious Diseases, Hamilton, MT*

GREGORY J. RAYMOND • *Laboratory of Persistent Viral Diseases, Rocky Mountain Laboratories, National Institutes of Allergies and Infectious Diseases, Hamilton, MT*

ROSALIND M. RIDLEY • *MRC Comparative Cognition Team, Department of Experimental Psychology, School of Clinical Veterinary Medicine, Cambridge, UK*

MARIO SALMONA • *Istituto di Ricerche Farmacologiche Mario Negri, Milano, Italy*

ANTHONY C. SCOTT • *Central Veterinary Laboratory, Surrey, UK*

JO R. SOLDAN • *Institute of Medical Genetics, University of Wales College of Medicine, Cardiff, UK*

MICHAEL J. STACK • *Central Veterinary Laboratory, Surrey, UK*

FABRIZIO TAGLIAVINI • *Istituto Nazionale Neurologico Carlo Besta, Milano, Italy*

DAVID M. TAYLOR • *BBSRC/MRC Neuropathogenesis Unit, Edinburgh, UK*

Kevin C. Taylor • *Ministry of Agriculture, Fisheries, and Food, Tolworth, UK*

David A. J. Tyrrell • *Centre for Applied Microbiology and Research, Salisbury, UK*

David Westaway • *Department of Molecular and Cellular Pathology and Centre for Research in Neurodegenerative Diseases, Faculty of Medicine, University of Toronto, Canada*

John W. Wilesmith • *Epidemiology Department, Central Veterinary Laboratory, Surrey, UK*

Robert G. Will • *National Creutzfeldt-Jakob Disease Surveillance Unit, Western General Hospital, Edinburgh, UK*

1

The Paradox of Prion Disease

Rosalind M. Ridley and Harry F. Baker

1. Introduction

The study of that group of diseases now collectively known as the prion diseases has always been a source of excitement and argument between scientists. These obscure diseases usually have been of extremely rare occurrence and have had little impact on the general public. When an epidemic occurs, however, as in the case of bovine spongiform encephalopathy (BSE) in the United Kingdom, the bizarreness of the prion diseases and the profound difference between them and any other "infectious" condition can (and did) lead to public consternation. Three features of prion diseases give them an apparently diabolical quality.

1.1. Their Occurrence Is Largely Unpredictable

1. In acquired cases there is a silent incubation period that can vary from a few weeks to up to 40 yr, depending on the species and the circumstances.
2. During the incubation period there are no symptoms and no detectable signs, e.g., there is no measurable immune response that will predict subsequent illness.
3. No mode of transmission could be established for the majority of human cases that appear to occur "out of the blue."

1.2. They Exhibit Certain Impossible Properties

In relation to established principles of microbiology, the prion diseases exhibit certain "impossible" properties.

1. "Infectious" disease seems to arise spontaneously.
2. Inherited cases give rise to a disease that is transmissible but acquired cases do not produce a heritable disease.

From: *Methods in Molecular Medicine: Prion Diseases*
Edited by: H. Baker and R. M. Ridley Humana Press Inc., Totowa, NJ

3. Procedures and chemical treatments that specifically destroy nucleic acids (which all replicating organisms possess) do not destroy the infectious agent.
4. The infectious agent can persist in the environment almost indefinitely.

1.3. They Are Invariably Fatal

1. No treatments have been found that will alter the progression of the disease.
2. In animal transmission studies, the duration of symptoms from onset to death can be as short as a few hours following an incubation period of many months, during which the animal appears to be completely healthy.

These three features have combined to strike fear into the general public, who regard prion diseases (perhaps with some justification) as uncontrollable, incomprehensible, indestructable, and incurable. This fear is probably only matched by an excessive dread of radioactivity, which is similarly imbued with great "strangeness." In sharp contrast, the public steadfastly refuses to worry about "ordinary" factors, such as diet, exercise, and safety, which are known to have an enormous effect on morbidity.

2. Scientific Investigation

In recent years the scientific investigation of prion diseases has expanded greatly, leading to strongly held convictions and arguments that have sometimes degenerated into acrimony. Scientists are particularly exercised by the fact that some accounts of the pathogenesis of prion disease seem to contradict basic tenets of molecular biology:

1. That proteins are made using information encoded in nucleic acid;
2. That different proteins are encoded by different nucleic acid sequences;
3. That protein synthesis cannot occur without nucleic acid; and
4. That any "agent" that increases in titer must involve replication of nucleic acid.

The central argument has been about "viruses vs genes (or proteins)." As in other great scientific debates, e.g., "particles vs waves" or "nature vs nurture," the resolution of the problem is unlikely to be that one view is correct and the other incorrect but that a new way of understanding the phenomenon allows the dichotomy to dissolve. For many years the body of data on prion diseases was like a Rubik cube; when one face was intact the others were not. Thus, it was easy enough on the basis of experimental transmission studies to make the case for disease being caused by a virus—except that it was not possible to explain how the majority of cases were "caught." The transmissible agent appeared to behave like a replicating, infectious agent—except that treatments that destroy nucleic acid did not reduce infectivity. Furthermore, some human cases clearly were inherited, whereas other cases clearly were not.

2.1. Misinterpretations

These difficulties in producing a coherent story for prion diseases were compounded by certain misinterpretations of the available evidence.

1. The demonstration of experimental transmissibility led to the presumption that *all* cases were acquired by infection.
2. The difficulty in demonstrating contact between cases of human prion disease led to exaggerated concern about the infectiousness of the disease. For example, the apparently higher incidence in cities (probably owing to sampling and demographic effects) could be interpreted as meaning that casual contact between strangers was all that was necessary to transmit disease.
3. The not infrequent occurrence of prion disease in another family member of the proband led to the fear that the disease was transmissible vertically through contamination *in utero* or through the incorporation of the agent (perhaps a retrovirus) in the genome of an acquired case. Where the disease occurred in a proband and a parent who had been separated for many years the suspicion grew that transmission of "agent" had occurred in childhood. Where disease occurred in a proband and a distant relative who had met for only the briefest of meetings it was supposed that genetically susceptible people were quite exquisitely vulnerable to infection.
4. The transmissibility of the disease and the apparent information-carrying property of the agent, as demonstrated by the phenomenon of strain of agent in experimental scrapie, lead to the insistence that failure to find the virus was only a temporary technical failure rather than evidence that there might not be such an independent organism in these diseases.

The basic phenomenon that the various hypotheses of the etiology of prion diseases have sought to address is experimental transmissibility, and advocates for each side draw on a number of facts to support their view.

2.2. Evidence in Favor of the Viral Hypothesis

1. The increase in infectivity titer in brain during disease progression is taken as evidence of a replicating agent.
2. No bacteria or other organism can be found so that, by default, the disease is attributed to a virus.
3. In experimental transmission the long incubation period followed by the rapid course of disease resembles the occurrence of diseases caused by lentiviruses.
4. Different isolates of prion disease "agent" appear to behave differently in terms of species specificity, incubation time, and lesion profile on experimental transmission to other animals leading to the concept of "strain variation."
5. The existence of different "strains of agent" is regarded as evidence of an informational, replicating molecule that is presumed to be composed of nucleic acid.
6. The familial occurrence of prion disease in humans is explained as a genetic susceptibility to an infectious agent in the environment.

2.3. Evidence in Favor of a Gene (Protein) Hypothesis

1. No virus particles are associated with infectivity.
2. No immune response to infection is found in an affected host.
3. No nucleic acid is associated specifically with infectivity.
4. Infectivity titer is associated with levels of the abnormal isoform of prion protein known as PrP^{Sc}.
5. PrP^{Sc} and the normal form of prion protein (PrP^C) do not differ in primary structure, thus negating the need for PrP^{Sc}-specific nucleic acid.
6. PrP^{Sc} has the primary structure dictated by the PrP gene of the host, thus negating the need for agent-specific nucleic acid in the production of PrP^{Sc}.
7. When prion disease is experimentally transmitted from one species to another the PrP^{Sc} in the recipient has the primary structure of the recipient species, not of the donor species or agent. Agent replication does not, therefore, need to occur for levels of PrP^{Sc} to increase.
8. Most (but not all) of the variation in species specificity, incubation time, and lesion profile can be accounted for by variations in the PrP protein of the host and the homology between that and the PrP protein of the agent (i.e., donor), thus reducing the variation attributable to "strain of agent."
9. Familial prion disease occurs in pedigrees with a mutation in the PrP gene.
10. Transgenic mice carrying mutations in the PrP gene produce a transmissible spongiform encephalopathy despite being kept in controlled conditions where accidental contamination with an exogenous agent can be avoided *(1)*.
11. PrP^{Sc} can be reversibly denatured with guanidine-chloride and renatured by diluting out the guanidine chloride. If radiolabeled PrP^C is added during the renaturation stage, proteinase K digestion results in the formation of radiolabeled proteinase K-resistant fragments of similar size to proteinase K-digested PrP^{Sc} from scrapie-affected animals *(2)* (and *see* Caughey et al., this volume). This suggests that PrP^C can be converted directly into PrP^{Sc} by interaction with other PrP^{Sc} molecules.

2.4. Resolution of the Dichotomy

A number of arguments are leading to a resolution of the dichotomy.

2.4.1. Acquired, Familial, and Sporadic Cases

It is now recognized that cases of prion disease arise in three different ways—acquired, familial, and sporadic, and that data from one type of case do not have to be applied to other cases. This resolves the argument between "inherited" and "acquired" factors in the etiology of different cases.

2.4.1.1. ACQUIRED CASES

Acquired cases of prion disease arise because a person or an animal has ingested or absorbed a quantity of the infectious agent. In humans, the majority of these acquired cases occurred in Papua New Guinea in the first half of this century. The disease was known as kuru and spread slowly around various related ethnic groups of the indigenous population. Epidemiological analysis implicated funerary practices and the people admitted to cannibalism, especially by the women and the young children in their care *(3)*. It is now supposed that the epidemic began with a sporadic case of prion disease. The epidemic has now almost run its course but the extremely long incubation periods that can occur, especially following low oral exposure to the agent, means that a few cases still occur each year.

In the West, prion disease in humans has not been associated with any dietary practice, but has occurred following the use of human tissue for medical purposes. The largest number of cases (currently approx 100 worldwide) has resulted from the prolonged use of human pituitary-derived growth hormone. A few further cases are associated with the use of human dura mater tissue in reconstructive surgery of the head and a small number of cases have occurred as a result of contamination during corneal grafting or neurosurgery. In the latter cases instruments were used that had previously been used in patients with prion disease and that probably were still contaminated despite sterilization treatments that were standard at that time. The tissues that have been implicated in these iatrogenic cases—pituitary gland, dura mater, and cornea—all come from close to the brain, which is known to have the highest levels of infectivity.

The largest epidemic of animal prion disease has affected cows in Britain. Bovine spongiform encephalopathy (BSE) was first recognized in 1986 and the number of cases has followed a large distribution with the peak (approx 3500 cases/mo) occurring in 1993. The total number of cows affected so far is approx 150,000 and the epidemic is expected to run beyond the year 2000. The epidemic is believed to have resulted from changes in the method of rendering carcasses (including ovine and bovine) for the production of animal food pellets such that the infectious agent associated with scrapie was not destroyed and affected the calf population. Subsequently, carcass material from BSE-affected cattle inevitably would have been incorporated in the calf food pellets, adding to the infectivity titer of this food preparation. The number of cases occurring in animals born after the feeding of ruminant-derived protein to ruminants was banned in 1988 has dropped dramatically, indicating that, apart from a possible very minor level of residual contamination in the system, the epidemic now is under control. The long incubation period that can occur, however, implies that there will be several years before the disease is eradicated. In addition to BSE, the contamination of animal feed has led to a small number of cases of prion disease in other animals, notably cats and exotic ungulates, and other food-related outbreaks have occurred in farmed mink and captive deer.

2.4.1.2. FAMILIAL CASES

About 15% of cases of human prion disease occur in families where at least one other family member has been affected and where it can be established that a large number of cases have occurred in one family, it can be seen that the disease runs in an autosomal dominant pattern. Linkage of disease to a mutation in the PrP gene open reading frame was first established by Hsiao et al. *(4)*. There are now known to be several point mutations and a number of expansions in an octapeptide repeat sequence within the PrP gene open reading frame *(5)* that are associated with disease.

2.4.1.3. SPORADIC CASES

Sporadic cases of human prion disease occur, by definition, without any known antecedent event through which the disease could have been acquired and without any known family history or mutation in the PrP gene. Some genetic influence is now being recognized, however, because most sporadic cases are homozygous for a common polymorphism in the PrP gene *(6)* although, because the disease is exceedingly rare (affecting approx one in a million people worldwide), the vast majority of people who share this genetic feature are not affected.

2.4.1.4. SCRAPIE: ACQUIRED, FAMILIAL, AND SPORADIC CASES

Scrapie in sheep occupies a particularly important position among the prion diseases because it is a relatively common, naturally occurring condition that shares features of the acquired, inherited, and sporadic etiologies. It is thought to be the original source of infection in the BSE epidemic, although this is difficult to prove. Persistent efforts to find an epidemiological or case-controlled association between scrapie and human prion disease, fortunately, have been unsuccessful.

In the 1960s natural scrapie was reported to occur in a pattern consistent with autosomal recessive inheritance in at least some breeds of sheep *(7)*. More recently it has been shown that in some breeds naturally occurring disease is linked to polymorphisms in the PrP gene that either are of partial dominance or recessive *(8)*, largely vindicating the original observation despite the many intervening years in which the experimental transmissibility of mouse-adapted scrapie blinded researchers to the possibility that the natural sheep disease might wholly be genetic. These PrP polymorphisms also are linked to susceptibility to experimental infection in sheep but it remains unproven whether linkage to naturally occurring disease is because of an influence of the polymorphisms on susceptibility to infection in the field or whether the majority of naturally occurring cases have a genetic disease, in the absence of any contact with agent in the environment. The more recent demonstration that the latter is the case in a proportion of human cases and that transgenic mice carrying mutations in the PrP gene *(1)* or additional copies of normal PrP alleles *(9)* can develop prion disease without contact with an exogenous agent make a purely genetic cause of natural scrapie, in at least some cases, more plausible than was believed until recently. The possibility that natural scrapie may arise without exogenous infection has, however, been recognized by some scientists for many years *(10)*. One of the arguments in favor of the infection hypothesis of the epidemiology of natural scrapie was the high coincidence of scrapie in dam and progeny that was thought to be indicative of maternal transmission *(11)*, but such a pattern clearly is compatible also with a recessive model, especially where the flock shares one ram, such that the variance is determined largely by the status of the dam. Furthermore, although scrapie has been seen in sheep fed with placentas from scrapie-affected sheep *(12)* this does not prove that oral transmission necessarily determines the natural incidence of disease. Indeed, it is not even clear that the placentas were the source of the disease, since the only sheep that can be guaranteed to be scrapie-free prior to exposure to infected tissue come from flocks that are genetically resistant to experimental infection and that therefore are largely unsuitable as recipients in this type of experiment. The recent demonstration that scrapie develops in genetically susceptible sheep born to unaffected, genetically unsusceptible dams following embryo transfer *(13)* greatly reduces the probable relevance of perinatal events in natural scrapie. In the original experiment, however, the embryo donors had been challenged with scrapie, therefore further experiments are required to establish whether this is relevant or whether genetically susceptible animals resulting from embryo transfer will develop scrapie without exposure of the embryo donor to the agent. The time is clearly right for a reevaluation of theories about the occurrence of natural sheep scrapie in the light of the finding of disease-specific polymorphisms in the PrP gene of sheep and other species and the demonstration of pathogenic mechanisms that do not require exposure to the infectious agent. "Naturally occurring scrapie in sheep is not the same as experimental scrapie in any species. It should be studied, in depth, as a unique disease" *(14)*.

2.4.2. Transmissible Disease in Genetic Cases

The recent demonstration that genetically manipulated mice spontaneously can develop neurological disease, which then can be experimentally transmitted to other rodents, has bridged the gap between the two opposing mechanisms of "inheritance" and "infection."

It is well established that disease can be transmitted experimentally to animals by intracerebral injection of brain tissue homogenate from affected members of those families in which typical prion disease is inherited as an autosomal dominant. It is also the case that in each family disease is associated with one of a number of different pathogenic mutations within the open reading frame of the PrP gene. Currently about 17 such mutations have been described, most of which appear to be fully penetrant. That an apparently genetic disease can involve a transmissible agent raises the question of whether the gene mutations are causing disease, *sui generis*, or whether they are predisposing individuals to develop disease as a result of infection with an exogenous agent. The argument that the genetic mutations are causing disease receives support from the work on transgenic mice expressing mutant prion protein. These animals spontaneously develop neurological disease in adulthood despite being protected from possible infection with exogenous agents. Furthermore, disease is then transmissible from these animals by serial passage to other animals *(1)*. These experiments suggest that attempts to classify at least some cases of prion disease as genetic or infectious may be inappropriate.

2.4.3. Conversion of PrP^C to PrP^{Sc}

Recent experimental findings have suggested that the fundamental feature of pathogenesis in this group of diseases consists of the conversion of a pre-existing molecule to a new shape or conformation rather than to *replication* of an agent.

2.4.3.1. THE RELATIONSHIP BETWEEN PRP^C AND PRP^{Sc}

The most important aspect of molecular pathogenesis is the relationship between PrP^C and PrP^{Sc}. A number of important features of this relationship have been established.

1. PrP^C and PrP^{Sc} are both encoded by the same gene and have the same amino acid sequence and relative mol wt of 33–35 kDa.
2. PrP^C and PrP^{Sc} are antigenically indistinguishable.
3. PrP^{Sc} is probably formed by the posttranslational modification of PrP^C.
4. PrP^C is destroyed completely by proteolytic enzymes but PrP^{Sc} is reduced to a protease-resistant core of mol wt 27–30 kDa, termed PrP27–30.
5. PrP^C is soluble in the presence of detergent, whereas PrP^{Sc} forms insoluble aggregates.
6. Turnover is rapid for PrP^C but very slow for PrP^{Sc}.
7. PrP^C is found in brain and some other tissues of all mammals, whereas PrP^{Sc} is found only in the brain and, to a lesser extent, other tissues of animals or humans with prion disease.
8. PrP^C levels are constant throughout disease, but PrP^{Sc} levels in brain rise during disease progression in cases of prion disease.
9. PrP^{Sc} aggregates have a fibrillar structure with the tinctorial properties of amyloid, whereas PrP^C does not aggregate in this way.
10. Levels of PrP^{Sc} are associated with infectivity titer, whereas no such relationship is seen for PrP^C.

2.4.3.2. WHY IS PRP^C CONVERTED TO PRP^{Sc}?

Since the conversion of PrP^C to PrP^{Sc} is irreversible, it would seem that this conversion consists of moving from a higher to lower energy state and it is only the intervening higher activation threshold that prevents the process happening more frequently. Four factors may contribute to the crossing of this barrier:

1. The presence of PrP^{Sc}. PrP seems to belong to a small group of proteins that can have one of several conformations, the presence of one form predisposing the other molecules to take up that shape *(15–17)*. Such a hypothesis can explain:
   a. Transmission of disease in the absence of agent-specific nucleic acid;
   b. The exponential accumulation of PrP^{Sc} during disease;
   c. The observation that, on transmission, the PrP^{Sc} in the recipient is host coded;
   d. The possibility that there is more than one self-perpetuating conformation allowing for the observation of "strains of agent;" and
   e. That where the PrP of the donor and host are different, the ability of the donor PrP to interact with the host PrP will be lessened, leading to the observation of a "species barrier."
2. The amino acid sequence of PrP. Mutations in the PrP gene may produce differences in the stability of the secondary structure of the PrP protein such that a spontaneous increase in the β-pleated sheet content is more probable. This could lead to the occurrence of genetic cases.
3. Availability of PrP^C. Transgenic mice with multiple copies of the normal PrP gene can develop spontaneous prion disease, those with only one copy show prolonged incubation times, and mice with no PrP gene cannot be infected. Factors such as an increase in general metabolic rate, which may *inter alia* affect the rate of production and breakdown of PrP^C, could occasionally tip the balance toward the production of PrP^{Sc} and therefore contribute to the occurrence of prion disease in sporadic cases.
4. Time. In a dynamic equilibrium, the probability that a rare event will have occurred increases with time. The amount of PrP^{Sc} required to initiate the massive production of PrP^{Sc} is not known, but it is likely to consist of a threshold quantity. Since PrP^{Sc} is not disposed of, the total amount present in one area will accrue with time. Naturally occurring prion disease has a characteristic age at onset, which usually is later in sporadic than inherited cases, but which is confined to middle age and beyond. Whether this reflects the effect of real time or the effect of age-related changes in the availability of PrP^C is not yet clear.

2.4.4. Prion Disease and Amyloidosis

The recognition that PrP^{Sc} has a propensity to take on the structure described as "amyloid" has allowed prion diseases to be compared to other amyloidoses rather than other infectious diseases.

PrP^C is a soluble protein containing about 40% α-helical domains and almost no β-pleated structure. PrP^{Sc} is largely insoluble with reduced α-helix content

(30%) and increased β-pleated structure (43%). The β-pleated content of PrP27–30 is even higher (>50%). Pathogenic mutations in the PrP gene are thought to alter the stability of the tertiary structural relations between the alpha helical structures in the PrP molecule, thereby altering the probability that this structural relationship will break down into a β-pleated sheet conformation *(18)*. Proteins with high β-pleated content have a propensity to form fibrils that themselves aggregate to form large deposits that have the tinctorial properties of "amyloid." For example, they exhibit birefringence when stained with Congo red and are viewed under polarized light.

The most important CNS amyloidosis occurs in Alzheimer disease (AD), which is characterized by the presence in brain of β-amyloid plaques, neurofibrillary tangles (NFTs), and often cerebral amyloid angiopathy (CAA), the deposition of β-amyloid in cerebral blood vessels. Similarities between AD and prion diseases include:

1. Approximately 15% of cases of AD and prion disease occur with an autosomal dominant pattern of inheritance. In familial cases, AD may be associated with one of several mutations in the gene that makes the precursor protein (amyloid precursor protein, APP) whose modification produces β-amyloid, whereas prion disease is associated with mutations in the PrP gene. (AD also may be associated with mutations in other genes that interact with the APP gene.)

2. AD and prion disease both have an age at onset largely confined to later middle age and beyond with onset being earlier in familial than in sporadic cases.

3. β-amyloid plaques are found in AD, whereas PrP-amyloid plaques are seen in about 20% of cases of prion disease. Even where PrP-amyloid plaques are not visible by light microscopy, electron microscopy reveals that the PrP^{Sc} exists in free fibrils in the neuropil, suggesting that prion disease involves amyloidosis even in cases where plaques are absent *(19)*.

4. In a few cases of prion disease the degree of CAA and β-amyloid either in separate plaques or integrated into PrP plaques is greater than would be expected by chance association, suggesting an interaction in the pathogenesis of both types of amyloidosis.

5. In AD, β-amyloid plaques are formed when one of two alternative processing pathways of APP results in the production of a truncated protein sequence that takes on a β-pleated sheet conformation. These molecules then polymerize into fibrils that aggregate into the characteristic form of a plaque. In prion disease, PrP-amyloid may consist of a truncated form of PrP^{Sc} *(20)*. It is, however, not yet clear whether the crucial difference between PrP^{C} and PrP^{Sc} that associates the latter with infectivity is the formation of PrP^{Sc} into amyloid fibrils or whether PrP^{Sc} exists in a pathogenic form prior to its subsequent breakdown into PrP-amyloid.

6. Although AD is not a transmissible disease, the laminar distribution of β-plaques in the cortex is consistent with the spread of pathology around the brain by a self-sustaining mechanism *(21)*. The injection of β-amyloid containing brain into the brains of monkeys has been found to result in the formation of β-amyloid plaques

in cortex *(22)*, suggesting that β-amyloidosis may be transmissible under experimental conditions.

3. Treatment and Prevention

Perhaps because prion disease in humans is so rare, the amount of effort devoted to consideration of treatment for prion disease is but a fraction of the interest and effort expended in understanding the nature of the disease. However, several strategies might be considered, some of which will be discussed in subsequent chapters.

3.1. Genetic Methods

1. The identification of mutations in the PrP gene opens up the possibility of genetic counseling for affected families, although counseling for adult-onset disorders is a complex issue. Where affected families request it, prenatal genetic testing can eliminate the disease from all subsequent progeny.
2. Selecting sheep that are resistant to scrapie on the basis of polymorphisms in the PrP gene can radically alter the incidence of scrapie in flocks.
3. Since mice that lack a PrP gene are not noticeably sick and are resistant to prion disease, a technique (if it existed) that blocked PrP gene expression might not have seriously deleterious effects and would prevent disease in disease-allele carriers and in those known to be at risk from acquired infection. Such a technique might also prevent disease progression but would not be expected to reverse the accumulated damage.

3.2. Biochemical Treatments

1. It might be speculated that treatments that decrease the availability of PrP^C, either by reducing PrP synthesis or by speeding up its metabolism, could slow down the progress of the disease. Alternatively, it should be borne in mind that centrally acting drugs that might be intended to relieve some of the symptoms of disease inadvertently might increase the availability of PrP^C and so increase the rate of disease progression.
2. Heterologous PrP molecules can interfere with the process of PrP polymerization and with the production of PrP^{Sc} in scrapie-infected cell cultures so, in theory, such molecules might be used to arrest the disease. Such a strategy, however, only is a remote possibility since it presents considerable difficulties of drug delivery and, if an inappropriate PrP molecule was used, it could accelerate the disease process.

3.3. Prevention

Acquired disease can be prevented by an understanding of the procedures that lead to transmission. For example, kuru has been controlled by a change in cultural habits. BSE has been reduced by banning the feeding of specified offals of all species and all ruminant-derived protein to ruminants, although the final eradication of the disease and prevention of its reoccurrence may require changes in the rendering process, reverting to a situation in which the agent from any source is destroyed. Iatrogenic human disease may be prevented by a better appreciation of the risk from brain and related tissue and the recognition that tissue from cases of human prion disease cannot be excluded from use on the basis of diagnosis alone because of the possible insidious onset and atypical presentation of the disease.

References

1. Hsiao, K. K., Groth, D., Scott, M., Yang, S.-L., Serban, H., Rapp, D., Foster, D., et al. (1994) Serial transmission in rodents of neurodegeneration from transgenic mice expressing mutant prion protein. *Proc. Natl. Acad. Sci. USA* **91,** 9126–9130.
2. Kocisko, D. A., Come, J. H., Priola, S. A., Chesebro, B., Raymond, G. J., Lansbury, P. T., et al. (1994) Cell-free formation of protease-resistant prion protein. *Nature* **370,** 471–474.
3. Glasse, R. and Lindenbaum, S. (1992) Fieldwork in the South Fore: the process of ethnographic inquiry, in *Prion Diseases of Humans and Animals* (Prusiner, S. B., Collinge, J., Powell, J., and Anderton, B., eds.), Ellis Horwood, London, pp. 77–91.
4. Hsiao, K., Baker, H. F., Crow, T. J., Poulter, M., Owen, F., Terwilliger, J. D., et al. (1989) Linkage of a prion protein missense variant to Gerstmann-Straussler syndrome. *Nature* **338,** 342–345.
5. Poulter, M., Baker, H. F., Frith, C. D., Leach, M., Lofthouse, R., Ridley, R. M., et al. (1992) Inherited prion disease with 144 base pair gene insertion. 1. Genealogical and molecular studies. *Brain* **115,** 675–685.
6. Palmer, M. S., Dryden, A. J., Hughes, J. T., and Collinge, J. (1991) Homozygous prion protein genotype predisposes to sporadic Creutzfeldt-Jakob disease. *Nature* **352,** 340–342.
7. Parry, H. B. (1983) *Scrapie Disease in Sheep: Historical, Clinical, Epidemiological and Practical Aspects of the Natural Disease,* Academic, London.
8. Goldman, W., Hunter, N., Foster, J. D., Salbaum, J. M., Beyreuther, K., and Hope, J. (1990) Two alleles of a neural protein gene linked to scrapie in sheep. *Proc. Natl. Acad. Sci. USA* **87,** 2476–2480.
9. Westaway, D., DeArmond, S. J., Cayetanocanlas, J., Groth, D., Foster, D., Yang, S. L., et al. (1994) Degeneration of skeletal muscle, peripheral nerves, and the central nervous system in transgenic mice overexpressing wild-type prion proteins. *Cell* **76,** 117–129.
10. Pattison, I. H. (1974) Scrapie in sheep selectively bred for high susceptibility. *Nature* **248,** 594–595.
11. Dickinson, A. G., Stamp, J. T., and Renwick, C. C. (1974) Maternal and lateral transmission of scrapie in sheep. *J. Comp. Pathol.* **84,** 19–25.
12. Pattison, I. H., Hoare, M. N., Jebbett, J. N., and Watson, W. A. (1972) Spread of scrapie to sheep and goats by oral dosing with foetal membranes from scrapie-affected sheep. *Vet. Rec.* **90,** 465–468.

13. Foster, J. D., McKelvey, W. A. C., Mylne, M. J. A., Williams, A., Hunter, N., Hope, J., et al. (1992) Studies on maternal transmission of scrapie in sheep by embryo transfer. *Vet. Rec.* **130,** 341–343.

14. Pattison, I. H. (1992) A sideways look at the scrapie saga, in *Prion Diseases of Humans and Animals* (Prusiner, S. B., Collinge, J., Powell, J., and Anderton, B., eds.), Ellis Horwood, London, pp. 16–22.

15. Caspar, D. L. D. (1991) Self-control of self-assembly. *Nature* **315,** 331–333.

16. Cheng, M. Y., Hartl, F.-U., and Horwich, A. L. (1990) The mitochondrial chaperonin hsp60 is required for its own assembly. *Nature* **348,** 455–458.

17. Milner, J. and Medcalf, E. A. (1991) Cotranslation of activated mutant p53 with wild type drives the wild type p53 protein into the mutant conformation. *Cell* **65,** 766–774.

18. Huang, Z., Gabriel, J.-M., Baldwin, M. A., Fletterick, R. J., Prusiner, S. B., and Cohen, F. E. (1994) Proposed three-dimensional structure for the cellular prion protein. *Proc. Natl. Acad. Sci. USA* **91,** 7139–7143.

19. Jeffrey, M., Goodsir, C. M., Bruce, M., McBride, P. A., Scott, J. R., and Halliday, W. G. (1994) Correlative light and electron microscopy studies of PrP localisation in 87V scrapie. *Brain Res.* **656,** 329–343.

20. Tagliavini, F., Prelli, F., Ghiso, J., Bugiani, O., Serban, D., Prusiner, S. B., et al. (1991) Amyloid protein of Gerstmann-Straussler-Scheinker disease (Indiana kindred) is an 11 kd fragment of prion protein with an N-terminal glycine at codon 58. *EMBO J.* **10,** 513–519.

21. Pearson, R. C. A., Esiri, M. M., Hiorns, R. W., Wilcock, G. K., and Powell, T. P. S. (1985) Anatomical correlates of the distribution of the pathological changes in the neocortex in Alzheimer disease. *Proc. Natl. Acad. Sci. USA* **82,** 4531–4534.

22. Baker, H. F., Ridley, R. M., Duchen, L. W., Crow, T. J., and Bruton, C. J. (1994) Induction of β-amyloid in primates by injection of Alzheimer's disease brain homogenate: comparison with transmission of spongiform encephalopathy. *Mol. Neurobiol.* **8,** 25–40.

2

Human Spongiform Encephalopathy

Clinical Presentation and Diagnostic Tests

Rajith de Silva

1. Introduction

The spectrum of human transmissible spongiform encephalopathies, or prion diseases, include sporadic Creutzfeldt-Jakob disease (CJD), familial CJD, iatrogenic CJD, and kuru. Although the disorders are rare and currently untreatable, establishing the diagnosis is of considerable importance for counseling relatives and in view of the ongoing epidemiological studies. The scientific study of these disorders has led to significant advances, particularly in the field of molecular biology. This progress has been accompanied by an expansion in the clinical "boundaries" of CJD, so that illnesses such as fatal familial insomnia (FFI) now are included within this group of disorders. In this chapter, the early descriptions of CJD are reviewed. The clinical phenotypes of sporadic CJD are described, and the frequency of common clinical signs is reviewed. Inherited CJD is discussed, in relation to classical descriptions of conditions such as Gerstmann-Sträussler syndrome (GSS), and more recent descriptions of conditions such as FFI. The clinical appearances of iatrogenic CJD cases are described. Diagnostic "pointers" to the condition are discussed. Data from the UK national CJD surveillance unit is presented then, and the differential diagnosis is discussed, based on suspect cases of CJD referred to the unit in which the final neuropathological diagnosis is not CJD. In conclusion, the common phenotypes of CJD are summarized, and possible novel diagnostic and therapeutic measures are discussed.

1.1. The Early Descriptions

H. G. Creutzfeldt is credited with the first description of the disorder, although by current diagnostic criteria his case would be highly atypical *(1)*. The patient

From: *Methods in Molecular Medicine: Prion Diseases*
Edited by: H. Baker and R. M. Ridley Humana Press Inc., Totowa, NJ

was 23 yr old at presentation, and there was a questionable family history of mental subnormality. In adolescence the patient was noted to be immature and had behavioral abnormalities. One year prior to presentation, she was treated for a skin rash and at that time spastic paraparesis was recorded. The latter improved thereafter, but at presentation there were features of psychiatric disturbance, cerebellar ataxia, and possibly dystonia. During the terminal hospitalization, her behavior continued to be disturbed, and there were a variety of cognitive deficits, pyramidal and cerebellar signs, and probably myoclonus. She died in status epilepticus, 2 mo after presentation.

A year later, in 1921, A. Jakob described four cases, at least two of which had clinical features suggestive of the entity we recognize as CJD *(2)*. A further case that Jakob considered similar was described in 1923 *(3)*. Over the next 40 yr there was considerable confusion over nomenclature with at least 12 synonyms being applied to this disorder. Of these, three descriptions are worthy of mention.

1.1.1. "Subacute Vascular Encephalopathy"

In 1954 and 1960, Nevin and colleagues in two elegant series described a total of 10 patients, all of whom had succumbed to a neurodegenerative process of subacute onset and rapid progression *(4,5)*. They drew attention to the combination of pyramidal and cerebellar disturbance, the presence of involuntary movements, especially myoclonus, the frequent occurrence of visual failure, the spectrum of speech disturbances (dysphasia, dyspraxia, and mutism), the recurrence of primitive reflexes, and the paratonic rigidity ("Passive movements were resisted…"). The association of the disorder with characteristic electroencephalogram (EEG) appearances was indicated, although it was stressed that the EEG abnormalities were not specific *(5)*. The authors assumed, on account of the spongiform change noted at neuropathology, that the etiology of the disorder was microvascular dysfunction. Despite the erroneous attribution of causation, these descriptions remain the most comprehensive accounts of the clinical course of sporadic CJD.

1.1.2. Heidenhain's Syndrome

In 1929 Heidenhain reported three cases of rapidly progressive dementia, two of which had blindness during the terminal phase *(6)*. In 1954, Meyer and colleagues reviewed Heidenhain's cases and reported a further case *(7)*. The latter was a man aged 38 yr who died 6 mo after the onset of progressive dementia accompanied by visual failure. On examination, he was severely demented, appeared to have a right homonymous hemianopia, had exaggerated limb reflexes on the left, and was ataxic. The cerebral spongiform change at postmortem was particularly marked in the occipital lobes. Meyer et al. felt that a variant form of CJD existed, characterized by rapidly progressive dementia and cortical blindness.

1.1.3. Ataxic CJD

In 1965 Brownell and Oppenheimer described four patients with pathologically confirmed CJD, all of whom had presented with cerebellar ataxia (as had six cases they reviewed from the literature) *(8)*. The authors drew attention to their fourth case, which had a slightly unusual course. A 60-yr-old woman was seen with a 6-mo history of difficulty walking. The key examination findings were gross limb tremor and gait ataxia. At initial assessment, she was "vague and forgetful" but fully orientated. Within the space of 1 mo she had become disoriented and demented. By then she was unable to stand or walk without support. Terminally, she had feeding difficulties and was doubly incontinent. At the time of death, the suspected diagnosis was (idiopathic) cerebellar degeneration (lumbar puncture and air encephalography had been normal). The total duration of illness had been 8 mo. The authors felt that this case of CJD was unusual in having a course that was for the most part dominated by cerebellar ataxia with dementia being a late feature.

1.2. Kuru

The epidemic of kuru that predominantly affected the Fore-speaking people of the highlands of Papua New Guinea was recognized as a progressive and eventually fatal neurodegenerative process mainly affecting the cerebellum *(9,10)*. In R. W. Hornabrook's 1979 review of the clinical features of this illness (based on his experience with 434 cases), the most salient observation was the remarkable uniformity of clinical features in affected patients *(11)*. ("A resemblance which could not be closer were they coined from the same mint" *[11]*.) During a prodrome of 12 mo or more, affected patients would show transient unsteadiness. At this time minor changes in personality and mood (mild euphoria, tendency toward fatuousness, and lack of insight) may have been present. The clinical illness itself was characterized by progressively worsening ataxia and the inability to maintain balance. In the last stages patients were unable to sit or perform any activity, were grossly dysarthric, and were unable to swallow. Death ensued 12–18 mo after the onset of the clinical illness. Dementia probably was present at the terminal stage of disease. Signs of extrapyramidal disease, rigidity, myoclonus, and seizures were absent. In children, the clinical course was more variable and of shorter duration. Brainstem and bulbar dysfunction appeared to occur more commonly than in adults.

2. Sporadic CJD

Much of the clinical data on sporadic CJD was gathered in the course of conducting large scale epidemiological studies on CJD all over the world in the 1970s. These used criteria modified from those put forward by Masters et al. *(12)*. On the whole, only cases that were neuropathologically confirmed (defined

Table 1
Comparison of Clinical Features
in Large Scale Epidemiological Studies of CJD

Clinical picture	Will and Matthews, %, (17)		Brown et al., %, (13)		Brown et al., %, (14)	
	Onset	Course	Onset	Course	Onset	Course
Cognitive impairment	21	100	64	100	69	100
Cerebellar dysfunction	19	42	43	61	33	71
Visual failure	9	13	17	42	19	42
Pyramidal disease	—	79[a]	2	43	2	62
Extrapyramidal disease	—	3[a]	2	67	0.5	56
Lower motor neuron signs	—	3	0.4	11	0.5	12
Seizures	?1[b]	9	0.4	8	0	19
Myoclonus	?5[c]	82	0	88	1	78

[a]Presence of rigidity alone was classified as "pyramidal."
[b]"Blackout attacks."
[c]"Involuntary movements."

as "definite"), or those with characteristic EEG appearances and appropriate clinical features (designated "probable") were included in these analyses.

2.1. Ages of Onset and Disease Durations

In Brown's 1986 study (13) the mean age of onset was 61.5 yr (range 19–83), and the mean disease duration was 7.6 mo (median 4.0). In his larger series of 1994 (based on 300 cases, all of whom had been experimentally transmitted to nonhuman primates) (14), the subgroup of sporadic cases (n = 234) had a mean age of onset of 60 yr (range 16–82, median 60) and a median disease duration of 4.5 mo (range 1–130, mean 8).

2.2. Clinical Courses

In a previous review of clinical characteristics, Cathala and Baron referred to a prodromal clinical stage in a third of patients, consisting of nonspecific features such as fatigue, sleeping difficulties, weight loss, headaches, malaise, and "sensations" (15). However, Knight has argued persuasively that these symptoms are "…common, nonspecific, and often noted later, when developments have lent them retrospective and possibly spurious significance" (16).

Symptoms and signs at the start and during the course of CJD in the larger series are summarized in Table 1. With respect to the debut of their illnesses, 30–40% of these patients had cognitive impairment alone, 30–40% had neurological dysfunction in isolation (most commonly cerebellar ataxia or cortical blindness), and 20–30% had mixed features. As can be observed in Table 1, inevitably there is some variation in the reported frequency of features, but several general conclusions can be drawn. First, dementia almost always was present during the course of illness. Second, myoclonus, which was rarely present at the onset, was frequently noted during the course of disease. Third, features of cerebellar, visual, pyramidal, and extrapyramidal dysfunction were noted regularly but were not universal. Finally, features of lower motor neuron dysfunction and convulsions were rare presenting features, and were unusual even during the course of illness. As the disease progresses, multifocal central nervous system (CNS) dysfunction is the norm, and is accompanied by the recurrence of primitive reflexes and, later, akinetic mutism. Terminally, the patients are rigid, mute, and unresponsive, and may have abnormal respiration.

It is appropriate at this juncture to consider some atypical, variant forms of sporadic CJD: cases with either very long or short durations of illness, so-called "amyotrophic" CJD, and "panencephalopathic" CJD.

2.2.1. Extreme Disease Durations

Will and Matthews described an "intermediate" group of patients with a mean disease duration of 33.4 mo *(17)*, who accounted for 6% of their series. These cases exhibited three distinct types of disease progression: a form of CJD in which there was slow but inexorable progression, a form in which a slow neurodegenerative process was followed by a rapid terminal phase, and (most rarely) a rapid early course that was followed by a protracted terminal phase in which there was little further decline. It is of note that two of the 12 patients considered by these authors had a family history of neurodegeneration. Brown and colleagues' seminal analysis of CJD of long duration in 1984 *(18)*, also is complicated by the inclusion of familial cases. The observation that these cases are difficult to differentiate from other chronic dementing processes (particularly Alzheimer's disease) nevertheless, is valid. In his later NIH series *(14)*, 4% of sporadic cases ($n = 9$) had illness durations longer than 2 yr. Of the previously identified subtypes, there were three each in the slow, slow-fast, and fast-slow groups.

In contrast, some CJD cases have a rapidly progressive course, with an onset and evolution resembling stroke. In a recent survey, approx 6% of cases in the UK CJD database (covering the period 1970–1994) were found to have this phenotype *(19)*. That a single neurodegenerative process can exhibit such diverse behavior is clinically fascinating, but merely may represent the extremes of biological behavior of this disorder in humans.

2.2.2. Amyotrophic CJD

As indicated in the previous epidemiological studies, signs of lower motor neuron dysfunction do occur in CJD, but they are uncommon and do not occur in the absence of more widespread cortical and cerebellar disease. Furthermore, their presence at the start of the clinical illness is extremely unusual. Salazar et al. reviewed this subject in 1983, and came to the conclusion that "...the great majority of cases involving syndromes of dementia and early onset of LMN (lower motor neuron) signs are clinically and pathologically distinct from the typical cases of CJD and do not represent transmissible disease caused by unconventional viruses as presently understood" *(20)*. They based their conclusion on the negative transmission studies they performed, and also on the atypical neuropathological appearances. The most consistent feature of the latter examinations was atrophy (with neuronal loss and gliosis) in the *frontotemporal cortex*. It is now recognized that frontotemporal atrophy is the pathological hallmark of the dementia that sometimes accompanies motor neuron disease *(21)*. It is our view that "amyotrophic CJD" is a misnomer, and that this entity in the majority of cases represents motor neuron disease with dementia. In the previous series of Salazar et al., only two out of 33 cases with dementia and early LMN signs transmitted: Both had chronic peripheral neuropathies and also typical CJD brain histology. They probably represented cases of CJD that had developed by chance in individuals with chronic neuropathies.

2.2.3. Panencephalopathic CJD

This pathological variant of CJD is described almost exclusively in the Japanese literature, and is characterized by extensive *white matter* degeneration in cases of CJD. There is probably no difference in the clinical presentations of these patients, although reports have implicated a longer duration of disease, and evidence of cortical atrophy and white matter disease (on CT and MRI) in these cases *(22,23)*. A further report has implicated the coexistence of amyloid plaques on neuropathological examination *(24)*.

3. Familial CJD

In 1928 Gerstmann described an unusual inherited cerebellar disorder. A 25-yr-old woman developed ataxia, dysarthria, and personality change, and on review a year later had more pronounced ataxia and dementia *(25)*. Over the next 5 yr (leading to death) the patient exhibited "pseudobulbar disturbance of swallowing," nystagmus (lateral and upgaze), limitation of upgaze, hypotonia, intention tremor, diminished reflexes, and bilaterally upgoing plantar responses. In a subsequent publication *(26)*, the detailed clinical and pathological features of this case along with the family history (there were seven other affected members) were described. The pathological appearances were dominated by *argyrophilic plaques throughout the brain*, and minimal spongiform change. In their landmark publication in 1981, Masters et al. reviewed Gerstmann's case along with nine other cases in the literature sharing similar clinical and pathological features *(27)*. With the inclusion of seven further cases with identical phenotypes referred to the authors' laboratory for transmission experiments, they concluded that the mean age of death in these cases was 48 yr (range 29–62) and that the mean disease duration was 59 mo (range 13–132). The key clinical features were characterized as cerebellar incoordination, pyramidal signs, and dementia, with myoclonus as an inconstant finding. The slow evolution of illness was felt to be the main differentiating feature of this group of patients from those with sporadic CJD.

Molecular biology has had an enormous impact on this form of CJD. After the demonstration of a pathogenicity-associated point mutation at codon 102 of the prion protein gene (PRNP) in patients with this phenotype *(28)*, the identical mutation was identified in Gerstmann's original family *(29)*. DNA extracted from archival material on the first described case of familial CJD ("Paul Backer," by Kirschbaum in 1924 *[30]*) recently has revealed a point mutation at codon 178 of PRNP *(31)*. To date, some 19 different point mutations and extra basepair insertions of the PRNP open reading frame have been associated with cases of familial CJD. Although specific clinical phenotypes are not consistently noted with individual mutations, some general observations can be made.

3.1. Gerstmann-Sträussler Syndrome

Although the clinical and pathological syndrome (described earlier) is classically associated with the codon 102 Pro-Leu change, individuals carrying this mutation even within the same pedigree will manifest highly variable phenotypes. For example, in the German "Sch" pedigree the clinical illness broadly was that of GSS, but individual cases exhibited features as diverse as anxiety and poor concentration, titubation, and myoclonus *(32)*. The evaluation of the impact of other sites within the genome on this phenotypic variability is in progress, but in a recent attempt the common polymorphism at codon 129 was not found to be influential *(33)*.

Confusingly, point mutations at codons 117, 198, and 217 are also associated with the pathological features of GSS. In the case of the latter two, the characteristic multicentric plaques are accompanied by neurofibrillary tangles ("the Indiana variant"), but probably are indistinct from GSS clinically *(34)*. The codon 117 mutation has also been described in association with a so-called "telencephalic" form of CJD, in which dementia is accompanied by pyramidal and extrapyramidal features, and variable cerebellar signs *(35)*.

3.2. Familial CJD Resembling Sporadic Disease

The point mutations at codons 200, 210, and 178 are associated with CJD that resembles sporadic CJD clinically. The last, however, is not associated with characteristic EEG appearances. Even in the case of the first (the explanation for the high incidence of CJD among Libyan-born Jews) perusal of the clinical features of a large pedigree revealed considerable phenotypic heterogeneity, including rarities (for CJD) such as demyelinating polyneuropathy *(36)*.

3.3. Fatal Familial Insomnia

The Asp-Asn change at codon 178 of the PRNP open reading frame has also been linked with an unusual neurodegenerative process *(37)*, first described in 1986 *(38)*. A 53-yr-old man first presented with progressive insomnia and dysautonomia. A previously sound sleeper, he could sleep only for 2–3 h at night. He became impotent and had loss of libido. There was episodic salivation, lacrimation, and rhinorrhoea, and he exhibited orthostatic diaphoresis, pyrexia, difficulties with micturition, and constipation. Two months later, he could only sleep for 1 h, and he was frequently disturbed by vivid dreams. He developed progressive dysarthria, intention tremor of his limbs, and gait ataxia. Examination at this stage revealed miosis, cerebellar signs, and brisk tendon reflexes. He was noted to lapse into a stuporose state if left alone, in which he performed complex and apparently purposeful gesturing and breathed noisily. He could be awakened quickly by light stimuli. As his condition progressed, there were oculomotor disturbances (limitation of upward gaze and saccadic movements), myoclonus, and irregular breathing patterns. Terminally he became confused and disoriented, had episodes of motor agitation, and exhibited severe truncal dystonias. The total duration of his illness was 9 mo. His EEG was never characteristic of CJD, but revealed diffuse slow waves and later became isoelectric. His dreamlike states coincided with EEG desynchronization, but physiological EEG patterns of sleep were absent. A pharmacological response to a short-acting benzodiazepine antagonist could be demonstrated both clinically and electrophysiologically. There were at least four other affected members in his pedigree, including two affected sisters. Study of the family in greater detail revealed that *insomnia, dysautonomia, dysarthria, ataxia, myoclonus, and pyramidal signs invariably were present* in affected members. Memory and attention deficits were minimal in the early stages, but tended to progress with time. During "sleep" there was loss of slow-wave and rapid-eye-movement phases. The mean age of onset was 49 yr, and the mean duration was 13 mo.

3.4. CJD Associated with Extra Repeat Insertions

Extra insertions of a variety of sizes have been described in the octapeptide repeat region of the PRNP open reading frame. The phenotypic descriptions of these cases resemble CJD, with two notable exceptions. First, in a large pedigree reported from the South East of England with a 144-basepair insertion, there was striking diversity in the clinical phenotypes of affected cases *(39)*.

Diagnoses, such as General Paralysis of the Insane, "spinal sclerosis," "cerebral softening," cerebral thrombosis, dementia praecox, Parkinsonism, Huntington disease, Pick disease, and Alzheimer disease, had been attached to affected members of the pedigree through the ages. This was accompanied by considerable variation in the neuropathological features of cases. With respect to the molecular biology, an important point of note was the stability of the expanded sequence over five successive generations. This contrasts with expansions of trinucleotide repeat sequences in successive generations, which are associated with the phenomenon of anticipation in associated neurodegenerative disorders, such as Huntington disease *(40)*. Anticipation is not a feature of inherited CJD. The age at death in this pedigree was affected by the common polymorphism at codon 129. The mutation was always carried on an allele encoding methionine at this site; cases homozygous for methionine had a significantly younger age at death than heterozygous cases *(41)*.

Second, in their report on a case of dementia associated with a 216-basepair insertion, Duchen et al. argued that the presence of neuritic plaques staining positively for β-amyloid protein and τ protein indicated that this case represented a transition (neuropathologically, at least) between CJD and Alzheimer disease *(42)*. From the clinical perspective, however, this patient's illness was entirely compatible with familial CJD.

3.5. Some Unusual Mutations, and Conclusions on Familial CJD

Two recent mutations described in cases of familial CJD from Japan are associated with a Pro-Leu change at codon 105, and an amber mutation at codon 145. Almost uniquely for cases of CJD (sporadic or familial), patients carrying the former mutation presented with spastic paraparesis *(43)*. Dementia eventually supervened, but there were no cerebellar features or myoclonic jerks. The latter mutation was associated (in a single patient) with a slowly progressive dementia of >10 yr duration *(44)*.

Familial CJD encompasses a wide variety of clinical syndromes, some of which have highly unusual features, such as dysautonomia and spastic paraparesis. There is considerable overlap of clinical features associated with individual PRNP mutations, and the variability of clinical features *within* pedigrees cannot be accounted for consistently by the common polymorphism at codon 129. Familial CJD may be associated with longer disease durations than sporadic cases. This would account for the higher prevalence of familial cases among series of CJD patients with long disease durations.

4. Iatrogenic CJD

Although rare, forms of iatrogenic CJD are of considerable clinical and public health importance. In particular, the disease phenotype appears to be influenced by the route by which the agent of infectivity gains entry into the host. Cases where "peripheral" (outside the CNS) inoculation has taken place (recipients of contaminated human pituitary-derived hormones) are characterized by a progressive cerebellar ataxia, with little or no cognitive impairment and the absence of characteristic EEG findings *(45)*. The disease phenotype in these individuals is remarkably homogeneous, and inevitable comparisons with the clinical course of kuru have been made. In addition to exhibiting relentlessly progressive truncal and limb ataxia, patients demonstrate a characteristic change in their personalities, appearing apathetic and unconcerned about their predicament. As their disease progresses, more CJD-like features emerge: limb rigidity, myoclonus, startle responses, and akinetic mutism are all apparent.

The clinical course of patients who have had the CJD agent inoculated centrally have had disease phenotypes consistent with sporadic CJD. However, in a recent report, four "Lyodura" recipients who had grafts placed in their posterior fossae were described *(46)*. They had presented with cerebellar syndromes (all four were ataxic at presentation, and three were dysarthric), possibly owing to the proximity of the cerebellum to the site of graft placement, but in our opinion the disease phenotypes were distinct from those with peripherally inoculated iatrogenic CJD. One patient had characteristic periodic sharp wave complexes on the EEG.

5. Diagnostic Tests

5.1. Electroencephalography

In the large epidemiological studies referred to previously, the EEG was found to be the most useful diagnostic test for CJD. However, because the inclusion criteria for these studies included characteristic EEG appearances, there has been a tendency for over-ascertainment of this feature. Also, given the subjective nature of EEG interpretation, there may have been some variation in the types of abnormalities included in the different series. Will and Matthews found characteristic EEG abnormalities in 84% of "subacute" cases *(17)*. In their "intermediate" group (with a longer disease duration) the EEG was typical in only two out of nine cases. Brown et al. report "periodic" EEG appearances in 60–80% and "triphasic 1 cycle/second" in 48–56% *(13,14)*. In our experience, an EEG showing periodic sharp wave complexes in the appropriate clinical setting is *virtually diagnostic* of CJD. Anecdotal reports of false positive diagnoses have usually arisen from the incorrect assessment of EEGs, and the inaccurate description of nonspecific EEG findings as "characteristic" of CJD. However, a proportion of CJD patients *never* manifest typical EEG appearances, reducing the sensitivity of this test.

5.2. Liver "Function" Tests

In the large epidemiological series, routine biochemistry and hematology usually were normal, with the exception of liver function tests (32 out of 80 cases in Will and Matthews' study). These amounted to mild elevations of hepatic enzymes, and overt liver failure was not observed. Although serial measurements rarely have been reported, the impression gained is that the elevations are transient *(47,48)*.

5.3. Neuroimaging

Computerized tomography in CJD is usually normal but sometimes atrophy is found, especially in cases with protracted illnesses. Radiology reports may overemphasize any degree of cortical atrophy in view of the clinical history of dementia. MRI abnormalities have been reported in CJD. The extensive white matter degeneration reported by Uchino et al. may be specific for the Japanese panencephalopathic variant of CJD *(49)*. In reports from Western countries, high T2-signal lesions in the basal ganglia have been described *(50,51)*. In our experience these abnormalities are not universal; nor are they necessarily associated with basal ganglia dysfunction. Magnetic resonance spectroscopy (for N-acetylaspartate) has been disappointing as an early diagnostic tool *(52)*.

5.4. Cerebrospinal Fluid

Examination of the cerebrospinal fluid (CSF) in patients with CJD is sometimes abnormal. Extrapolating from Will and Matthews' data, the CSF protein was >0.4 g/L in 45% of cases where it was examined. The protein content rarely exceeds 1.0 g/L, and a leukocyte response is absent. A pair of novel proteins in the CSF, identified by two-dimensional electrophoresis, has been described in CJD *(53,54)*, but may not be specific for this condition (*see* note added in proof).

6. Data from the UK National Surveillance Unit

Surveillance of CJD has been ongoing in the UK since 1990, and affords an opportunity for the systematic collection of clinical data on patients with this disorder. Cases designated "definite" and "probable" using criteria modified from Masters et al. are included in the survey *(55)*. During the first 4 yr of the study, 144 sporadic and 14 familial cases of CJD were identified. The ages of onset and disease durations of these patients are tabulated in Table 2. As can be seen, patients with familial disease tend to present approx 10 yr earlier than sporadic cases, and have a disease duration approximately twice as long.

The frequency of clinical features during the course of illness in sporadic cases was entirely consistent with the previous surveys. The detailed clinical questionnaire, however, enabled a closer scrutiny of patients' *presenting* features. It emerges that at presentation, approx 40% of patients with CJD have some aspect of cognitive dysfunction in isolation, 30% present with cerebellar ataxia alone, 10% have a combination of cognitive and cerebellar dysfunction, and 10%

Table 2
Ages at Onset and Disease Durations of Sporadic and Familial CJD Cases

	Age at onset, yr				Disease duration, mo			
	Mean	Minimum	Maximum	Median	Mean	Minimum	Maximum	Median
Sporadic	65[a]	43	86	65	7[b]	1	62	4
	(n = 143/144)				(n = 139/144)			
Familial	52[a]	35	67	55	19[b]	1	112	8
	(n = 14/14)				(n = 12/14)			

[a]z-4.1270, p < 0.0001, corrected for ties.
[b]z-2.2424, p < 0.0249, corrected for ties.

have cortical blindness (Heidenhain syndrome). These presentations account for 90% of cases, and alternative presenting features, such as expressive dysphasia, motor dyspraxia, and pyramidal signs, are all much less common. Patients with the Brownell-Oppenheimer variant of CJD (with progressive cerebellar ataxia and no cognitive impairment until late in the illness) were unusual, accounting for no more than 4% of sporadic cases.

6.1. Differential Diagnosis

Cases of suspect CJD referred to the surveillance unit but whose neuropathological examination revealed an alternative neurodegenerative process are listed in Table 3. It can be seen that the majority of these (58%) had Alzheimer disease. These patients had a mean disease duration of 31 mo (median 16; range 1–151), which was significantly longer than that in sporadic cases (Mann-Whitney U-test, z [corrected for ties] –4.8675, p < 0.0001).

Comparison of clinical features during the course of illness between CJD and non-CJD cases enabled the estimation of their relative sensitivities and specificities. These are tabulated in Table 4. Most of the clinical criteria on which the diagnosis of CJD is traditionally based emerge with high sensitivities, with the exceptions of Parkinsonism and (particularly) neurogenic muscle wasting. Disorientation is found by this analysis to have a poor specificity, but this is owing to the inclusion of a large number of patients with dementia in the non-CJD group.

7. Conclusions

In this chapter, the clinical characteristics of the human transmissible spongiform encephalopathies have been reviewed. These disorders are summarized in Table 5. The diagnosis essentially is based on a high index of clinical suspicion backed up by neuropathological confirmation, usually at postmortem. Brown's NIH study gives systematic data on brain biopsy: The

Table 3
Cases with (Pathologically Confirmed) Non-CJD

Diagnosis	Frequency
Alzheimer-type dementia (ATD)	17
ATD + multi-infarct disease (MID)	5
MID	3
Diffuse Lewy body disease (DLBD)	1
ATD + DLBD	1
Motor neuron disease	2
Cerebrovascular disease	1
Cerebellar degeneration	1
Pick disease	1
Progressive supranuclear palsy	1
Multiple system atrophy	1
Corticobasal degeneration	1
Viral encephalomyelitis	1
Metastatic carcinoma	1
Hypoxia	1
Epilepsy	1
No abnormality found	1

Table 4
Sensitivities and Specificities of Clinical Features in CJD

Sign	Sensitivity	Specificity
Cognitive impairment		
Personality change	0.5223	0.6389
Behavior change	0.6242	0.5000
Memory loss	0.6369	0.4117
Disorientation	0.7898	0.2500
Myoclonus	0.8535	0.4444
Cortical blindness	0.5096	0.8333
Pyramidal signs	0.5924	0.4444
Parkinsonism	0.3312	0.6944
Cerebellar incoordination	0.8535	0.6111
Akinetic mutism	0.7261	0.7778
Muscle wasting	0.1656	0.8333
Characteristic electroencephalogram	0.4552	1.0000

procedure achieved a high rate of diagnosis, with 52 out of 55 autopsy-verified cases revealing spongiform change *(14)*. Against this has to be weighed the attendant risks of accidental transmission to other patients and staff *(56)*. In

Table 5
Summary of the Human Spongiform Encephalopathies

Sporadic CJD	
Classical	Presenting with cognitive dysfunction alone (40%)
	Presenting with cerebellar dysfunction alone (30%)
	Presenting with cognitive and cerebellar dysfunction (10%)
	Other presenting features (10%)
Heidenhain variant	Presenting with occipital blindness (10%)
Brownell-Oppenheimer variant	Progressive cerebellar syndrome (rare)
Panencephalopathic	(Rare)

Familial CJD

Phenotypes linked with 19 separate PRNP open reading frame mutations, including:
  Phenotype resembling sporadic CJD;
  Gerstmann-Sträussler syndrome;
  "Telencephalic" CJD;
  Fatal familial insomnia;
  Familial CJD with spastic paraparesis; and
  Phenotype resembling Alzheimer disease.
However, there is considerable overlap and the phenotypes are not mutation-specific.

Iatrogenic CJD	
Central inoculation	Resembles sporadic CJD
Peripheral inoculation	Progressive cerebellar syndrome
Kuru	

familial cases PRNP genome analysis clearly is invaluable. Since a family history of neurodegeneration is not available always; this should perhaps be considered routinely in suspect CJD cases, especially in cases presenting at a young age and appearing to run a long course. The role of the common polymorphism at codon 129 in the diagnosis of CJD is more uncertain. Although most sporadic CJD cases are likely to be methionine homozygous at this site *(57,58)*, approx 39% of the normal Caucasian British population are also of the same genotype *(59)*. Extrapolating from our own data (based on the genotypes of 68 sporadic CJD cases, and 16 non-CJD cases), the relative sensitivity and specificity of methionine homozygosity at codon 129 for the diagnosis of CJD are 0.8382 and 0.5000, respectively.

CJD is finally an incurable disease. Despite their early promise, both Amantidine and Amphotericin have little or no effect on the course of illness *(60,61)*. However, sulfated glycosaminoglycans *(62)* and even antisense oligonucleotides may offer hope in the future as therapies. At present, patients only can be treated supportively. Depending on the certainty of the diagnosis and the relatives' wishes, hydration and artificial feeding may be instituted. If the patient is perceived to be in pain, opiate analgesics should be administered. Myoclonus (which can be particularly distressing for relatives to watch) usually responds to Clonazepam or Valproate.

Note Added in Proof

Zerr et al. have recently reported elevated neuron-specific enolase concentrations in the CSF of patients with CJD *(63)*. The enzyme is known to be elevated in CSF in a variety of neurological disorders, and may lack diagnostic specificity.

Acknowledgments

The author is grateful to R. Will for reviewing the text. The clinical data presented was collected by R. Will, T. Esmonde, M. Zeidler, and the author. The UK National Surveillance Unit receives funding from the Department of Health and the Scottish Home and Health Department.

References

1. Creutzfeldt, H. G. (1920) Über eine eigenartige herdförmige Erkrankung des Zentralnervensystems. *Zeitschrift für die gesamte Neurologie und Psychiatrie* **57**, 1–18.
2. Jakob, A. (1921) Über eine der multiplen Sklerose klinisch nahestehende Erkrankung des Zentralnervensystems (spastische Pseudosklerose) mit bemerkenswertem anatomischem Befunde. Mitteilung eines vierten Falles. *Med. Klin.* **17**, 372–376.
3. Jakob, A. (1923) *Die Extrapyramidalen Erkrankungen*, Springer, Berlin, pp. 218–245.
4. Jones, D. P. and Nevin, S. (1954) Rapidly progressive cerebral degeneration (subacute vascular encephalopathy) with mental disorder, focal disturbances, and myolonic epilepsy. *J. Neurol. Neurosurg. Psychiat.* **17**, 148–159.
5. Nevin, S., McMenemey, W. H., Behrman, S., and Jones, D. P. (1960) Subacute spongiform encephalopathy—a subacute form of encephalopathy attributable to vascular dysfunction (spongiform cerebral atrophy). *Brain* **83**, 519–564.
6. Heidenhain, A. (1929) Klinische und Anatomische Untersuchungen über eine eigenartige Erkrankung des Zentralnervensystems im Praesenium. *Z. Ges. Neurol. Psychiat.* **118**, 49.
7. Meyer, A., Leigh, D., and Bagg, C. E. (1954) A rare presenile dementia associated with cortical blindness (Heidenhain's syndrome). *J. Neurol. Neurosurg. Psychiat.* **17**, 129–133.
8. Brownell, B. and Oppenheimer, D. R. (1965) An ataxic form of subacute presenile polioencephalopathy (Creutzfeldt-Jakob disease). *J. Neurol. Neurosurg. Psychiat.* **28**, 350–361.
9. Gajdusek, D. C. and Zigas, V. (1959) Kuru. Clinical, pathological and epidemiological study of an acute progressive degenerative disease of the central nervous system among natives of the Eastern Highlands of New Guinea. *Am. J. Med.* **26**, 442–469.
10. Simpson, D. A., Lander, H., and Robson, H. N. (1959) Observations on kuru: II. Clinical features. *Australas. Ann. Med.* **8**, 8–15.

11. Hornabrook, R. W. (1979) Kuru and clinical neurology, in *Slow Transmissible Diseases of the Nervous System*, vol. 1 (Prusiner, S. B. and Hadlow, W. J., eds.), Academic, New York, pp. 37–66.
12. Masters, C. L., Harris, J. O., Gajdusek, C., Gibbs, C. J., Jr., Bernoulli, C., and Asher, D. M. (1979) Creutzfeldt-Jakob disease: patterns of worldwide occurrence and the significance of familial and sporadic clustering. *Ann. Neurol.* **5,** 177–188.
13. Brown, P., Cathala, F., Castaigne, P., and Gajdusek, D. C. (1986) Creutzfeldt-Jakob disease: clinical analysis of a consecutive series of 230 neuropathologically verified cases. *Ann. Neurol.* **20,** 597–602.
14. Brown, P., Gibbs, C. J., Jr., Rodgers-Johnson, P., Asher, D. M., Sulima, M. P., Bacote, A., et al. (1994) Human spongiform encephalopathy: the National Institutes of Health series of 300 cases of experimentally transmitted disease. *Ann. Neurol.* **35,** 513–529.
15. Cathala, F. and Baron, H. (1987) Clinical aspects of Creutzfeldt-Jakob disease, in *Prions: Novel Infectious Pathogens Causing Scrapie and Creutzfeldt-Jakob Disease* (Prusiner, S. B. and McKinley, M. P., eds.), Academic, San Diego, pp. 467–509.
16. Knight, R. (1987) Transmissible dementia: clinical aspects, in *Degenerative Neurological Disease in the Elderly* (Griffiths, R. A. and McCarthy, S. T., eds.), Wright, Bristol, pp. 109–118.
17. Will, R. G. and Matthews, W. B. (1984) A retrospective study of Creutzfeldt-Jakob disease in England and Wales 1970–79 I: clinical features. *J. Neurol. Neurosurg. Psychiat.* **47,** 134–140.
18. Brown, P., Rodgers-Johnson, P., Cathala, F., Gibbs, C. J., Jr., and Gajdusek, D. C. (1984) Creutzfeldt-Jakob disease of long duration: clinicopathological characteristics, transmissibility, and differential diagnosis. *Ann. Neurol.* **16,** 295–304.
19. McNaughton, H. and Will, R. (1994) Creutzfeldt-Jakob disease presenting as stroke: an analysis of 30 cases. *Ann. Neurol.* **36,** 313.
20. Salazar, A. M., Masters, C. L., Gajdusek, D. C., and Gibbs, C. J., Jr. (1983) Syndromes of amyotrophic lateral sclerosis and dementia: relation to transmissible Creutzfeldt-Jakob disease. *Ann. Neurol.* **14(1),** 17–26.
21. Neary, D. (1990) Non Alzheimer's disease forms of cerebral atrophy. *J. Neurol. Neurosurg. Psychiat.* **53,** 929–931.
22. Kitagawa, Y., Gotoh, F., Koto, A., Ebihara, S., Okayasu, H., Ishii, T., et al. (1983) Creutzfeldt-Jakob disease: a case with extensive white matter degeneration and optic atrophy. *J. Neurol.* **229,** 97–101.
23. Macchi, G., Abbamondi, A. L., Di Trapani, G., and Sbriccoli, A. (1984) On the white matter lesions of Creutzfeldt-Jakob disease. Can a new subentity be recognised in man? *J. Neurol. Sci.* **63,** 197–206.
24. Kawata, A., Suga, M., Oda, M., Hayashi, H., and Tanabe, H. (1992) Creutzfeldt-Jakob disease with congophilic kuru plaques: CT and pathological findings of the cerebral white matter. *J. Neurol. Neurosurg. Psychiat.* **55,** 849–851.
25. Gerstmann, J. (1928) Über ein noch nicht beschriebenes Reflexphänomen bei einer Erkrankung des zerebellaren Systems. *Weiner Medizinische Wochenschrift* **78,** 906–908.

26. Gerstmann, J., Sträussler, E., and Scheinker, I. (1936) Über eine eigenartige hereditär-familiäre Erkrankung des Zentralnervensystems. Zugleich ein Beitrag zur Frage des vorzeitigen lokalen Alterns. *Zeitschrift für Neurologie* **154,** 736–762.

27. Masters, C. L., Gajdusek, D. C., and Gibbs, C. J., Jr. (1981) Creutzfeldt-Jakob disease virus isolations from the Gerstmann-Sträussler syndrome with an analysis of the various forms of amyloid plaque deposition in the virus-induced spongiform encephalopathies. *Brain* **104,** 559–588.

28. Hsiao, K., Baker, H. F., Crow, T. J., Poulter, M., Owen, F., Terwilliger, J. D., et al. (1989) Linkage of a prion protein missense variant to Gerstmann-Sträussler syndrome. *Nature* **338,** 342–345.

29. Kretzschmar, H. A., Honold, G., Seitelberger, F., Feucht, M., Wessely, P., et al. (1991) Prion protein mutation in family first reported by Gerstmann, Sträussler, and Scheinker. *Lancet* **337,** 1160.

30. Kirschbaum, W. R. (1924) Zwei eigenartige Erkrankungen des Zentralnervensystems nach Art der spastischen Pseudosclerose (Jakob). *Z. Ges. Neurol. Psychiat.* **92,** 175–202.

31. Brown, P., Cervenáková, L., Boellaard, J. W., Stavrou, D., Goldfarb, L., and Gajdusek, D. C. (1994) Identification of a PRNP genemutation in Jakob's original Creutzfeldt-Jakob disease family. *Lancet* **344,** 130–131.

32. Brown, P., Goldfarb, L. G., Brown, W. T., Goldgaber, D., Rubenstein, R., Kascsak, R. J., et al. (1991) Clinical and molecular genetic study of a large German kindred with Gerstmann-Sträussler-Scheinker syndrome. *Neurology* **41,** 375–379.

33. Barbanti, P., Fabbrini, G., Salvatore, M., Petraroli, R., Pocchiari, M., Macchi, G., et al. (1994) Phenotypic heterogeneity in Gerstmann-Sträussler-Scheinker syndrome with codon 102 mutation of the prion protein gene is not related to codon 129 polymorphism. *Ann. Neurol.* **36,** 309.

34. Ghetti, B., Tagliavini, F., Hsiao, K., Dlouhy, S. R., Yee, R. D., Giaccone, G., et al. (1992) Indiana variant of Gerstmann-Sträussler-Scheinker disease, in *Prion Diseases of Humans and Animals* (Prusiner, S. B., Collinge, J. Powell, J., and Anderton, B., eds.), Ellis Horwood, London, pp. 154–167.

35. Hsiao, K. K., Cass, C., Schellenberg, G. D., Bird, T., Devine-Gage, E., Wisiniewski, H., et al. (1991) A prion protein variant in a family with the telencephalic form of Gerstmann-Sträussler-Scheinker syndrome. *Neurology* **41,** 681–684.

36. Chapman, J., Brown, P., Goldfarb, L. G., Arlazoroff, A., Gajdusek, D. C., and Korczyn, A. D. (1993) Clinical heterogeneity and unusual presentations of Creutzfeldt-Jakob disease in Jewish patients with the PRNP codon 200 mutation. *J. Neurol. Neurosurg. Psychiat.* **56,** 1109–1112.

37. Medori, R., Tritschler, H. J., LeBlanc, A., Villare, F., Manetto, V., Chen, H. Y., et al. (1992) Fatal familial insomnia, a prion disease with a mutation at codon 178 of the prion protein gene. *N. Engl. J. Med.* **326,** 444–449.

38. Lugaresi, E., Medori, R., Montagna, P., Baruzzi, A., Cortelli, P., Lugaresi, A., et al. (1986) Fatal familial insomnia and dysautonomia with selective degeneration of thalamic nuclei. *N. Engl. J. Med.* **315(16),** 997–1003.

39. Collinge, J., Brown, J., Hardy, J., Mullan, M., Rossor, M. N., Baker, H., et al. (1992) Inherited prion disease with 144 base pair gene insertion 2. Clinical and pathological features. *Brain* **115**, 687–710.

40. La Spada, A. R., Paulson, H. L. and Fischbeck, K. H. (1994) Trinucleotide repeat expansion in neurological disease. *Ann. Neurol.* **36**, 814–821.

41. Poulter, M., Baker, H. F., Frith, C. D., Leach, M., Lofthouse, R., Ridley, R. M., et al. (1992) Inherited prion disease with 144 base pair gene insertion 1. Genealogical and molecular studies. *Brain* **115**, 675–685.

42. Duchen, L. W., Poulter, M., and Harding, A. E. (1993) Dementia associated with a 216 base pair insertion in the prion protein gene. *Brain* **116**, 555–567.

43. Kitamoto, T., Amano, N., Terao, Y., Nakazato, Y., Isshiki, T., Mizutani, T., et al. (1993) A new inherited prion disease (PrP-P105L mutation) showing spastic paraparesis. *Ann. Neurol.* **34**, 808–813.

44. Kitamoto, T., Iizuka, R., and Tateishi, J. (1993) An amber mutation of prion protein in Gerstmann-Sträussler syndrome with mutant PrP plaques. *Biochem. Biophys. Res. Commun.* **192**, 525–531.

45. Brown, P. (1988) The clinical neurology and epidemiology of Creutzfeldt-Jakob disease, with special reference to iatrogenic cases, in *Novel Infectious Agents and the Central Nervous System* (Bock, G. and Marsh, J., eds.), Wiley, Chichester, pp. 3–23.

46. Martínez-lage, J. F., Poza, M., Sola, J., Tortosa, J. G., Brown, P., Cervenáková, L., et al. (1994) Accidental transmission of Creutzfeldt-Jakob disease by dural cadaveric grafts. *J. Neurol. Neurosurg. Psychiat.* **57**, 1091–1094.

47. Tanaka, M., Iizuka, O., and Yuasa, T. (1992) Hepatic dysfunction in Creutzfeldt-Jakob disease. *Neurology* **42**, 1249.

48. Tanaka, M. (1993) Liver in Creutzfeldt-Jakob disease. *Neurology* **43**, 457.

49. Uchino, A., Yoshinaga, M., Shiokawa, O., Hata, H., and Ohno, M. (1991) Serial MR imaging in Creutzfeldt-Jakob disease. *Neuroradiology* **33**, 364–367.

50. Milton, W. J., Atlas, S. W., Lavi, E., and Mollman, J. E. (1991) Magnetic resonance imaging of Creutzfeldt-Jakob disease. *Ann. Neurol.* **29**, 438–440.

51. Onofrj, M., Fulgente, T., Gambi, D., and Macchi, G. (1993) Early MRI findings in Creutzfeldt-Jakob disease. *J. Neurol.* **240**, 423–426.

52. Graham, G. D., Petroff, O. A. C., Blamire, A. M., Rajkowska, G., Goldman-Rakic, P., and Prichard, J. W. (1993) Proton magnetic resonance spectroscopy in Creutzfeldt-Jakob disease. *Neurology* **43**, 2065–2068.

53. Harrington, M. G., Merril, C. R., Asher, D. M., and Gajdusek, D. C. (1986) Abnormal proteins in the cerebrospinal fluid of patients with Creutzfeldt-Jakob disease. *N. Engl. J. Med.* **315**, 279–283.

54. Blisard, K. S., Davis, L. E., Harrington, M. G., Lovell, J. K., Kornfeld, M., and Berger, M. L. (1990) Pre-mortem diagnosis of Creutzfeldt-Jakob disease by detection of abnormal cerebrospinal fluid proteins. *J. Neurol. Sci.* **99**, 75–81.

55. Esmonde, T. G. and Will, R. G. (1992) Creutzfeldt-Jakob disease in Scotland and Northern Ireland. *Scot. Med. J.* **37**, 181–184.

56. Advisory Committee on Dangerous Pathogens (1994) Precautions for work with human and animal transmissible spongiform encephalopathies. HMSO, London.

57. Palmer, M. S., Dryden, A., Hughes, J. T., and Collinge, J. (1991) Homozygous prion protein genotype predisposes to sporadic Creutzfeldt-Jakob disease. *Nature* **352,** 340–342.

58. Laplanche, J.-L., Delasnerie-Lauprêtre, N., Brandel, J. P., Chatelain, J., Beaudry, P., Alpérovitch, A., et al. (1994) Molecular genetics of prion diseases in France. *Neurology* **44,** 2347–2351.

59. Collinge, J. and Palmer, M. (1991) CJD discrepancy. *Nature* **353,** 802.

60. Sanders, W. L. and Dunn, T. L. (1973) Creutzfeldt-Jakob disease treated with amantidine. A report of two cases. *J. Neurol. Neurosurg. Psychiat.* **36,** 581–584.

61. Masullo, C., Macchi, G., Xi, Y. G., and Pocchiari, M. (1992) Failure to ameliorate Creutzfeldt-Jakob disease with amphotericin B therapy. *J. Infect. Dis.* **165,** 784–785.

62. Caughey, B. (1994) Scrapie-associated PrP accumulation and agent replication: effects of sulphated glycosaminoglycan analogues. *Phil. Trans. R. Soc. Lond. B* **343,** 399–404.

63. Zerr, I., Bodemer, M., Räcker, S., Grosche, S., Poser, S., Kretzchmar, H. A., and Weber, T. (1995) Cerebrospinal fluid concentration of neuron-specific enolase in diagnosis of Creutzfeldt-Jakob disease. *Lancet* **345,** 1609,1610.

3

Neuropathological Diagnosis
of Human Prion Disease

Morphological Studies

James W. Ironside

1. Introduction

The neuropathology of the classical human prion diseases, Creutzfeldt-Jakob disease (CJD), Gerstmann-Sträussler-Scheinker syndrome (GSS), and kuru, is characterized by four main features: spongiform change, neuronal loss, gliosis, and amyloid plaque formation *(1–3)*. These features are shared with prion diseases in animals; the recognition of these similarities prompted the first attempts to transmit a human prion disease (kuru) to a primate in 1966 *(4)*, followed by CJD in 1968 *(5)* and GSS in 1981 *(6)*. These neuropathological features have formed the basis of the morphological diagnosis of human prion diseases for many years, although it was recognized that these changes are enormously variable both from case to case and within the central nervous system (CNS) in individual cases *(7)*. In this respect, it is interesting to note that the original case reported by Creutzfeldt *(8,9)* and two of the original cases reported by Jakob *(10–12)* do not show any of these characteristic neuropathological features; the diagnosis in these cases remains uncertain on review *(7,13)*. However, at least two of Jakob's original cases show typical neuropathological changes and other cases subsequently reported from his laboratory (including the members of the Backer family) also exhibited classical histological features *(12,14)*. It is also of interest to note that prion protein (PrP) gene analysis has recently been performed on one of the Backer family cases, showing a codon 178 Asn mutation with met/val at codon 129 *(15)*. This genotype has been described in other familial forms of human prion disease *(16)* *(see* Chapter 2).

From: *Methods in Molecular Medicine: Prion Diseases*
Edited by: H. Baker and R. M. Ridley Humana Press Inc., Totowa, NJ

Table 1
Classification of Human Prion Diseases

Creutzfeldt-Jakob disease	Sporadic
	Familial
	Iatrogenic
Gerstmann-Sträussler-Scheinker disease	Classical
	Variants with neurofibrillary tangles
Kuru	
Fatal familial insomnia	
Atypical prion dementia	

Early clinical and neuropathological reports on human prion diseases suffered from a confusion of nomenclature, in which the significance of the diagnostic feature of spongiform change occasionally was overlooked *(14)*. The subsequent demonstration that human prion diseases were transmissible reinforced the importance of spongiform change as a diagnostic neuropathological feature, reflected in the use of the term "spongiform encephalopathy" for this group of disorders *(7,13)*. Recent substantial advances in the understanding of the infectious agent, in particular the central role of prion protein in relation to transmissibility *(17–19)* along with increasing knowledge on the pathogenetic significance of mutations and polymorphisms in the human prion protein gene *(20)* have prompted a re-evaluation of classical neuropathology in this group of diseases, and a tendency to use the generic term "prion disease" rather than spongiform encephalopathy *(21)*.

Neuropathological assessment of the structural changes in the CNS has been the mainstay in diagnosis of human prion diseases for many years *(1,3)*. A new range of investigative techniques, including PrP gene analysis, PrP immunocytochemistry and detection by the Western blot, histoblot, and immunoblot techniques, prion rod/SAF detection by electronmicroscopy, and transmissibility to both wild-type and transgenic laboratory animals all now have diagnostic applications *(17)*. In the laboratory investigation of human prion diseases a combined morphological, immunocytochemical, and molecular genetic approach is desirable. However, many cases can be diagnosed on morphological assessment alone, including the vast majority of cases of sporadic CJD *(1,2)*.

Historically, human prion diseases comprised Creutzfeldt-Jakob disease in sporadic and familial forms, GSS, a rare inherited disease, and kuru, which was confined to the Fore tribes in New Guinea. A modified classification for human prion diseases to incorporate recent entities recognized by clinical, immunocytochemical, and molecular biological studies in addition to classical neuropathology is given in Table 1. This classification will be employed when discussing morphological aspects of diagnosis.

2. Neuropathological Diagnosis

2.1. Autopsy Sampling

In the vast majority of cases the neuropathological diagnosis of human prion diseases is based on the examination of the fixed brain and spinal cord following removal at autopsy. Cortical biopsies are now rarely performed for the diagnosis of dementia; current guidelines in the United Kingdom require neurosurgical instruments to be destroyed after use on a suspected case of CJD *(22)*. Furthermore, the marked variability in the distribution of the cortical lesions in CJD makes for difficulty in choosing a biopsy site, since a negative result could be potentially misleading. There is also evidence to suggest that patients with CJD experience a significant clinical deterioration following brain biopsy *(23)*.

Autopsy procedures for cases of human prion disease must take into account the information known on the transmissibility of the agents responsible for these disorders and the specialized techniques required for decontamination of the mortuary environment and mortuary equipment following autopsy. In the United Kingdom, the agent responsible for human prion diseases has recently been reclassified as a Category 3 pathogen *(24)*. Provided adequate mortuary facilities exist, a full autopsy may be performed in these cases. However, when facilities and/or staff are limited, then autopsy may be restricted to the removal of the brain *(25)*. This can be accomplished with minimal contamination to the immediate environment using tools that can be decontaminated according to recommended standards *(24)* (*see* Table 2).

2.1.1. Brain Fixation, Examination, and Sampling

After removal at autopsy, the unfixed brain should be sampled if fresh frozen tissue is required for molecular biological or transmission studies. Because of the variability in distribution of the lesions in the CNS, it is advisable to sample at least several cortical areas, the cerebellum, and brainstem. The brain should then be fixed in 10% formalin (or equivalent) for a period of 2–3 wk prior to dissection. In the past, it has been recommended that phenol be added to the formalin solution to help inactivate the transmissible agent. Phenol is now known to be ineffective against the agents responsible for human prion diseases, and can frequently impair tissue fixation and morphology, thereby hindering diagnosis *(24)*. Fixation in formalin will substantially reduce but not abolish the potential infectivity of CNS tissues in prion diseases, so the fixed brain should be dissected in a Class 1 safety cabinet to minimize contamination of the environment *(24)*. Since the greatest risk of infectivity results from inoculation of CNS tissues, it is recommended that dissection of the brain be performed while wearing gloves that will offer protection against cuts and

Table 2
Autopsy and Decontamination Procedures for Human Prion Diseases

Autopsy procedures
    Access to a "high-risk" autopsy room is desirable but not essential
    Fully trained medical and technical staff required
    Restricted access to autopsy room during the procedure
    Disposable protective clothing to be worn
    Open body bag recommended to minimize contamination of the autopsy room
    All disposable clothing and disposable instruments to be incinerated on completion
    Nondisposable instruments should be autoclaved or decontaminated *(see the
        following)* on completion
    Working surfaces to be decontaminated with sodium hypochlorite containing
        20,000 ppm chlorine *(see the following)* on completion
Decontamination
    Heat-stable equipment and nondisposable protective clothing
    Porous load autoclave
        A single cycle 134°C (+4/0) (30 lb psi) 18 min (holding time at temperature)
        Six temperature cycles 134°C (+4/0) (30 lb psi) 3 min (holding time at temperature)
    Nondisposable nonheat stable equipment and work surfaces
        At least 1 h exposure to sodium hypochlorite containing 20,000 ppm available
           chlorine, with repeated wetting
        Exposure to 2*M* sodium hydroxide with repeated wetting
    All disposable equipment, temporary bench coverings, contaminated fluids, and
        tissues to be incinerated

inoculation. Fine chain-mail gloves are available for this purpose *(25)*; these have the additional advantage of being readily autoclaved at the required temperature for decontamination.

Macroscopic examination of the brain in human prion diseases will usually yield little in the way of specific findings *(1,3)*. Most cases of CJD exhibit cortical atrophy in a global distribution, which is often accompanied by cerebellar atrophy, particularly in the superior vermis. Cases of GSS and iatrogenic (pituitary hormone-associated) CJD may show disproportionate atrophy of the cerebellar hemispheres in addition to the vermis, often with relative sparing of the cerebral cortex *(2)*. Similar changes have been described in kuru *(26)*. In CJD, the brain may exhibit a range of age-associated abnormalities, which include meningeal fibrosis and atheroma affecting the circle of Willis. A small but significant percentage of CJD cases occur in more elderly individuals, in whom the brain may exhibit a more severe degree of age-related changes both on external inspection and on microscopy *(see* Table 3).

On coronal sectioning, the deep gray matter structures (basal ganglia and thalamus) may also appear atrophied, although this is also a variable finding

Table 3
Age-Related Changes in the CNS

Macroscopic	Meningeal thickening and fibrosis
	Atheroma/arteriosclerosis in the circle of Willis
	Mild cerebral cortical and hippocampal atrophy
	Depigmentation of substantia nigra and locus ceruleus
	Ventricular dilatation
	Enlarged perivascular spaces in deep gray and white matter
Microscopic	Variable cerebral cortical and hippocampal neuronal loss
	Hippocampal Aβ plaques and neurofibrillary tangles
	Occasional Lewy bodies in substantia nigra and locus ceruleus
	Perivascular lacunes in deep gray matter
	Cerebral arteriolosclerosis and calcification

(1,3). The white matter of the cerebral hemispheres often appears normal, but will be reduced in bulk if severe cerebral cortical atrophy is present; these features will also be accompanied by ventricular dilatation. Other age-related changes may be noted in sections of the cerebrum and brainstem, including a loss of pigment in the substantia nigra and locus ceruleus, hippocampal atrophy, and enlargement of the perivascular space around small blood vessels within the thalamus, central white matter, and basal ganglia (*see* Table 3).

The markedly variable distribution of the characteristic pathological features in human prion diseases necessitates extensive and comprehensive histological sampling. Representative material should be examined from the frontal lobes (including the enterorhinal cortex), temporal lobes and insula, parietal parasaggital and convexity regions, and occipital lobes to include the visual cortex. The hippocampi should be examined, along with representative regions of the basal ganglia (to include the caudate nucleus, putamen, globus pallidus, amygdala, and basal nucleus), thalamus, and hypothalamus. Brainstem sections should include the midbrain, pons (including the locus ceruleus), and medulla. Cerebellar blocks should be taken to include the vermis and both hemispheres (comprising the cortex, white matter, and dentate nucleus). In most cases of CJD the characteristic neuropathological changes are present bilaterally, although not always in a symmetrical fashion *(1,2)*, so it is necessary to ensure that both cerebral and cerebellar hemispheres are sampled.

The spinal cord in human prion diseases usually shows no significant external abnormality. The dorsal root ganglia, dura mater, and spinal nerve roots usually appear normal and no consistent abnormalities are present on the external surface of the spinal cord. On cross-section, the spinal cord also appears unremarkable in most cases. Blocks for histology should be taken from representative regions of the spinal cord, dorsal root ganglia, spinal nerve roots, and cauda equina.

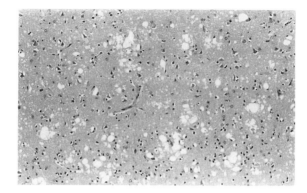

Fig. 1. Spongiform change in CJD consists of numerous rounded vacuoles within the neuropil that occur both singly and in confluent groups, distorting the cortical cytoarchitecture. Hematoxylin and eosin.

Only by performing extensive histological sampling will the full nature and extent of any histological abnormalities be identified. This is particularly important for clinicopathological correlation and will also facilitate differential diagnosis. The most common disorders that can be confused clinically with CJD include Alzheimer's disease, Pick's disease, and diffuse Lewy body disease *(27)*. Using the appropriate histological techniques, the mentioned sampling protocol will allow a full assessment of these possibilities.

2.2. Histological Investigation

2.2.1. Sporadic CJD

Most neuropathological studies in human prion diseases are performed on paraffin-embedded tissues. Tissue blocks from the CNS and other organs can be decontaminated in 96% formic acid for 1 h *(28)* prior to processing into paraffin wax. Sections are then cut at 5 μm in thickness and stained for routine analysis with hematoxylin and eosin. Spongiform change is most easily recognized at this tissue section thickness; the use of thicker (10 μm or more) sections carries the danger of misinterpretation of other sponge-like changes in the cerebral cortex *(see the following)*.

In most cases of human prion diseases the histological features are distinctive and will allow a diagnosis to be reached without undue difficulty. In CJD, the most consistent histological abnormality is spongiform change *(29)*, which is characterized by a fine vacuole-like appearance in the neuropil, with vacuoles varying from 20–200 μm in diameter (Fig. 1). These vacuoles can appear in any layer of the cerebral cortex and may become confluent, resulting in large vacuoles that substantially distort the cortical cytoarchitecture. Vacuola-

Fig. 2. Spongiform change in the cerebellum comprises multiple small vacuoles in the molecular layer that usually do not appear confluent, with relatively little distortion of the tissue architecture. Hematoxylin and eosin.

tion may also be seen within the cytoplasm of larger neurons, particularly in layers 3 and 5 within the cortex. Cortical involvement is detectable in most cases of CJD, and is usually accompanied by spongiform change in the basal ganglia, thalamus, and cerebellar cortex. Cerebellar involvement is present in most cases, although the severity and distribution of the spongiform change within the cerebellum is markedly variable *(29)*. Confluent spongiform change is unusual in the cerebellum, which may, however, exhibit a widespread microvacuolar change with smaller vacuoles 20–50 μm in diameter in the molecular layer (Fig. 2).

In long-standing cases, the neuronal loss and spongiform change maybe so severe as to result in status spongiosus *(13)*, where widespread coarse vacuolation is present throughout the cerebral cortex, resulting in collapse of the cortical cytoarchitecture, leaving an irregular distorted rim of gliotic tissue containing few remaining neurons (Fig. 3). The basal ganglia and thalamus also may exhibit severe neuronal loss with gliosis and atrophy, and in the cerebellum there is often an irregular loss of neurons in the granular cell and Purkinje cell populations, with proliferation of Bergmann and radial glia. Spongiform change in most brain regions is accompanied by neuronal loss and gliosis involving both astrocytes and microglia (Figs. 4 and 5). Microglial hypertrophy and hyperplasia occur in a widespread distribution within the CNS in CJD *(30)*, but no evidence of a classical inflammatory or immune response occurs. Microglia are also implicated in the pathogenesis of PrP plaques *(see the following) (31)*.

A range of clinical and neuropathological variants in sporadic CJD have been described *(7,32,33)*, the most striking of which (the panencephalic vari-

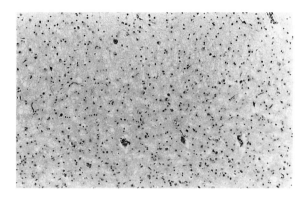

Fig. 3. Status spongiosis in a case of long-standing CJD shows extensive neuronal loss in the cerebral cortex with a widespread mesh-like distribution of irregular vacuoles, accompanied by reactive gliosis. The cortical cytoarchitecture has collapsed (compare with Fig. 1). Hematoxylin and eosin.

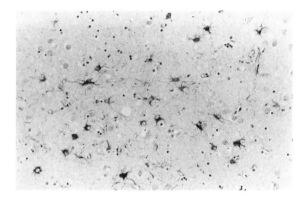

Fig. 4. Astrocytosis is widespread in CJD, with increased numbers of enlarged astrocytes in the cerebral cortex. Astrocyte processes extend around spongiform vacuoles and degenerate neurons. Immunocytochemistry for glial fibrillary acidic protein.

ant) *(34)* contains extensive necrotizing lesions in the white matter. The histological features in these rare variants have been described largely on the basis of traditional neuropathological staining techniques; further immunocytochemical studies for PrP are required in order to more fully characterize these unusual findings.

Around 10% of CJD cases contain PrP plaques that are usually visible as rounded eosinophilic structures *(1,3)*. These are most frequently observed in the cerebellum, where they usually occur as unicentric plaques with a hyaline eosi-

Fig. 5. Increased numbers of microglia within the cerebral cortex in CJD are visualized as irregular dark structures on immunocytochemistry for CD68. Microglial cell bodies are evenly distributed throughout the cortex, with cell processes occurring adjacent to small spongiform vacuoles.

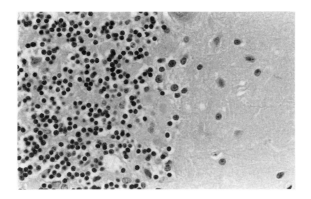

Fig. 6. Cerebellar amyloid plaques in sporadic CJD consist of a highline core surrounded by a pale halo (center). These structures are more easily observed on immunocytochemistry for PrP. Hematoxylin and eosin.

nophilic core and a paler halo (Fig. 6). In kuru, these plaques often showed a peripheral margin of radiating fibrils *(26)*, and similar changes can be identified within plaques in sporadic *(2,35)* and familial CJD *(20)*. Plaques are identified out with the cerebellum in a minority of these cases in the thalamus, basal ganglia, or cerebral cortex. The occurrence of plaques is related to PrP genotype and is associated with codon 129 polymorphism (being more common in met/val or val/val genotypes) *(35)*; PrP plaques have also been reported in familial CJD in association with several different PrP gene mutations *(20)*.

Ultrastructural and immunocytochemical studies in both human and animal prion diseases have demonstrated that microglial cells are intimately involved in PrP plaque formation, and may perhaps play a role in the processing of PrP into an amyloid structure, as is thought to occur with Aβ protein in Alzheimer's plaques *(31,36)*. PrP plaques share other structural similarities to Alzheimer's Aβ plaques, including the formation of neuritic processes in the plaque periphery *(2)*. Furthermore, Aβ protein has been detected immunocytochemically in the periphery of some PrP plaques, although the mechanisms of this deposition are as yet unclear *(37,38)*.

2.2.2. Familial CJD

Molecular biological studies on human prion diseases have revealed an ever-increasing number of pathogenic abnormalities in the PrP gene on chromosome 20 *(19,20)*. Some of these genetic abnormalities are associated with a clinical phenotype indistinguishable from sporadic CJD, whereas others exhibit more distinct clinical features, including predominantly cerebellar and extrapyramidal signs and symptoms. Many of these genetic abnormalities are also associated with characteristic pathological features, in particular the presence of PrP plaques, although considerable pathological and clinical variation may occur within affected families *(20)* (*see* Table 4).

2.2.3. Iatrogenic CJD

The neuropathological changes in iatrogenic CJD are variable in their character and distribution; this variability may depend on the route of inoculation of the agent. In patients who have acquired CJD following a dura mater graft, or implantation of intracerebral electrodes, the distribution of lesions in the CNS is very similar to those described herein for sporadic CJD, with predominantly cerebral cortical pathology *(39)*. However, in patients who develop CJD as a consequence of peripheral inoculation with pituitary hormone preparations a different pattern of pathology is observed, which in some cases bears more similarity to kuru than sporadic CJD *(40–42)*. In these patients, the cerebellum bears the brunt of neuropathological changes and often is markedly atrophic with correspondingly severe neuronal loss and gliosis (Fig. 7). PrP plaques are frequently found in this region, along with a more widespread deposition of PrP within the granular layer *(42)* (*see* Chapter 4). Although spongiform change may be observed in the cerebral cortex in such cases, it is usually not as widespread as in sporadic CJD. Spongiform change in the basal ganglia, however, is common and usually of marked severity. Examination of the spinal cord in pituitary hormone-associated cases often shows plaque-like accumulations of PrP *(43)*, with PrP deposition in the substantia gelatinosa and ascending pathways (*see* Chapter 4). However, it must be emphasized that these

Table 4
Relationship Between Pathology
and PrP Genotype in Human Prion Diseases

Pathology	PrP genotype
CJD	No mutation in open reading frame
	129 Met/Met, Met/Val
	178 Asp > Asn, 129 Met/Val
	180 Val > Ile
	200 Glu > Lys
	210 Val > Ile
	219 Glu > Lys
	232 Met > Arg
	Octapeptide repeat region
	(codons 51–91)
	48 bp insert
	96 bp insert
	120 bp insert
	144 bp insert
	168 bp insert
	192 bp insert
	216 bp insert
CJD with PrP plaques	129 Met/Val, Val/Val
GSS	102 Pro > Leu
	105 Pro > Leu
	117 Ala > Val
	145 Tyr > stop (TAG)
GSS with neurofibrillary tangles	198 Phe > Ser
	217 Gln > Arg
FFI	178 Asp > Asn, 129 Met/Met
Atypical prion dementia	Octapeptide repeat region
	144 bp insert

spinal cord abnormalities are not specific for iatrogenic CJD cases and similar patterns of PrP accumulation have been observed in some cases of sporadic CJD *(43)*.

2.2.4. Kuru

In kuru, PrP plaque formation was the most striking abnormality, and was recorded as being present within the cerebellum in at least 70% of cases *(26)*. The typical kuru plaque is described in Section 2.2.1. and although not confined to the cerebellum was most easily identified at this site. Spongiform change was not identified in the earliest cases of kuru *(44)*, but subsequent

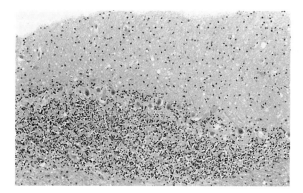

Fig. 7. Iatrogenic CJD occurring in human growth hormone recipients shows severe neuronal loss in the cerebellum involving the granular and Purkinje cell layers, with widespread gliosis and severe spongiform change in the molecular layer (compare with Fig. 2). Hematoxylin and eosin.

investigations showed it to be present in a variable distribution in most cases *(26,45)*. It remains uncertain as to whether this observation was a true reflection of a change in the spectrum of pathology occurring in kuru.

2.2.5. GSS

In GSS, the characteristic abnormality is the multicentric PrP plaque, which differs in structure from the smaller unicentric or kuru plaque *(46,47)*. As their name implies, multicentric plaques are larger, more diffuse structures that have several core-like areas that give a positive staining reaction with Congo red, Sirius red, and other tinctorial stains *(48)*. These plaques are present in large numbers in the cerebellum, within the molecular layer, granular layer, and occasionally in the white matter. However, they are widespread also in the cerebral hemispheres involving the cerebral cortex, basal ganglia, thalamus, and hypothalamus. Spongiform change occurs to a variable degree in GSS; the cerebellar molecular layer is often devoid of spongiform change, but it has been described in the cerebral cortex, basal ganglia, and thalamus in some GSS cases *(47)*. There is an interesting relationship between the presence of spongiform change and transmissibility in GSS; cases that show little or no spongiform change (although PrP plaques are present) appear to transmit to animals with difficulty or not at all, whereas cases with spongiform change are more readily transmissible *(49)*. The cerebellum in GSS often shows severe neuronal loss and gliosis, which is disproportionate to the neuronal loss occurring in the cerebrum or spinal cord. In the Indiana variant of GSS, neurofibrillary tangles are present within a wide distribution in the cerebral cortex *(47,50)*.

This striking abnormality usually occurs in the absence of any other features suggestive of Alzheimer's disease.

2.2.6. Fatal Familial Insomnia

Recent studies employing molecular genetic and immunocytochemical investigations have identified a spectrum of human prion diseases that is broader than hitherto suspected, resulting in the description of novel disease entities, including fatal familial insomnia *(50,51)* (FFI) and atypical prion dementia *(52)*. FFI is characterized clinically by progressive insomnia and extrapyramidal neurological abnormalities, often in the absence of overt dementia *(51,53)*. This disease is associated with a unique PrP genotype, codon 178Asn, and met/met at codon 129. The neuropathology of this entity is characterized by thalamic gliosis, particularly in the dorsomedial nuclei, often in the absence of spongiform change and in disproportionate severity to neuronal loss. Spongiform change has been described in the cerebral cortex in occasional patients with FFI, although this is not a prominent feature *(53)*. PrP plaque deposition has not been described in any case to date, although cerebellar atrophy and neuronal loss with gliosis in the inferior olivary nuclei have been reported in addition to other minor and inconstant histological abnormalities *(51)*. Thalamic gliosis in FFI is the most characteristic neuropathological feature, although the distribution of lesions within the thalamus and associated structures does not readily correlate with the predominant clinical features in each case. It is also interesting to note that PrP deposition in this disorder is most evident within the thalamus, both on Western blot studies and immunocytochemistry *(54)* *(see* Chapter 4). Attempts to transmit disease to animals from cases of FFI have recently been successful *(55)*, further confirming this entity as a prion disease.

2.2.7. Atypical Prion Diseases

The term atypical prion dementia was employed to describe dementing illnesses occurring in young adults within a large family with an inherited prion disorder characterized by a 144-bp insert in the PrP gene *(52)*. Initial descriptions of the neuropathology in these individuals indicated that no significant abnormalities were present, although there was a generalized mild cerebral atrophy with variable neuronal loss in the cerebrum and cerebellum. Spongiform change was not a prominent feature, but immunocytochemistry for PrP yielded a striking pattern of PrP accumulation in the cerebellum *(56)* *(see* Chapter 4). Other members of this large kindred exhibited classical neuropathological features of CJD, reinforcing the phenotypic variability that may occur in human prion diseases. Although it has been suggested that atypical prion dementia may account for many cases of dementia that are not readily catego-

Table 5
Histological Changes in Human Prion Disease

Classical changes	Spongiform change
	Neuronal loss
	Astrocytosis
	PrP plaques
Other changes	Status spongiosus
	Swollen neurons
	Abnormal neuritic dendrites
	Neuritic dystrophy around PrP plaques
	White matter necrosis and cavitation
	Microgliosis
	Aβ protein amyloid angiopathy

rized into precise clinicopathological entities *(57)*, subsequent investigations on a range of other neurodegenerative disorders have failed to provide any evidence to support this claim. Since cases of atypical prion dementia appear to occur exclusively within the context of inherited human prion disease, it would seen unlikely that similar cases account for a large proportion of human dementias.

2.3. Other Histological Features in Human Prion Diseases

A range of other histological abnormalities have been described in human prion diseases, some of which relate to the effect of aging on the human brain and have no specific association with this group of disorders. These abnormalities include swollen cortical neurons *(58,59)* and amyloid angiopathy *(60,61)*; other changes are summarized in Table 5. It is critically important to be aware of age-associated histological abnormalities in the human brain and not to interpret these as being indicative of a coexisting CNS disorder (*see* Table 3). Alzheimer disease and CJD been described concurrently, although this appears to be an exceptional event *(62)*. Other more apparently specific abnormalities have been described in cases of CJD that have been studied by ubiquitin immunocytochemistry *(63)*. These abnormalities include dot-like ubiquitinated structures in the neuropil, within neurons and around PrP plaques. The latter probably represent dystrophic neurites, whereas the intracellular lesions may represent lysosomal structures, as suggested by animal scrapie models *(64)*.

2.4. Other Conditions Associated with Spongiform-Like Change in the CNS

The importance of spongiform change as one of the histological hallmarks of human prion diseases is widely accepted, and clear distinctions have been

Table 6
Pathological Features Similar
to Spongiform Change Seen in Other Neurodegenerative Diseases

CNS disorders with focal spongiform change	Alzheimer's disease
	Pick's disease
	Diffuse Lewy body disease
	Dementia in motor neuron disease
	Dementia of frontal lobe type
Sponge-like changes in other CNS disorders	
Gray matter	Status spongiosis
	Edema
	Metabolic encephalopathies
	Neuronal storage disorders
	Tissue fixation and processing
	Artifacts
White matter	Edema
	Ischemia
	Metabolic encephalopathies
	Canavan disease
	Spongy degeneration of the white matter in infancy
	Tissue fixation and processing artifacts

made between spongiform change and status spongiosis. Status spongiosis can result from any neurodegenerative disorder that results in widespread neuronal death and collapse of the cerebral cortical cytoarchitecture. It is encountered commonly in Pick's disease and may also occur in Alzheimer's disease, cortical ischemia, and as a consequence of viral encephalitis. Spongiform change *per se* is not pathonomonic for human prion diseases. Appearances identical to spongiform change have been described in other neurodegenerative disorders (*see* Table 6), particularly in Alzheimer's disease and diffuse Lewy body disease *(65)* (Fig. 8). In these disorders spongiform change is usually confined to layer 2 of the cerebral cortex and is present in a restricted distribution in the frontal lobes, cingulate gyrus, temporal poles, and inferior temporal cortex. The basal ganglia, thalamus, hypothalamus, cerebellum, and spinal cord in these disorders do not show spongiform change. In occasional cases of Alzheimer's disease and diffuse Lewy body disease the spongiform change may be particularly conspicuous and necessitate widespread histological sampling along with additional investigative techniques, including PrP immunocytochemistry, to investigate the possibility of human prion disease. It should also be recalled that other neurodegenerative disorders resulting in focal

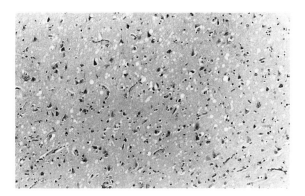

Fig. 8. Spongiform change in diffuse Lewy body disease tends to occur in a characteristic distribution in the inferior frontal and temporal lobes. Numerous small vacuoles are present throughout the cortex, but there is usually less evidence of confluent vacuolation in this condition than in sporadic CJD (compare with Fig. 1). Hematoxylin and eosin.

spongiform-like change are associated with distinctive lesions that help clarify diagnosis. Other disorders that may cause a sponge-like appearance in the CNS are listed in Table 6. In these disorders, confusion with spongiform change in human prion disease is not a major difficulty because of the clinical and pathological context in which these changes occur (Figs. 9 and 10).

3. Relation of Neuropathology to Host Genotype

As mentioned earlier, the molecular biological studies on human prion disorders have revealed a wide range of abnormalities in the PrP gene in familial prion disorders. The importance of the influence of PrP genotype on both clinical and pathological features is well illustrated by the effect of the codon 129 genotype in patients with a codon 178Asn mutation, since this influences presentation either as familial CJD (met/val genotype) or FFI (met/met genotype) *(54)*. Even in the absence of PrP gene mutations, codon 129 genotype appears to confer disease susceptibility for both sporadic and iatrogenic CJD *(66,67)*; codon 129 genotype also influences PrP plaque formation in sporadic CJD, which occurs most frequently in patients with a met/val or val/val genotype *(67)*. Other genes may also influence clinical and pathological features in human prion diseases, including the apolipoprotein E gene on chromosome 19 *(35)*. As in Alzheimer disease, the ApoE e4 genotype is more common in cases of sporadic CJD than in age-matched normal control cases, and in affected individuals the ApoE e2 allele appears to be associated with a more prolonged clinical course, independent of any PrP gene abnormalities *(68)*. Whether these

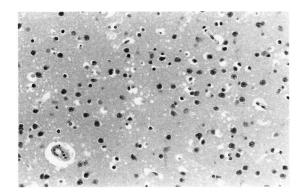

Fig. 9. Cerebral edema can occasionally result in multiple irregular vacuoles within the neuropil of the cerebral cortex, accompanied by a marked retraction artifact around blood vessels and individual cell bodies. This usually can be distinguished easily from spongiform change (compare with Fig. 1). Hematoxylin and eosin.

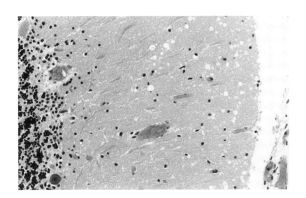

Fig. 10. Cerebellar edema occurring with hypoxic brain damage occasionally results in the appearance of multiple small vacuoles in the molecular layer, particularly in its superficial aspect (compare with Figs. 2 and 7). Hematoxylin and eosin.

findings are relevant to disease susceptibility is uncertain, and the influence of ApoE genotype on neuropathology has not yet been fully evaluated.

4. Recent Developments in Neuropathology

Neuropathological studies in human prion diseases have been greatly facilitated by the development of reliable techniques for PrP immunocytochemistry *(69)* (*see* Chapter 4). This technique has revealed an entirely unsuspected spectrum of pathology, although the relationship between PrP deposition and other classical neuropathological features is as yet unclear. In order to address this question, and to help provide a more precise analysis of neuropathological changes in CJD, computer-based image analysis techniques have been developed to evaluate the neuropathology of human prion diseases in a subjective manner (70). To date, successful programs have been developed to quantitatively assess spongiform change, PrP deposition as revealed by immunocytochemistry, and astrocytosis. Preliminary application of these validated techniques to human prion diseases has demonstrated a variation in the relative involvement of the cerebral cortex and basal ganglia in terms of PrP deposition, spongiform change, and astrocytosis in relation to PrP genotype. One major advantage of this automated quantitative approach is the ability to analyze numerous large sections of the human brain, allowing a representative survey of neuropathology without observer bias or fatigue.

Other recent neuropathological investigations in human prion diseases have included the use of confocal scanning laser microscopy (71) to investigate the complex structural changes occurring in relation to spongiform change and PrP plaque formation. The use of confocal laser microscopy with immunocytochemistry, and the ability to serially reconstruct three-dimensional images from thick (60-µm) brain sections will allow a more detailed understanding of cellular interactions in the development of classical neuropathological abnormalities. Electron microscopy in human prion diseases has not been employed as an investigative tool to a large extent (particularly in comparison with studies on the neuropathology of animal scrapie models) partly because of difficulties encountered with postmortem autolysis and the relative unavailability of rapidly fixed cortical biopsy specimens. However, immunocytochemical techniques for PrP in the murine scrapie model have been developed at the ultrastructural level (64) (see Chapter 4), which should be applied to the fine structural changes occurring in human prion disease in relation to PrP deposition. Ultrastructural investigations of animal and human prion diseases have reported the presence of tubulovesicular bodies (72,73), which are apparently specific for this group of disorders, although their precise nature and significance in relation to the transmissible agents responsible for these diseases remains uncertain.

5. Conclusion

The enormous increase in knowledge of the protein chemistry, biochemistry, and molecular biology of prion diseases in humans has been accompanied by a renewed interest in the neuropathology of this fascinating group of diseases. The broadening spectrum of human prion diseases encompasses a wider range of clinicopathological entities than suspected a decade ago, whereas the refinement of immunocytochemical techniques for PrP and other proteins in the brain has allowed a more detailed study of cellular reactions and structural abnormalities in the CNS. The development of quantitative techniques to analyze these neuropathological abnormalities will facilitate a further understanding of the relationship between genotype, clinical features, and neuropathology in sporadic, iatrogenic, and familial forms of human prion disease. The recognition that many of the neuropathological abnormalities in human prion diseases are not exclusive to this group of disorders, but may occur in other diseases *(21,74)*, has reinforced the concept of common mechanisms operating in a range of neurodegenerative disorders and has also prompted efforts to develop more specific neuropathological diagnostic investigations.

Acknowledgments

I am grateful to J. E. Bell, R. Will, R. de Silva, I. Goodbrand, and K. Sutherland for helpful discussion; to L. McCardle, C. Barrie, and D. Nicolson for technical assistance; and S. Honeyman for preparing the manuscript. The National CJD Surveillance Unit is supported by the Department of Health, BBSRC (AG 15/610) and MRC.

References

1. Bastian, F. O. (ed.) (1991) *Creutzfeldt-Jakob Disease and Other Transmissible Human Spongiform Encephalopathies.* Mosby Year Book, St. Louis, MO.
2. Bell, J. E. and Ironside, J. W. (1993) Neuropathology of spongiform encephalopathies in humans. *Br. Med. Bull.* **49**, 738–777.
3. Tomlinson, B. E. (1992) Ageing and the dementias, in *Greenfield's Neuropathology*, 5th ed. (Adams, J. H. and Duchen, L. W., eds.), Edward Arnold, London, pp. 1284–1410.
4. Gajdusek, D. C., Gibbs, C. J., and Alpers, M. P. (1966) Experimental transmission of a kuru-like syndrome in chimpanzees. *Nature* **209**, 794–796.
5. Gibbs, C. J., Gajdusek, D. C., Asher, D. M., Alpers, M. P., Beck, E., Daniel, P. M., et al. (1968) Creutzfeldt-Jakob disease (subacute spongiform encephalopathy): transmission to the chimpanzee. *Science* **161**, 388–389.
6. Masters, C. L., Gajdusek, D. C., and Gibbs, C. J. (1981). Creutzfeldt-Jakob disease: virus isolations from the Gerstmann-Sträussler syndrome. *Brain* **104**, 559–588.
7. Richardson, E. P., Jr. and Masters, C. L. (1995) The nosology of Creutzfeldt-Jakob disease and conditions related to the accumulation of PrP[CJD] in the nervous system. *Brain Pathol.* **5**, 33–41.
8. Creutzfeldt, H. G. (1920) Über eine eigenartige herdförmige Erkrankung des Zentralnervensystems. *Z. ges Neurol. Psychiat.* **57**, 1–18.
9. Creutzfeldt, H. G. (1921) Über eine eigeartige herdförmige Erkrankung des Zentralnervensystems, in *Histologische und Histopathologische Arbeiten über die Grosshirnrinde* (Nissl, F. and Alzheimer, A., eds.), Gustav Fischer, Jena, pp. 1–48.
10. Jakob, A. (1921) Über eigenartige Erkrankungen des Zentralnervensystems mit bemerkenswertem anatomischen Befunde. *Z. ges Neurol. Psychiat.* **64**, 147–228.
11. Jakob, A. (1921) Über eine der multiplen sklerose klinische nahestehende Erkrankung des Zentralnervensystems (spastische Pseudosklerose) mit bemerkenswertem anatomischem Befunde. *Med. Klinik* **17**, 372–376.

12. Jakob, A. (1923) Spastische Pseudosklerose, in *Monographien aus dem Gesamtgebiete der Neurologie und Psychiatrie, Die Extrapyramidalen Erkrankungen.* (Foester, O. and Wilmanns, K., eds.), Springer, Berlin, **37**, 212–245.

13. Masters, C. L., and Gajdusek, D. C. (1982) The spectrum of Creutzfeldt-Jakob disease and the virus-induced spongiform encephalopathies, in *Recent Advances in Neuropathology*, vol. 2 (Smith, W. T. and Canavagh, J. B., eds.), Churchill Livingstone, Edinburgh.

14. Meggendorfer, F. (1930) Klinische und genealogische beobachtungen bei einem Fall von spastischer Pseudosklerose Jakobs. *Z. ges Neurol. Psychiat.* **128**, 337–341.

15. Brown, P., Cervenakova, L., Boellaard, J. W., Stavrou, D., Goldfarb, L. G., and Gajdusek, D. C. (1994) Identification of a PRNP gene mutation in Jakob's original Creutzfeldt-Jakob disease family. *Lancet* **344**, 130–131.

16. Nieto, A., Goldfarb, L. G., Brown, P., McCombie, W. R., Trapp, S., Asher, D. M., et al. (1991) Codon 178 mutation in ethnically diverse Creutzfeldt-Jakob disease families. *Lancet* **337**, 622–623.

17. Prusiner, S. B. and DeArmond, S. J. (1991) Molecular biology and pathology of scrapie and the prion diseases of humans. *Brain Pathol.* **1**, 297–309.

18. DeArmond, S. L. (1993) Overview of the transmissible spongiform encephalopathies: prion protein disorders. *Br. Med. Bull.* **49**, 725–737.

19. Baker, H. F. and Ridley, R. M. (1992) The genetics and transmissibility of human spongiform encephalopathy. *Neurodegeneration* **1**, 3–16.

20. Tateishi, J. and Kitamoto, T. (1995) Inherited prion diseases and transmission to rodents. *Brain Pathol.* **5**, 53–59.

21. Ridley, R. M. (1994) Perceptions of prion disease. *J. Clin. Pathol.* **47**, 876–879.

22. Metters, J. S. (1992) Neuro and ophthalmic surgery procedures in patients with, or suspected to have, or at risk of developing CJD or GSS. *Letter PL(92)CO/4 to Consultants and Health Managers.* Department of Health, London.

23. Brown, P. and Cathala, F. (1979) Creutzfeldt-Jakob disease in France: I. Retrospective study of the Paris area during the ten year period 1968–1977. *Ann. Neurol.* **5**, 189–192.

24. Advisory Committee on Dangerous Pathogens (1994) *Precautions for Work with Human and Animal Transmissible Spongiform Encephalopathies.* HMSO, London.

25. Bell, J. E. and Ironside, J. W. (1993) How to tackle a possible Creutzfeldt-Jakob disease necropsy. *J. Clin. Pathol.* **46**, 193–197.

26. Klatzo, I., Gajdusek, D. C., and Vigas, V. (1959) Pathology of kuru. *Lab. Invest.* **8**, 799–847.

27. Will, R. G. (1991) The spongiform encephalopathies. *J. Neurol. Neurosurg. Psychiat.* **54**, 761–763.

28. Brown, P., Wolff, A., and Gajdusek, D. C. (1990) A simple and effective method for inactivating virus infectivity in formalin-fixed tissue samples from patients with Creutzfeldt-Jakob disease. *Neurology* **40**, 887–890.

29. Masters, C. L. and Richardson, E. P. (1978) Subacute spongiform encephalopathy (Creutzfeldt-Jakob disease). The nature and progression of spongiform change. *Brain* **101**, 333–334.

30. Sasaki, A., Hirato, J., and Nakazato, Y. (1993) Immunohistochemical study of microglia in the Creutzfeldt-Jakob diseased brain. *Acta Neuropathol.* **86**, 337–344.
31. Miyazono, M., Iwaki, T., Kitamoto, T., Kaneko, Y., Doh-ura, K., and Tateishi, J. (1991) A comparative immunohistochemical study of kuru and senile plaques with a special reference to glial reactions at various stages of amyloid plaque formation. *Am. J. Pathol.* **139**, 589–598.
32. Will, R. G. and Matthews, W. B. (1984) A retrospective study of Creutzfeldt-Jakob disease in England and Wales 1970–79. I: Clinical features. *J. Neurol. Neurosurg. Psychiat.* **47**, 134–140.
33. Mizutani, T. and Shiraki, H. (eds.) (1985) *Clinicopathological Aspects of Creutzfeldt-Jakob Disease.* Elsevier, Amsterdam.
34. Mori, S., Hamada, C., Kumanishi, T., Fukuhara, N., Ichihashi, Y., Ikuta, F., et al. (1989) A Creutzfeldt-Jakob disease agent (Echigo-1 strain) recovered from brain tissue showing the 'panencephalopathic type' disease. *Neurology* **39**, 1337–1342.
35. Pickering-Brown, S., Mann, D. M. A., Owen, F., Ironside, J. W., de Silva, R., Roberts, D. A., et al. (1995) Allelic variations in apolipoprotein E and prion protein genotype related to plaque formation and age of onset in sporadic Creutzfeldt-Jakob disease. *Neurosci. Lett.* **187**, 127–129.
36. Frackowiak, J., Wisniewski, H. M., Wegiel, J., Merz, G. S., Iqbal, K., and Wang, K. C. (1992) Ultrastructure of the microglia that phagocytose amyloid and the microglia that produce β-amyloid fibrils. *Acta Neuropathol.* **84**, 225–233.
37. Bugiani, O., Giaccone, G., Verga, L., Pollo, B., Frangione, B., Farlow, M. R., et al. (1993) βPP participates in PrP-amyloid plaques of Gerstmann-Straussler-Scheinker disease, Indiana kindred. *J. Neuropathol. Exp. Neurol.* **52**, 66–70.
38. Miyazono, M., Kitamoto, T., Iwaki, T., and Tateishi, J. (1992) Colocalization of prion protein and A4/β protein in the same amyloid plaques in patients with Gerstmann-Sträussler syndrome. *Acta Neuropathol.* **83**, 333–339.
39. Martinez-Lage, J. F., Poza, M., Sola, J., Tortosa, J. G., Brown, P., Cervenakova, L., et al. (1994) Accidental transmission of Creutzfeldt-Jakob disease by dural cadaveric grafts. *J. Neurol. Neurosurg. Psychiat.* **57**, 1091–1094.
40. Weller, R. O., Steart, P. V., and Powell-Jackson, J. D. (1986) Pathology of Creutzfeldt-Jakob disease associated with pituitary-derived human growth hormone administration. *Neuropathol. Appl. Neurobiol.* **12**, 117–129.
41. Gibbs, C. J., Joy, A., Heffner, R., Franko, M., Miyazaki, M., Asher, D., et al. (1985) Clinical and pathological features and laboratory confirmation of Creutzfeldt-Jakob disease in a recipient of pituitary-derived human growth hormone. *N. Engl. J. Med.* **313**, 734–738.
42. Billette de Villemeur, T., Gelott, A., Deslys, J. P., Dormont, D., Duyckaerts, C., Jardin, L., et al. (1994) Iatrogenic Creutzfeldt-Jakob disease in three growth hormone recipients: a neuropathological study. *Neuropathol. Appl. Neurobiol.* **20**, 111–117.
43. Goodbrand, I. A., Ironside, J. W., Nicolson, D., and Bell, J. E. (1995) Prion protein accumulation in the spinal cords of patients with sporadic and growth hormone associated Cruetzfeldt-Jakob disease. *Neurosci. Lett.* **183**, 127–130.

44. Gajdusek, D. C. and Zigas, V. (1957) Degenerative disease of the central nervous system in New Guinea. The endemic occurrence of "kuru" in the native population. *N. Engl. J. Med.* **257,** 974–978.
45. Neumann, M. A., Gajdusek, D. C., and Zigas, V. (1965) Neuropathologic findings in exotic neurologic disorders among natives of the highlands of New Guinea. *J. Neuropathol. Exp. Neurol.* **18,** 486–507.
46. Gerstmann, J., Sträussler, E., and Scheinker, I. (1936) Über eine eigenartige hereditar-familiare Erkrankung des Zentralnervensystems, zugleich ein Beitrag zur Frage des vorzeitigen lokalen Alterns. *Z. ges Neurol. Psychiat.* **154,** 736–762.
47. Ghetti, B., Dlouhy, S. R., Giaccone, G., Bugiani, O., Frangione, B., Farlow, M. R., et al. (1995) Gerstmann-Sträussler-Scheinker disease and the Indiana kindred. *Brain Pathol.* **5,** 61–75.
48. Guiroy, D. C., Yanagihara, R., and Gajdusek, D. C. (1991) Localisation of amyloidogenic proteins and sulfated glycosaminoglycans in nontransmissible and transmissible cerebral amyloidoses. *Acta Neuropathol.* **82,** 87–92.
49. Brown, P., Gibbs, C. J., Jr., Rogers-Johnson, P., Asher, D. M., Sulima, M. P., Bacote, A., et al. (1994) Human spongiform encephalopathy: the National Institutes of Health series of 300 experimentally transmitted disease. *Ann. Neurol.* **35,** 513–529.
50. Ghetti, B., Tagliavina, F., Masters, C. L., Beyreuther, K., Giaccone, G., Verga, L., et al. (1989) Gerstmann-Sträussler-Scheinker disease. II. Neurofibrillary tangles and plaques with PrP amyloid coexist in an affected family. *Neurology* **39,** 1453–1461.
51. Medori, R., Tritschler, J. H., LeBlanc, A., Villare, F., Manetto, V., Ghen, H. Y., et al. (1992) Fatal familial insomnia, a prion disease with a mutation at codon 178 of the prion protein gene. *N. Engl. J. Med.* **7,** 444–449.
52. Collinge, J., Owen, F., Poulter, M., Leach, M., Crow, T. J., Rossor, M. N., et al. (1990) Prion dementia without characteristic pathology. *Lancet* **336,** 7–9.
53. Manetto, V., Medori, R., Cortelli, P., Montagna, P., Tinuper, P., Baruzzi, A., et al. (1992) Fatal familial insomnia: clinical and pathological study of five new cases. *Neurology* **42,** 312–319.
54. Gambetti, P., Parchi, P., Petersen, R. B., Chen, S. G., and Lugaresi, E. (1995) Fatal familial insomnia and familial Creutzfeldt-Jakob disease: clinical, pathological and molecular features. *Brain Pathol.* **5,** 43–51.
55. Collinge, J., Palmer, M. S., Sidle, K. C., Gowland, I., Medori, R., Ironside, J., and Lantos, P. (1995) Transmission of fatal familial insomnia to laboratory animals. *Lancet* **346,** 569,570.
56. Lantos, P. L., McGill, I. S., Janota, I., Doey, L. J., Collinge, J., Bruce, M., et al. (1992) Prion protein immunocytochemistry helps to establish the true incidence of prion diseases. *Neurosci. Lett.* **147,** 67–71.
57. Editorial (1990) Prion disease—spongiform encephalopathies unveiled. *Lancet* **336,** 21–22.
58. Nakazato, Y., Hirato, J., Ishida, Y., Hoshi, S., Hasegawa, M., and Fukuda, T. (1990) Swollen cortical neurons in Creutzfeldt-Jakob disease contain a phosphorylated neurofilament epitope. *J. Neuropathol. Exp. Neurol.* **49,** 197–205.

59. Kato, S., Hirano, A., Umahara, T., Llena, J. F., Herz, F., and Ohama, E. (1992) Ultrastructural and immunohistological studies of ballooned cortical neurons in Creutzfeldt-Jakob disease: expression of αB-crystallin, ubiquitin and stress-response protein 27. *Acta Neuropathol.* **84,** 443–448.

60. Tateishi, J., Kitamoto, T., Doh-Ura, K., Boellaard, J. W., and Peiffer, J. (1992) Creutzfeldt-Jakob disease with amyloid angiopathy: diagnosis by immunologic analysis and transmission experiments. *Acta Neuropathol.* **83,** 559–563.

61. Gray, F., Chretien, F., Cesaro, P., Chatelain, J., Beaudry, P., Laplanche, J. L., et al. (1994) Creutzfeldt-Jakob disease and cerebral amyloid angiopathy. *Acta Neuropathol.* **88,** 106–111.

62. Powers, J. M., Liu, Y., Hair, L. S., Kascsack, R. J., Lewis, L. D., and Levy, L. A. (1991) Concomitant Creutzfeldt and Alzheimer diseases. *Acta Neuropathol.* **83,** 95–98.

63. Ironside, J. W., McCardle, L., Hayward, P. A. R., and Bell, J. E. (1993) Ubiquitin immunocytochemistry in human spongiform encephalopathies. *Neuropathol. Appl. Neurobiol.* **19,** 134–140.

64. Laszlo, L., Lowe, J., Self, T., Kenward, N., Landon, M., McBride, T., et al. (1992) Lysosomes as key organelles in the pathogenesis of prion encephalopthies. *J. Pathol.* **166,** 333–341.

65. Hansen, L. A., Masliah, E., Terry, R. D., and Mirra, S. S. (1989) A neuropathological subset of Alzheimer's disease with concomitant Lewy body disease and spongiform change. *Acta Neuropathol.* **78,** 194–201.

66. Collinge, J., Palmer, M. S., and Dryden, A. J. (1991) Genetic predisposition to iatrogenic Creutzfeldt-Jakob disease. *Lancet* **337,** 1441,1442.

67. de Silva, R., Ironside, J. W., McCardle, L., Esmonde, T., Bell, J. E., Will, R. G., et al. (1994) Neuropathological phenotype and "prion protein" genotype correlation in sporadic Creutzfeldt-Jakob disease. *Neurosci. Lett.* **179,** 50–52.

68. Amouyel, P., Vidal, O., Launay, J. M., and Laplanche, J. L. (1994) The apolipoprotein E alleles as major susceptibility factors for Creuztfeldt-Jakob disease. *Lancet* **344,** 1315–1318.

69. Hayward, P. A. R., Bell, J. E., and Ironside, J. W. (1994) Prion protein immunocytochemistry: the development of reliable protocols for the investigation of Creutzfeldt-Jakob disease. *Neuropathol. Appl. Neurobiol.* **20,** 375–383.

70. Sutherland, K., Barrie, C., and Ironside, J. W. (1994) Automatic quantification of amyloid plaque formation in human spongiform encephalopathy. *Neurodegeneration* **3,** 293–300.

71. Turner, C., Bell, J. E., and Ironside, J. W. (1993) Localisation of microglia in CNS amyloid plaques: an immunocytochemical and confocal microscopic study. *J. Pathol.* **170,** 401.

72. Liberski, P. P., Budka, H., Sluga, E., and Barcikowska Kwiecinski, H. (1992) Tubulovesicular structures in Creutzfeldt-Jakob disease. *Acta Neuropathol.* **84,** 238–243.

73. Liberski, P. P., Yanagihara, R., Gibbs, C. J., and Gajdusek, D. C. (1990) Appearance of tubulovesicular structures in experimental Creutzfeldt-Jakob disease and scrapie precedes the onset of clinical disease. *Acta Neuropathol.* **79,** 349–354.

74. DeArmond, S. J. (1993) Alzheimer's disease and Creutzfeldt-Jakob disease: overlap of pathogenic mechanisms. *Curr. Opin. Neurol.* **6,** 872–881.

4

Neuropathological Diagnosis of Human Prion Disease

PrP Immunocytochemical Techniques

Jeanne E. Bell

1. Introduction

Routine histological examination of cases of Creutzfeldt-Jakob disease (CJD) reveals a spectrum of pathological involvement, particularly with regard to the extent and localization of spongiform change. Although there is no difficulty in confirming a diagnosis of CJD in cases with widespread and severe spongiform change involving the cortical gray matter or cerebellar molecular layer, all too often the cortex may be only minimally or focally affected, and in some cases pathological changes are largely confined to the basal ganglia or cerebellum. Unless wide-ranging pathological examination is undertaken, the diagnosis of CJD may be missed in such cases. This is a particular problem in the interpretation of cortical biopsies in cases of suspected CJD and is one reason why biopsy in such cases is now discouraged *(1)*. In addition, there are other conditions that give rise to spongy appearances in the gray matter *(2)* and the difficulties of achieving a firm diagnosis of CJD, based on conventional histological appearances, have been outlined in Chapter 3. Occasional reports of the coexistence of CJD with other dementing illnesses, particularly Alzheimer disease, also add to the diagnostic difficulties *(3)*. CJD is a rare disease (0.5 cases/million of the population/yr) and it is likely that individual experience of the possible range of CJD pathological changes will be limited. For all of these reasons, it would be useful to develop some additional method of confirming the diagnosis of CJD in tissue sections. Since it is now known that CJD and related diseases are associated with accumulation in the brain of a protease-resistant protein known as prion protein *(4)*, the demonstration of

From: *Methods in Molecular Medicine: Prion Diseases*
Edited by: H. Baker and R. M. Ridley Humana Press Inc., Totowa, NJ

prion presence in suspected cases would represent such a development. Recently, immunocytochemical techniques have been devised that accurately localize disease-related prion protein (PrP^{CJD}) in human cases *(5–13)*. The potential value of such additional techniques in clarifying the diagnosis in difficult and subtle cases of CJD is obvious. However, the implementation of PrP^{CJD} immunocytochemistry may have other benefits. Since PrP^{CJD} is closely associated, if not identical, with the transmissible agent known to be present in the central nervous system (CNS), its localization clearly is a matter of scientific interest and may help to throw light on the spread of the agent from the point of entry, at least in those cases in which the disease has an exogenous origin, such as patients treated with human pituitary-derived hormones *(14)*. All the major human forms of prion disease, including sporadic *(2,15)* and familial CJD *(2,16)*, Gerstmann-Sträussler-Scheinker syndrome (GSS) *(16)*, and iatrogenic CJD (dura mater and human pituitary hormone cases) *(17)* show deposits of PrP^{CJD} in different CNS locations. Similar accumulations occur in animal prion diseases, notably scrapie *(18)* and bovine spongiform encephalopathy (BSE) *(19)*. This chapter is concerned with the methodology for immunolocalization of PrP^{CJD} within the human CNS, and with the dilemmas of interpretation associated with this technique.

2. PrP^{CJD}

PrP^{CJD} is an insoluble, protease-resistant protein that accumulates in the brains of patients with CJD and other related diseases *(20)*. It is deposited within the brain parenchyma in part as amyloid that is detectable by periodic acid Schiff and congo red staining if present in sufficient quantity to form the plaques that are seen in some cases of CJD and more commonly in other prion diseases, particularly GSS *(2,20,21)*. It is believed that PrP^{CJD} is a perverted form of a normal soluble cellular protein, PrP^C, which is encoded by a gene on chromosome 20 and is expressed in many cells (including non-CNS cells) but particularly on neuronal membranes. The function of PrP^C is not known. In CJD and other related diseases, PrP^C is posttranslationally altered to PrP^{CJD}, which, because it is protease resistant, is capable of accumulating in tissues in which it is generated *(20)*. It is known that a number of different mutations in the PrP gene predispose to this posttranslational modification with consequent accumulation of PrP^{CJD} *(22)*. The association of PrP mutations with different forms of prion disease is reviewed elsewhere in this volume. The relationship between PrP genotype and phenotypic expression of prion disease is complex, and currently is the focus of intense research activity.

PrP^C and PrP^{CJD} are closely similar and antibodies raised against one will crossreact against the other *(2)*. The gene involved is highly conserved across species *(20)* and is expressed in mammals and in some birds. Antibodies have

Table 1
Pretreatments Used
in PrP Immunostaining Protocols

Proteinase K 10 μg/mL 15 min 37°C *(6)*
Pepsin 10% 30 min 37°C *(8)*
80–100% formic acid 5–60 min *(5,7,9,10,13)*[a]
30% formic acid 1 min in microwave oven *(11)*
4M guanidine thiocyanate 2 h *(13,23)*[a]
Hydrated autoclaving distilled water *(13)*[a]
Hydrolytic autoclaving 2.5 mM HCl 10 min 121°C *(24)*

[a]Including those currently used in Edinburgh.

been raised both to the disease-related forms of PrP, such as scrapie-associated fibrils, and more recently to synthetic peptides following the successful sequencing of the PrP gene. A number of different polyclonal and monoclonal antibodies are available (but not yet commercially) and many of these have been used in immunocytochemical studies *(5–13)*. Antibodies raised against scrapie fibrils readily produce positive staining in human CJD cases, thus demonstrating crossreactivity between different species *(2,13)*.

The fact that these antibodies cannot clearly distinguish between PrP^C and PrP^{CJD} raises potential problems of interpretation. In order to circumvent these difficulties, strategies have been developed that aim to eliminate PrP^C from tissue sections before exposure to the primary antibody, with the aim of visualizing only disease-related, protease-resistant PrP^{CJD} (Table 1). A number of different enhancing treatments are also currently in use that allow the primary antibody to be used at much higher dilution *(5–13,23,24)*. As a result, it now seems unlikely that the visualization of PrP^C is a significant confounding factor in sections stained for PrP^{CJD}. Nevertheless, it is even more important than usual to include immunocytochemical checks and controls in any protocol for the demonstration of PrP^{CJD}. The use of negative and positive control case material, and negative sections in which the primary antibody is omitted, are mandatory in each test run. Other checks include comparison of results using a panel of different PrP antibodies, including antibodies preabsorbed against the specific antigen.

Protocols for PrP^{CJD} immunocytochemistry (Table 2) are now reasonably well validated for CNS tissue sections. In our laboratory, PrP immunostaining is part of the diagnostic procedure, used in conjunction with routine histopathology of the CNS, and to demonstrate the topographical cellular localization of PrP^{CJD}. Western blotting and immunoblotting *(25)* represent alternative investigatory procedures for the presence of PrP^{CJD} but neither allows localization at the cellular level in the way that immunocytochemistry does.

Table 2
Protocol for PrP Immunocytochemistry

1. 5-μm sections floated on to slides coated with Vectabond
2. Sections to water
3. Picric acid 15 min (to remove formalin pigment)
4. Water
5. Hydrogen peroxide 30 min (to block endogenous peroxide)
6. Water
7. Autoclave in distilled water 121°C for 10 min
8. Water
9. 96% Formic acid 5 min
10. Water
11. 4M guanidine thiocyanate 2 h at 4°C
12. Water followed by Tris buffered saline
13. Blocking serum 10 min
14. Primary PrP antibody: most are used overnight (Table 3)
15. Wash in Tris buffered saline
16. Secondary antibody (1:200) for 30 min
17. Wash in Tris buffered saline
18. Avidin biotin for 30 min
19. Wash in Tris buffered saline
20. Visualize with diaminobenzidine
21. Wash in water
22. Counterstain lightly with hemotoxylin
23. Sections dehydrated, cleared, and mounted in Pertex
24. Mounted slides (after drying) further decontaminated by immersion in 96% formic acid for 5 min

Just as the correlation between PrP genotype, clinical phenotype, and classical pathology is under intense investigation at present, so also is the distribution of PrP^{CJD} in different manifestations of prion disease. It remains to be seen whether tightly consistent patterns of PrP^{CJD} deposition will be demonstrated but considerable progress is being made in this field.

3. Protocols for PrP Immunocytochemistry

Based on the published experience of PrP immunostaining, we introduced this procedure in the UK National CJD Surveillance Unit in Edinburgh in 1992, experimentally at first in order to explore its potential role in diagnostic confirmation of CJD. Since that time we have modified the protocol in accordance with recently published work and with our own experience in order to further enhance PrP^{CJD} localization and to reduce the risk of PrP^C detection. In 1993, our Unit hosted an MRC-funded workshop to which all the UK centers undertaking PrP immunocytochemistry were invited. This workshop was aimed at exploring the then current dilemmas and difficulties in PrP immunostaining. A five-center study was set up in which duplicate slides were circulated for PrP immunostaining to all the centers, each center using its own protocol and antibodies. Subsequently, all the stained slides have been circulated for assessment of staining results to each of the five centers in turn and a consensus report is being drawn up. It has emerged that despite some slight variation in the intensity of staining, similar patterns of PrP deposition were noted in each case regardless of the antibody used or of slight variations in protocol (each center used at least one of the pretreatment steps described, usually exposure to formic acid). Control cases, such as age matched, nondemented subjects, and cases of Alzheimer disease, did not show PrP staining.

Pretreatments that have been advocated in the literature include those listed in Table 1. The protocol that is currently in use in our laboratory is shown in Table 2. In summary, the pretreatments that we use before exposure to the primary antibody include hydrated autoclaving, formic acid for 5 min, and guanidine thiocyanate for 2 h (Table 2). Although in our experience guanidine thiocyanate has proved useful, the results of the five-center workshop study suggest that this pretreatment could be excluded with no significant detriment to PrP^{CJD} visualization. In contrast, the formic acid and hydrated autoclaving pretreatments are extremely useful, leading to consistent and unequivocal patterns of PrP^{CJD} immunolocalization. In particular, sections that are slightly positive with other pretreatments always show more widespread and specific staining following the autoclave pretreatment. We have experimented with other pretreatments (Table 1), including protease K and pepsin (but have found that they are extremely detrimental to tissue preservation); hydrolytic autoclaving (in 2.5 mM HCl as compared with distilled water; this appears to confer no advantage over hydrated autoclaving); and microwave enhancement (1 min in formic acid). Our experience with over 150 cases of prion disease suggests that the current immunocytochemical protocol and pretreatments provide the most reliable results. In most cases a panel of three or four anti-PrP antibodies is used for diagnostic confirmation (Table 3) *(2,13)*.

4. PrP Immunostaining in Sporadic CJD Cases

Given the range of pathological changes seen with routine staining methods in cases of sporadic CJD it is not, perhaps, surprising that PrP^{CJD} immunolocalization also varies from case to case. The patterns of PrP^{CJD} deposition include:

1. Cortex: In those cases with severe confluent spongiform change in the cortical gray matter, irregular diffuse deposits of PrP^{CJD} are seen in relation to the spongiform vacuoles (Figs. 1 and 2). However, when the spongiform change is represented only by small discrete parenchymal vacuoles, PrP^{CJD} staining may

Table 3
PrP Antibodies Currently in Use in Edinburgh

Antibody	Clonality	Antigen	Dilution	Source
1A8	Polyclonal	Scrapie fibrils	1 in 5000 o/n[a]	J. Hope, Edinburgh
KG9	Monoclonal	Recombinant bovine PrP	1 in 200 o/n	C. Birkett, Compton
3F4	Monoclonal	Synthetic peptide (residues 108–111 of human PrP^C)	1 in 2000 o/n	R. Kascsak, New York
SP40	Polyclonal	Synthetic peptide (partial sequence sheep PrP)	1 in 1500 o/n	B. Anderton, London

[a]o/n = overnight.

Fig. 1. Immunolocalization of PrP^{CJD} in the cortical gray matter from a patient with sporadic CJD. Darkly stained PrP^{CJD} deposits are seen in association with confluent spongiform change through the full thickness of the cortical ribbon from cortical surface above to the white matter below.

be apparently absent in gray matter unless specifically enhanced with autoclaving pretreatment, which reveals diffuse granular staining of the gray matter in such cases *(13)*. In some cases, distinct perineuronal staining is seen particularly in layers 3 and 5 of the cortex (Fig. 3). Astrocytes may display granular cytoplasmic staining either as an isolated feature or in company with vacuolar deposits of PrP^{CJD} (Fig. 4). Plaques are not usually visible in the cortex, although βA4 plaques

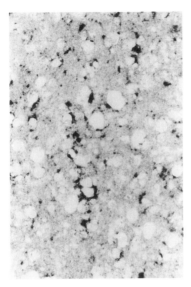

Fig. 2. PrP^{CJD} immunolocalization in association with severe spongiform change in the cortical gray matter from a case of sporadic CJD.

Fig. 3. Laminar neuronal staining of membranes and processes with antibody KG9. Microcystic spongiform change is seen in the adjacent unstained cortex.

may be present in small number as an aging phenomenon or, rarely, as a sign of concurrent presence of Alzheimer disease. The cornu ammonis of the temporal cortex is frequently spared both in terms of spongiform change and of PrP^{CJD}.

2. Basal ganglia: PrP^{CJD} immunostaining is sometimes visible in the basal ganglia, particularly in the form of small plaques in the white matter when the gray matter shows spongiform change (Fig. 5). Fine granular staining may be visible in the areas of gray matter.

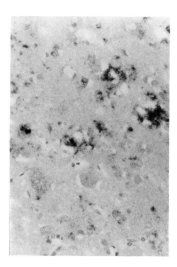

Fig. 4. Focal deposits of PrP^{CJD} associated with spongiform vacuoles in the cortex of a case of sporadic CJD. Neighboring neurons (arrow) are unstained. Occasional astrocytes show granular cytoplasmic staining for PrP^{CJD} (closed arrow).

3. Cerebellum: In a minority of cases of sporadic CJD small PrP^{CJD} positive plaques may be present in any of the layers of the cortex and in the white matter but are particularly prominent in the Purkinje cell layer and in the granular layer (Fig. 6). The plaques are circular in profile and may have a pale center, a dark circumferential layer, and radiating spicules from the outer surface—the so-called kuru plaques *(26)* (Fig. 7). The majority of cases do not have well-defined plaques and show more diffuse positivity in the granular layer and focally in the molecular layer (Fig. 8). The white matter is unstained in such cases. The presence of plaques in the cerebellum bears no obvious relationship to the presence of spongiform change in the molecular layer but does correspond with the presence of cerebellar atrophy. The Purkinje cells are always PrP^{CJD} negative *(2,13)* (Fig. 8). Perineuronal staining may be visible in the dentate nucleus.

4. Brain stem: A minority of cases show scattered small plaques in the brainstem, particularly in the peripheral white matter tracts (Fig. 9). More typically, in one-third of cases of sporadic CJD granular positivity is seen in the gray matter of the periaqueductal region, tectum, substantia nigra, and pontine nuclei. Staining associated with neurons may be very obvious in some of these nuclei, particularly in the olive (Figs. 9 and 10). The corticospinal tracts are always negative.

5. Spinal cord: In these cases that show positivity in the brainstem, PrP^{CJD} immunopositivity is seen in the substantia gelatinosa and dorsal horns of the spinal cord gray matter *(27)* (Fig. 11). These cases also show small plaques within the peripheral white matter tracts and linear PrP^{CJD} deposits extending into the white matter (Fig. 12).

Fig. 5. Section of basal ganglia from a case of sporadic CJD showing spongiform change in areas of gray matter at the top and bottom of the figure. PrP^{CJD}-positive plaques are present in the intervening white matter as well as in the gray matter. Most of the spongiform gray matter is apparently unstained for PrP^{CJD}.

It is interesting to note that, whereas in some instances, PrP^{CJD} accumulates in those areas that show the most profound pathological change (such as the cerebral cortex in the presence of confluent spongiform change as shown in Figs. 1 and 2) or that are associated with clinical signs and symptoms (cerebellar ataxia with cerebellar atrophy, displaying plaques in the granular layer as shown in Fig. 6), there are also considerable discrepancies between the localization of classical pathology (as revealed by routine staining) and PrP^{CJD} immunolocalization *(2)*. Thus, quite strong positivity for PrP^{CJD} may be evident in parts of the CNS, such as spinal cord and brain stem (Figs. 9–12), which rarely show recognizable cellular changes or spongiform degeneration in human cases, and conversely, the cerebellar molecular layer may show conspicuous microcystic spongiform change without colocalizing evidence of PrP^{CJD} *(13)*. Similarly, we have not thus far been successful in demonstrating PrP^{CJD} by immunocytochemical techniques in dura mater or in the pituitary gland, both of which have been implicated in the transmission of prion diseases from one human case to another. Further work needs to be undertaken in this field since there are several possible explanations for this finding. Our initial studies with spinal cord tissue proved negative immunocytochemically, but with improving techniques and a wider range of material, we have now

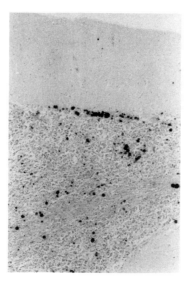

Fig. 6. Section of cerebellum from a case of sporadic CJD displaying PrP^{CJD}-positive plaques. The plaques are most numerous in the granular layer but also show focal concentrations in the Purkinje cell layer at the interface between the largely unstained molecular layer and the granular layer.

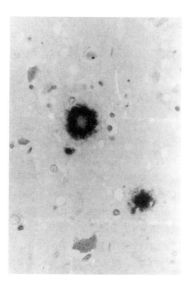

Fig. 7. High power view of a plaque in the molecular layer of the cerebellar cortex in a patient with sporadic CJD. The larger plaque has a pale center, a dark edge, and radiating outer fibrils, displaying the classic kuru plaque appearance.

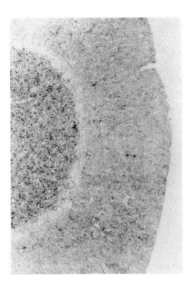

Fig. 8. Cerebellar cortex in a case of sporadic CJD without plaques. There is diffuse granular staining, which is focally accentuated, in the granular and molecular layers. The Purkinje cell layer is unstained.

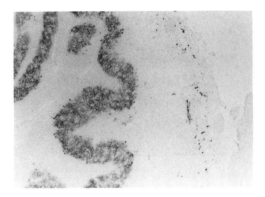

Fig. 9. Section of the medulla in sporadic CJD showing immunolocalization of PrP^{CJD} in the olivary nucleus and in the subpial white matter where small plaques are present.

described immunopositivity for PrP^{CJD} in the spinal cord in 30% of cases of sporadic CJD.

In many respects, the PrP^{CJD} immunostaining patterns in human cases are similar to those described in mice with transmitted scrapie agent *(18)* although the localization of PrP^{CJD} is slightly different in the two species.

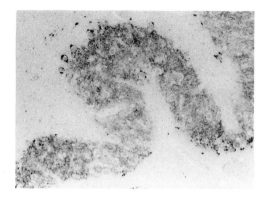

Fig. 10. High power view of the olivary nucleus shown in Fig. 9. Perineuronal and diffuse granular staining is seen in the gray matter.

5. PrP^CJD Staining Patterns in Other Prion Dementias

In other prion dementias, the patterns of PrP^{CJD} immunopositivity sometimes resemble, and are sometimes quite widely divergent from, those seen in the CNS of patients with sporadic CJD as described. Although these differing patterns sometimes segregate with the particular genotype of the individual or with the mode of acquisition of the transmissible agent, there is also a diversity of phenotype within individual genetic abnormalities as well as among cases of sporadic disease *(2,27–30)*.

5.1. Inherited Prion Diseases Including Familial CJD and GSS

These cases have a different clinical presentation from sporadic CJD and may display a longer duration of illness with a predominantly cerebellar symptomatic presentation. Quite a number of different PrP gene mutations and inserts have now been described in these families and to a certain extent, the findings are linked to particular genetic abnormalities. As might be expected from the clinical presentation, the major pathological changes are found in the cerebellum, which shows predominantly plaques in familial disorders. In cases of GSS, large multicentric plaques, more complex than the compact kuru-type plaque, are seen scattered in the molecular layer and in the granular layer of the cerebellum. These plaques are readily seen with routine stains (Fig. 13) and those that bind to amyloid, but are very strongly positive for PrP^{CJD} (Fig. 14). The cerebellum in GSS has a heavier burden of plaques (Fig. 14) than in those cases of CJD (Fig. 6) that also display cerebellar plaques. The plaques may have associated microglial and neuritic abnormalities and in this way resemble those seen in Alzheimer disease *(31)*. Antibodies for βA4 amyloid may colocalize in these large complex plaques *(32)*. Spongiform change may be

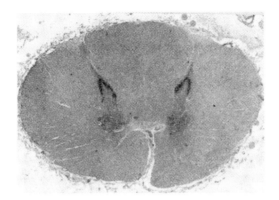

Fig. 11. Computer image of PrP^{CJD} immunolocalization in the spinal cord of a growth hormone recipient who developed CJD. The strongest positivity is noted in the substantia gelatinosa and dorsal horns.

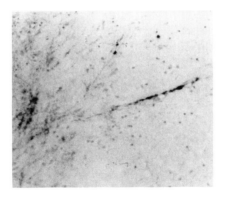

Fig. 12. Section of thoracic spinal cord showing the junctional zone between gray and white matter. Immunopositivity for PrP^{CJD} extends along cell processes from gray matter into the white matter.

absent from the cerebellar molecular layer and may be very focal in the cerebrum, which generally shows less conspicuous changes than in cases of sporadic CJD. In one particular GSS kindred, very numerous neurofibrillary tangles have been described *(29)* and when combined with the presence of amyloid plaques, represent a possible source of diagnostic confusion and may be mistakenly interpreted as the changes of familial Alzheimer disease. Further study of these interesting families presents a useful opportunity for linking particular neurodegenerative manifestations with identified genetic abnormalities.

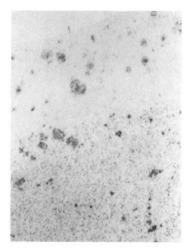

Fig. 13. Routinely stained section of the cerebellar cortex in a case of GSS. Eosinophilic plaques are conspicuous in the molecular later (above) and are also present within the granular layer (below).

5.2. Acquired Prion Diseases

5.2.1. Kuru

This virtually extinct condition, which was associated with cannibalistic rituals in the natives of the Fore tribe in Papua New Guinea, presented initially with cerebellar signs and symptoms and progressed to dementia only at a late stage. In 70% of cases that were examined, the cerebellum displayed a significant number of small plaques *(26)* with a dense center when stained for amyloid, but with a dense peripheral rim on staining for PrP^{CJD}. A corona of radiating fibrils typically is present in these classical "kuru-type plaques." The plaques have been demonstrated in all the layers of the cerebellum, particularly in the granular layer. They also are present elsewhere in the CNS. It is of interest that some cases of apparently sporadic CJD closely resemble kuru cases with regard to the type and distribution of plaques within the cerebellum and elsewhere (Fig. 7).

5.2.2. Iatrogenic Cases

The development of CJD in some patients previously treated with human hormones extracted from cadaver-derived pituitary glands *(14,17)*, or in occasional patients who have received corneal or dura mater grafts *(33,34)*, has provided a tragic opportunity to study the pathology and pattern of PrP^{CJD} deposits in iatrogenic cases in whom the site of entry of the transmissible agent,

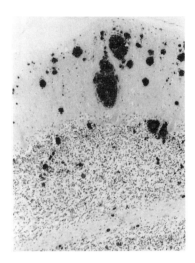

Fig. 14. Section from the cerebellar cortex (same case as Fig. 13) showing immunolocalization of PrP^{CJD} in large and complex plaques in the molecular layer. Smaller plaques are present within the granular layer and several large plaques are present at the interface between molecular and granular layers. The plaque burden is much greater in this case than in the plaque-bearing case of sporadic CJD shown in Fig. 6.

and the latent period before development of signs and symptoms of CJD, can be accurately identified. In patients treated with dural grafts, the pattern of PrP^{CJD} deposition resembles that seen in classical sporadic CJD, with predominantly cortical involvement and with a clinical presentation of dementia *(2)*. In contrast, patients treated with peripheral injections of cadaver-derived pituitary hormones first display cerebellar signs and symptoms *(35)* (which may be misdiagnosed as paraneoplastic syndrome or in some instances as recurrence of the primary CNS tumor that originally had led to their treatment with pituitary hormones), and immunocytochemical studies show predominantly cerebellar positivity with accumulation of kuru-type plaques (similar to Fig. 6) or more commonly, diffuse PrP^{CJD} deposition in the granular layer (similar to Fig. 8) *(2,17,35,36)*. There may be very limited PrP^{CJD} deposition in the cortical gray matter in patients treated with pituitary hormone injections. It is of interest to note that the gray matter of the spinal cord shows heavy PrP^{CJD} deposition (Fig. 15) in the anterior and posterior horns and Clarke's column (autonomic interneurons of the thoracic spinal cord) *(27)*. These cases of iatrogenic prion disease with spinal cord involvement resemble a subset of sporadic CJD cases (Fig. 11) that also show spinal cord deposits of PrP^{CJD} *(27)*.

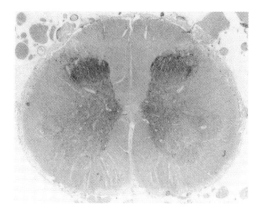

Fig. 15. Computer image of the upper lumbar spinal cord from a case of growth-hormone-induced CJD. Immunopositivity for PrP^{CJD} is present in the gray matter and particularly pronounced in the dorsal horn and substantia gelatinosa on each side. Positivity is also seen in Clarke's column, indicated by the arrow.

5.3. Fatal Familial Insomnia

This genetic disorder is associated with a point mutation at codon 178 *(37)*. The pathological substrate of the clinical signs and symptoms appears to be thalamic gliosis (Fig. 16) and there may be very little in the way of spongiform change in this prion disorder. No plaques are seen. PrP^{CJD} is demonstrated by immunocytochemical means rather inconsistently in the thalamus and predominantly associated with neurons (Fig. 17). Reactive astrocytes are PrP^{CJD}-negative even when associated with profound neuronal loss as in the inferior olive but focal positivity for PrP^{CJD} may be present in the granular layer of the cerebellum (Fig. 18).

5.4. The Atypical Prion Dementias

In these familial cases, there are abnormalities of the PrP genotype but no very characteristic neuropathological abnormalities (*see* Chapter 3). The brain is atrophic and there may be cortical loss and gliosis, but spongiform change is not a feature of these diseases *(10,38)*. The cerebellar cortex may appear atrophic but routine staining does not demonstrate any particular abnormality in the molecular layer (Fig. 19). However, immunostaining with antibodies to PrP^{CJD} reveals striking and consistent linear deposits confined to the molecular layer (Fig. 20).

Some prion dementias may be particularly difficult to diagnose on routine staining and the benefits of reliable PrP^{CJD} immunolocalization clearly relate not only to diagnostic clarification but also to judging the extent of pathological change.

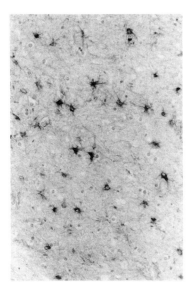

Fig. 16. Case of fatal familial insomnia showing thalamic gliosis. Astrocytes are identified by strong immunoreactivity for glial fibrillary acidic protein. Neuronal loss is evident but spongiform change is not a noticeable feature in this section.

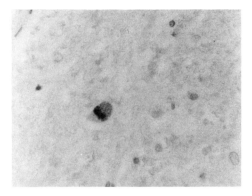

Fig. 17. Occasional neurons in the thalamus in fatal familial insomnia show strong positivity for PrP^{CJD}.

6. Differentiation from Other Dementias

Although there should be no difficulty in differentiating classical CJD cases from other dementias, clearly there is some neuropathological overlap and diagnostic difficulty in separating subtle cases of prion diseases from the range of

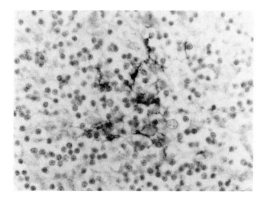

Fig. 18. Cerebellar cortex from a case of fatal familial insomnia showing focal positivity for PrP^{CJD} in the granular layer.

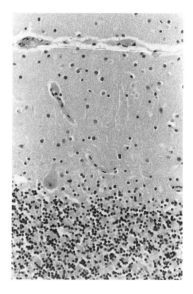

Fig. 19. Section of the cerebellar cortex from a case of atypical prion dementia with a mutation in the prion gene but no evidence of spongiform change. In routine staining, the cerebellum appears somewhat atrophic but otherwise unremarkable.

other dementias in which spongiform change, amyloid plaques, or neurofibrillary tangles, or simply neuronal loss and extensive gliosis, are common features *(2,3,5,8,9,32,39–41)*. These difficulties have been alluded to both in Chapter 3 and in this chapter. Very occasional cases have been reported in which CJD does appear to coexist with other diseases *(3,5,41)* and we have had

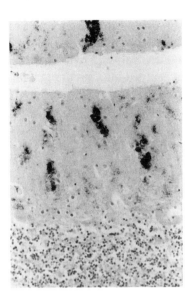

Fig. 20. Section from the case of atypical prion dementia shown in Fig. 19, stained for PrP^CJD. Linear deposits are clearly identified in the molecular layer, whereas other layers of the cortex are negative. Sections from this case, and permission to photograph the staining results (Figs. 19 and 20), were kindly provided by P. Lantos.

two cases of this type in the national surveillance project in the last 4 yr. There are a number of factors that may be taken into consideration when trying to finalize a diagnosis. These include the clinical and family history, EEG and neuroimaging findings, the general pathology and neuropathology findings, and the results of immunocytochemical investigation. In many cases information regarding the PrP genotype will be a useful adjunct in achieving diagnosis. The advent of antibodies that will specifically detect different forms of amyloid, particularly βA4 amyloid and PrP^CJD, has been very helpful. Transmissibility would also help to separate prion diseases from other dementias *(22)*, but until recently it has not been clear that all the prion diseases are clearly transmissible. In any case this is not a practical possibility in all cases because of the cost and time involved in transmission experiments. In this regard, reliable PrP^CJD immunolocalization is a most useful development.

7. Role of PrP Immunocytochemistry in Diagnosis

Confidence in the reliability of PrP^CJD immunolocalization has come slowly in view of the potential pitfalls in its interpretation *(see the following)*. In cases currently under examination that have been fixed in formalin for <4 wk and with routine processing of tissue blocks after immersion in 96% formic acid for 1 h, there is no difficulty in demonstrating PrP^{CJD} by immunocytochemistry. Occasional difficulty is encountered in archival cases or in cases referred from elsewhere, in which classical CJD neuropathology, including confluent spongiform change, is widespread but immunocytochemical results are negative. This we have attributed to variation in fixation time or processing schedules. Nevertheless, the consistency with which the abnormal protein is localized in nearly all disease cases; the positivity with all the different antibodies and the abolition of positivity when preabsorbed antibodies are used; the high level of conformity between reports from different studies and the negativity of control cases, all suggest that PrP immunocytochemistry has now graduated to take its place as a useful diagnostic procedure. The topographical discrepancies between conventional pathology of prion diseases and the PrP^{CJD} deposition raise some interesting questions but do not interfere with the diagnostic usefulness of the technique, provided these discrepancies are recognized. Reliable PrP^{CJD} immunolocalization will allow us to better define the incidence of these interesting diseases *(10)*, particularly those diseases that are very rare or that show atypical neuropathological changes, and also has a useful role in separating prion diseases from other neurodegenerative diseases. The interesting relationships between the deposition of PrP^{CJD}, clinical signs, and symptoms, and PrP genotype require further study before the diagnostic applications of PrP^{CJD} immunocytochemistry are apparent in the full range of prion diseases.

8. Role of PrP Immunocytochemistry in Furthering the Understanding of Prion Diseases

If PrP^{CJD} can be localized accurately within CNS tissues, the first step is to delineate the cell types in which it is present at light microscope and ultrastructural level. Clearly some of the deposits are extracellular and these are best exemplified by the presence of plaques. Whether these deposits commence as intracellular accumulations that are actively extruded or become extracellular as a result of cell breakdown is not clear *(42)*. However, current immunolocalization studies suggest that at least some PrP^{CJD} deposits are intracellular or closely pericellular *(2,13)*. We are currently investigating the ultrastructural immunolocalization of PrP^{CJD}. Double immunostaining at the light microscope level will also help to identify the cells in which, or around which, PrP^{CJD} deposition occurs.

The mapping of PrP^{CJD} localization may throw light on the spread of PrP^{CJD} within the CNS. The finding that the spinal cord is involved in all cases of iatrogenic CJD, in which the agent has been introduced peripherally, and in some cases of sporadic CJD, suggests that the spinal cord may be one point of initial entry and accumulation of PrP^{CJD} in the CNS *(27)*. Current interest is focusing on possible accumulation of PrP^{CJD} in peripheral nerves and in autonomic nerves. Although the results to date are negative in these tissues, it may simply reflect the fact that the abnormal PrP is transported along nerves and accumulates only once it reaches the CNS. Accumulation of abnormal PrP in peripheral tissues has been demonstrated in muscle cells in cases of inclusion body myositis *(43)*. How this condition relates to PrP mutations and other abnormalities in muscle tissue is not clear. It is of interest that some transgenic mice who express high levels of PrP transgenes do develop a severe generalized skeletal muscle disorder that on pathological examination proves to be necrotizing and associated with PrP accumulation *(44)*. Whether PrP^{CJD} can be demonstrated in spleen and lymphoid tissues, as has been reported in animals *(24)*, awaits further investigation. It is too early to say whether patterns of PrP^{CJD} deposition can yet be linked to the putative route of entry of the agent to the CNS, but the similarity of some cases of sporadic CJD to kuru raises interesting questions. PrP^{CJD} immunolocalization adds an extra dimension to the study of disease distribution within the CNS that is particularly important in the investigation of human cases for any possible links with the outbreak of BSE. If BSE has had a significant effect on human disease, it may be that the pattern of pathology in human cases will come to resemble that seen in BSE *(19)* and that there may be a significant shift of pathology, and associated PrP^{CJD} immunolocalization, to the brainstem.

9. Image Analysis Studies PrP^{CJD} Immunolocalization

In order to achieve an objective assessment of PrP^{CJD} load and distribution, we have actively pursued image analysis techniques that allow observer-independent assessment of the correlation of PrP^{CJD} deposits with the degree of spongiform change and with other parameters, such as the extent of colocalizing gliosis *(45)*. This is achieved by semi-interactive or automatic assessment of immunocytochemically stained slides. Figures 11 and 15 represent computer images of PrP^{CJD} immunolocalization in the spinal cord in cases of sporadic and growth-hormone-associated CJD. These methods show potential for mapping large numbers of sections from a range of cases in the different subsets of prion diseases and may add further useful information on the pathological variants to be linked to the different genotypes of prion diseases.

10. Pitfalls in PrP^{CJD} Immunocytochemistry

Most of these have already been alluded to in this chapter. To summarize, the major difficulties surround the crossreactivity between antibodies that do not differentiate between PrP^{C} and PrP^{CJD}. Strategies to minimize the risks of immunolocalizing PrP^{C} and of restricting the reaction to PrP^{CJD} are employed in most published studies and are included in our protocols (Table 2). In practice, there does not seem to be a difficulty with crossreactivity when CNS tissue sections are examined in adult subjects. However, the application of these techniques to non-CNS tissues, or to embryonic and fetal tissues, is still at an experimental stage and will benefit from further understanding of the normal function of PrP^C. The discordance that is sometimes present between pathology changes and PrP^{CJD} localization may be genuine or may be a problem of sensitivity. It is certain that autoclave pretreatment significantly increases visualization of PrP^{CJD} distribution in disease cases. However, accumulation of abnormal PrP has already occurred in symptomatic prion diseases and sensitivity of techniques may be an issue in presymptomatic or short duration cases since the temporal relationship between PrP^{CJD} accumulation, onset of cellular neurodegenerative changes, and clinical signs and symptoms is not understood. The fact that some long-fixed archival material does not respond positively with immunocytochemical techniques for PrP^{CJD} does suggest that laboratory processing of tissues has some effect on the likelihood of visualization. Some very practical difficulties are associated with the previous practice of immersion of tissue in phenol that was undertaken in an attempt to decontaminate prion disease cases *(46)*. Phenol has a deleterious interaction with formic acid and the two pretreatments should not be used in conjunction. This is largely historical since phenol immersion has been shown to be ineffective in decontamination. However, pretreatment with formic acid leads to some technical problems in its own right; sections taken from tissue that has been exposed to formic acid for any length of time are difficult to handle in the laboratory since their adherence to glass slides is poor unless the slides are pretreated with very effective bonding agents. However, the advantages of formic acid in enhancing not only the visualization of PrP^{CJD} but also some other antigens outweigh this disadvantage.

Occasional false positive staining, or what is assumed to be false positive staining, is observed in blood vessel walls both in the subarachnoid space and within CNS tissue. This may be present with one or other PrP antibody and is an inconsistent finding.

Despite these difficulties, PrP immunocytochemistry has assumed a credible place in the diagnostic and scientific investigation of human cases of prion disease. Consensus in the hands of many investigators lends weight to the reliability of these techniques and when the results are correlated with classical histopathology, clinical findings, genetic analysis, and other techniques, such as Western blotting and immunoblotting, they may contribute to significant progress in our understanding of prion disease.

Acknowledgments

The development of reliable PrP immunostaining was supported by the Medical Research Council (SPG9119619) and the CJD surveillance project is supported by the Department of Health. The skilled assistance of L. McArdle, C. Barrie, D.

Nicolson, P. Hayward, and I. Goodbrand is gratefully acknowledged. K. Sutherland generated the computer images of PrP^{CJD} immunolocalization in the spinal cord and R. Melrose prepared the manuscript. I am indebted to R. Will for helpful discussion and this work was done in collaboration with J. W. Ironside.

References

1. Metters, J. S. (1992) Neuro and ophthalmic surgery procedures in patients with, or suspected to have, or at risk of developing CJD or GSS. *Letter PL(92)CO/4 to Consultants and Health Managers.* Department of Health, London.
2. Bell, J. E. and Ironside, J. W. (1993) Neuropathology of spongiform encephalopathies in humans. *Br. Med. Bull.* **49,** 738–777.
3. Hansen, L. A., Masliah, E., Terry, R. D., and Mirra, S. S. (1989) A neuropathological subset of Alzheimer's disease with concomitant Lewy body disease and spongiform change. *Acta Neuropathol.* **78,** 194–201.
4. Bockman, J. M., Kingsbury, D. I., McKinley, M. P., Bendheim, P. E., and Pruisiner, S. B. (1991) Creutzfeldt-Jakob disease prion proteins in human brains. *N. Engl. J. Med.* **312,** 73–78.
5. Powers, J. M., Lui, Y., Hair, L. S., Kascsack, R. J., Lewis, L. D., and Levy, L. A. (1991) Concomitant Creutzfeldt-Jakob and Alzheimer diseases. *Acta Neuropathol.* **83,** 95–98.
6. Piccardo, P., Safar, J., Ceroni, M., Gajdusek, D. C., and Gibbs, C. J. (1990) Immunohistochemical localization of prion protein in spongiform encephalopathies and normal brain tissue. *Neurology* **40,** 518–522.
7. Kitamoto, T. and Tateishi, J. (1988) Immunohistochemical confirmation of Creutzfeldt-Jakob disease with a long clinical course with amyloid plaque core antibodies. *Am. J. Pathol.* **131,** 435–443.
8. Guiroy, D. C., Yanagihara, R., and Gajdusek, D. C. (1991) Localisation of amyloidogenic proteins and sulfated glycosaminoglycans in nontransmissible and transmissible cerebral amyloidoses. *Acta Neuropathol.* **82,** 87–92.
9. Bugiani, O., Giaccone, G., Verga, L., Pollo, B., Frangione, B., Farlow, M. R., et al. (1993) βPP participates in PrP-amyloid plaques of Gerstmann Sträussler-Scheinker disease, Indiana kindred. *J. Neuropathol. Exp. Neurol.* **52,** 64–70.
10. Lantos, P. L., McGill, I. S., Janota, I., Doey, L. J., Collinge, J., Bruce, M., et al. (1992) Prion protein immunocytochemistry helps to establish the true incidence of prion diseases. *Neurosci. Lett.* **147,** 67–71.
11. Hashimoto, K., Mannen, T., and Nobuyuki, N. (1992) Immunohistochemical study of kuru plaques using antibodies against synthetic prion protein peptides. *Acta Neuropathol.* **83,** 613–617.
12. Kitamoto, T., Shin, R., Doh-Ura, K., Tomokane, N., Miyazano, M., Muramoto, T., et al. (1992) Abnormal isoform of prion proteins accumulates in the synaptic structures of the CNS in patients with Creutzfeldt-Jakob disease. *Am. J. Pathol.* **140,** 1285–1294.
13. Hayward, P. A. R., Bell, J. E., and Ironside, J. W. (1994) Prion protein immunocytochemistry: the development of reliable protocols for the investigation of Creutzfeldt-Jakob disease. *Neuropathol. Appl. Neurobiol.* **20,** 375–383.

14. Brown, P., Gajdusek, D. C., Gibbs, C. J., and Asher, D. M. (1985) Potential epidemic of Creutzfeldt-Jakob disease from human growth hormone therapy. *N. Engl. J. Med.* **313**, 728–731.

15. Pickering-Brown, S., Mann, D. M. A., Owen, F., Ironside, J. W., de Silva, R., Roberts, D. A., et al. (1995) Allelic variations in apolipoprotein E and prion protein genotype related to plaque formation and age of onset in sporadic Creutzfeldt-Jakob disease. *Neurosci. Lett.* **187**, 127–129.

16. Tateishi, J. and Kitamoto, T. (1995) Inherited prion diseases and transmission to rodents. *Brain Pathol.* **5**, 53–59.

17. Billette de Villemeur, T., Gelott, A., Deslys, J. P., Dormont, D., Duyckaerts, C., Jardin, L., et al. (1994) Iatrogenic Creutzfeldt-Jakob disease in three growth hormone recipients: a neuropathological study. *Neuropathol. Appl. Neurobiol.* **20**, 111–117.

18. Fraser, H. (1993) Diversity in the neuropathology of scrapie-like disease in animals. *Br. Med. Bull.* **49**, 792–809.

19. Wells, G. A. H., Wilesmith, J. W., and McGill, I. S. (1992) Bovine spongiform encephalopathy. *Brain Pathol.* **1**, 69–78.

20. Prusiner, S. B. and DeArmond, S. J. (1991) Molecular biology and pathology of scrapie and the prion diseases of humans. *Brain Pathol.* **1**, 297–309.

21. DeArmond, S. L. (1993) Overview of the transmissible spongiform encephalopathies: prion protein disorders. *Br. Med. Bull.* **49**, 725–737.

22. Baker, H. F. and Ridley, R. M. (1992) The genetics and transmissibility of human spongiform encephalopathy. *Neurodegeneration* **1**, 3–16.

23. Serban, D., Taraboulos, A., DeArmond, S. J., and Prusiner, S. B. (1990) Rapid detection of Creutzfeldt-Jakob disease and scrapie prion proteins. *J. Neuropathol. Exp. Neurol.* **49**, 290.

24. Kitamoto, T., Muramoto, T., Mohri, S., Doh-Ura, K., and Tateishi, J. (1991) Abnormal isoform of prion protein accumulates in follicular dendritic cells in mice with Creutzfeldt-Jakob disease. *J. Virol.* **65**, 6292–6295.

25. Tateishi, J. and Kitamoto, T. (1993) Developments in diagnosis for prion diseases. *Br. Med. Bull.* **49**, 971–979.

26. Klatzo, I., Gajdusek, D. C., and Zigas, V. (1959) Pathology of kuru. *Lab. Invest.* **8**, 799–847.

27. Goodbrand, I. A., Ironside, J. W., Nicolson, D., and Bell, J. E. (1995) Prion protein accumulation in the spinal cords of patients with sporadic and growth hormone associated Cruetzfeldt-Jakob disease. *Neurosci. Lett.* **183**, 127–130.

28. Gambetti, P., Parchi, P., Petersen, R. B., Chen, S. G., and Lugaresi, E. (1995) Fatal familial insomnia and familial Creutzfeldt-Jakob disease: clinical, pathological and molecular features. *Brain Pathol.* **5**, 43–51.

29. Ghetti, B., Tagliavini, F., Masters, C. L., Beyreuther, K., Giaccone, G., Verga, L., et al. (1989) Gerstmann-Straussler-Scheinker disease. II. Neurofibrillary tangles and plaques with PrP amyloid coexist in an affected family. *Neurology* **39**, 1453–1461.

30. de Silva, R., Ironside, J. W., McCardle, L., Esmonde, T., Bell, J. E., Will, R. G., et al. (1994) Neuropathological phenotype and "prion protein" genotype correlation in sporadic Creutzfeldt-Jakob disease. *Neurosci. Lett.* **179**, 50–52.

31. Turner, C., Bell, J. E., and Ironside, J. W. (1993) Localisation of microglia in CNS amyloid plaques: an immunocytochemical and confocal microscopic study. *J. Pathol.* **170**, 401.

32. Miyazono, M., Kitamoto, T., Iwaki, T., and Tateishi, J. (1992) Colocalization of prion protein and A4/β protein in the same amyloid plaques in patients with Gerstmann-Straussler syndrome. *Acta Neuropathol.* **83**, 333–339.

33. Martinez-Lage, J. F., Poza, M., Sola, J., Tortosa, J. G., Brown, P., Cervenakova, L., et al. (1994) Accidental transmission of Creutzfeldt-Jakob disease by dural cadaveric grafts. *J. Neurol. Neurosurg. Psychiat.* **57**, 1091–1094.

34. Will, R. G. and Matthews, W. B. (1982) Evidence for case-to-case transmission of Creutzfeldt-Jakob disease. *J. Neurol. Neurosurg. Psychiat.* **45**, 235–238.

35. Gibbs, C. J., Joy, A., Heffner, R., Franko, M., Miyazaki, M., Asher, D., et al. (1985) Clinical and pathological features and laboratory confirmation of Creutzfeldt-Jakob disease in a recipient of pituitary-derived human growth hormone. *N. Engl. J. Med.* **313**, 734–738.

36. Weller, R. O., Steart, P. V., and Powell-Jackson, J. D. (1986) Pathology of Creutzfeldt-Jakob disease associated with pituitary-derived human growth hormone administration. *Neuropathol. Appl. Neurobiol.* **12**, 117–129.

37. Medori, R., Tritschler, J. H., LeBlanc, A., Villare, F., Manetto, V., Ghen, H. Y., et al. (1992) Fatal familial insomnia, a prion disease with a mutation at codon 178 of the prion protein gene. *N. Engl. J. Med.* **7**, 444–449.

38. Collinge, J., Owen, F., Poulter, M., Leach, M., Crow, T. J., Rossor, M. N., et al. (1990) Prion dementia without characteristic pathology. *Lancet* **336**, 7–9.

39. Tomlinson, B. E. (1992) Ageing and the dementias, in *Greenfield's Neuropathology*, 5th ed. (Adams, J. H. and Duchen, L. W., eds.), Edward Arnold, London, pp. 1284–1410.

40. Tateishi, J., Kitamoto, T., Doh-ura, K., Boellaard, J. W., and Peiffer, J. (1992) Creutzfeldt-Jakob disease with amyloid angiopathy: diagnosis by immunologic analysis and transmission experiments. *Acta Neuropathol.* **83**, 559–563.

41. Gray, F., Chretien, F., Cesaro, P., Chatelain, J., Beaudry, P., Laplanche, J. L., et al. (1994) Creutzfeldt-Jakob disease and cerebral amyloid angiopathy. *Acta Neuropathol.* **88**, 106–111.

42. Laszlo, L., Lowe, J., Self, T., Kenward, N., Landon, M., McBride, T., et al. (1992) Lysosomes as key organelles in the pathogenesis of prion encephalopathies. *J. Pathol.* **166**, 333–341.

43. Askanas, V., Bilak, M., Engel, W. K., Alvarez, R. B., Tome, F., and Leclerc, A. (1993) Prion protein is abnormally accumulated in inclusion-body myositis. *Neuroreport* **5**, 25–28.

44. Westaway, D., DeArmond, S. J., Cayetano-Canlas, J., Groth, D., Foster, D., Yang, S., et al. (1994) Degeneration of skeletal muscle, peripheral nerves and the central nervous system in transgenic mice overexpressing wild-type prion proteins. *Cell* **76**, 117–129.

45. Sutherland, K., Barrie, C., and Ironside, J. W. (1994) Automatic quantification of amyloid plaque formation in human spongiform encephalopathy. *Neurodegeneration* **3**, 293–300.

46. Advisory Committee on Dangerous Pathogens (1994) *Precautions for Work with Human and Animal Transmissible Spongiform Encephalopathies*. HMSO, London.

5

The Diagnosis
of Bovine Spongiform Encephalopathy
and Scrapie by the Detection of Fibrils
and the Abnormal Protein Isoform

Michael J. Stack, Paula Keyes, and Anthony C. Scott

1. Introduction

Traditionally, confirmation of Transmissible Spongiform Encephalopathy (TSE) disease in humans or animals is by conventional light microscopy of stained tissue sections prepared from specific sites of formalin-fixed tissue after embedding in paraffin wax. This is the statutory method of confirmation of Bovine Spongiform Encephalphathy (BSE) in the United Kingdom and it is also used in other Member States of the European Community. However, two additional diagnostic criteria for TSE, the detection of disease-specific fibrils by transmission electron microscopy and the detection of the main constituent of the fibrils, an abnormal protein by immunoblotting, have also been reported. The former is the current method of confirmation for the statutory diagnosis of scrapie in the United Kingdom.

The purpose of this chapter is to provide a clear account of easy to use and effective protocols, including reagent preparation, interpretation of results, and equipment required, for the extraction and detection of fibrils and the abnormal protein from central nervous tissue of clinically suspect cases of BSE in cattle or scrapie in sheep. The protocols also can be applied to other species (such as goats, moufflon, captive wild *Bovidae*, domestic cats, or wild *Felidae*) in which TSE is suspected, but experience of the reliability of the methods in such species is less extensive.

1.1. Electron Microscopy

Disease-specific structures, called Scrapie-Associated Fibrils (SAF), were first reported in brain extracts from affected mice and hamsters experimentally

From: *Methods in Molecular Medicine: Prion Diseases*
Edited by: H. Baker and R. M. Ridley Humana Press Inc., Totowa, NJ

infected with scrapie by the use of negative contrast Transmission Electron Microscopy (TEM) *(1,2)*. Rod-like structures, designated "prion rods," were also reported in extracts from scrapie-infected hamster brains *(3)*. Further studies also showed fibrils to be present in the human TSE diseases; Creutzfeldt-Jakob disease, Gerstmann-Sträussler-Scheinker syndrome, and kuru *(4)*. The role of SAF detection as a diagnostic aid for natural infections of scrapie in sheep has been well established *(5–8)* demonstrating a close correlation (90–100%) between fibril detection and histopathological diagnosis. Fibrils closely resembling those detected from natural sheep scrapie-affected brains have also been extracted from the brains of BSE-affected cattle *(9–12)*. Similar fibrils have also been reported in a domestic cat *(13)*, a captive cheetah *(14)*, a greater kudu antelope *(15)*, and in Rocky Mountain elk *(16)*.

Although biochemically similar, electron microscopy of SAF and prion rods does show morphological differences in fibril size, orientation, and clarity of substructure. It has been suggested that these morphological differences are not only owing to the differences in preparation method used but may be affected by the centrifugal sedimentation rate of the species of the brain tissue tested, or by the strain of scrapie agent used for transmission into laboratory rodents *(17)*. For convenience the term "scrapie-associated fibrils" or SAF is used throughout this chapter, irrespective of the species from which they are derived.

1.2. Immunoblotting

Biochemically, the main constituent of SAF and prion rods is a protease-resistant neuronal membrane glycoprotein, which is disease specific, and is an abnormal isoform of a host protein. The normal isoform is found in a number of tissues, including the central nervous system (CNS), but is distinguishable from the abnormal isoform because the latter forms disease-specific fibrils and is relatively resistant to protease digestion.

Immunoblotting is used to detect the purified, protease-resistant, abnormal protein by its molecular weight and reaction with specific antibodies. At present, the test only can be regarded as consistently effective when applied to CNS tissue where the abnormal isoform accumulates in particularly large quantities. Further technical developments may allow the reliable and consistent detection of much smaller amounts of the protease-resistant protein in lymphoreticular or other more accessible tissues.

Several different terminologies for the normal and abnormal isoforms of the protein are used in the scientific literature. Treatment of infected brain extracts with Sarkosyl (a detergent), followed by centrifugation and proteinase K (PK) digestion, can be used to separate the two protein isoforms that have been termed "prion protein" *(18)* or "protease-resistant protein" (PrP) *(19)*. The normal cellular protein isoform (PrP^C) is soluble and susceptible to proteinase K digestion, whereas the abnormal scrapie isoform (PrP^{Sc}) is sedimentable and resistant, but treatment with detergents other than Sarkosyl, followed by centrifugation, may separate the PrP isoforms differently. For example, PrP^C may be found in either or both the sedimentable or soluble fractions, depending on the experimental conditions. The type of protease used is also an important variable in distinguishing between the normal and abnormal isoforms of PrP.

1.3. Safety

There would appear to be little risk to humans from the pathogens causing animal TSE under normal conditions, but there may be risks associated with very high exposure via parenteral routes. Therefore, the guiding safety principle is to minimize or confine the practices that may result in high exposure of the pathogens to humans. It should also be remembered that samples sent in for testing might contain conventional pathogens, some of which may cause disease in humans.

There are strict codes of practice for working with TSE-infected material, with several documents used as references. The safety situation is complex, with different hazard levels being assigned according to the species being studied, the tissue locale, and the type of procedures being carried out. The safety precautions presently recommended are formulated by the Advisory Committee on Dangerous Pathogens (ACDP) *(20)* and there are some published guidelines by the European Commission of Agriculture *(21)*. Safety is an important issue and each laboratory should formulate its own guidelines based on the most recent credible evidence and guidance publications from the ACDP. It is particularly important to recognize that the procedures employed in the protocols described in this chapter will serve to concentrate infectivity as purification proceeds.

Ideally all potentially hazardous procedures associated with possible operator exposure via parenteral routes (i.e., homogenization), should be carried out in a Class 1 Safety Cabinet (BS 5726 or equivalent) with latex gloves worn at all times. The chemicals marked **Care: Toxic** in the reagent lists are hazardous to health. Local risk assessments should be carried out and appropriate safety precautions should be taken throughout. All glassware, gloves, plasticware, and instruments used in the following protocols must be decontaminated before disposal, or reuse, by one of three procedures:

1. Porous load autoclaving at a holding temperature of 134–138°C for 18 min;
2. Soaking overnight at 20°C in sodium hypochlorite to provide a final concentration of 2% (20,000 ppm) of available chlorine;

3. Soaking overnight at 20°C in sodium hydroxide to provide a final concentration of 2N NaOH.

2. Materials

2.1. Tissue Collection

For adequate diagnosis of TSE there generally would be a priority to collect a portion of tissue from the medulla brain region (dissected at the obex) to assess any spongiform change by histopathological examination. If the brain at necropsy is considered to be adequately preserved, then the sites specified below should be sampled for SAF or PrP^{Sc} detection. However, if the brain is determined to be autolysed at necropsy and, in the opinion of the pathologist, is unsuitable for histopathology, it is recommended that the brainstem regions be sampled, including the medulla region usually retained for histopathological examination.

The source tissue for both SAF detection by TEM and PrP^{Sc} by Western immunoblotting must be untreated and especially not fixed in formalin or other fixative. Ideally it should be processed as soon after death as practical but some autolysis will not invalidate the examination *(22,23)*. It is recommended that if a significant delay is likely between death and the initiation of extraction of SAF and PrP^{Sc} (e.g., more than 12 h) then samples of brain and spinal cord should be frozen, ideally at –70°, but –20°C is acceptable. It has been recommended that tissue from more than one CNS site be sampled, particularly for natural scrapie cases *(24)*. Samples for processing should maximize the gray matter content by dissecting off white matter. Thus, for spinal cord only the central core would be used and for tissue from cerebral hemispheres and cerebellum only gray matter should be used for the extractions. If BSE brain is autolysed and unsuitable for histopathological examination, the basal nuclei and midbrain can be used as additional sites *(25)*.

2.1.1. Tissue Requirements
for Electron Microscopy (EM) and Immunoblotting (IMB)

1. Scrapie (sheep, goats, and moufflon). For both fresh and autolysed tissue the requirement is: 1 g (EM), 4 g (IMB) of spinal cord (cervical vertebrae segments C1 and C2) plus 1 g (EM), 4 g (IMB) of brainstem region.
2. BSE (cattle or other bovidae). For fresh tissue the requirement is: 1 g (EM), 4 g (IMB) of spinal cord (cervical vetebrae segments C1 and C2) plus 1 g (EM), 4 g (IMB) of caudal medulla, avoiding damage to the obex.
3. For autolysed BSE tissue the requirement is: 1 g (EM), 4 g (IMB) of brainstem, to include the medulla at obex.

These are the minimum tissue requirements for one test but collecting larger quantities does enable retests to be conducted if necessary.

2.1.2. Reagents Used in the Methods Described

2.1.2.1. REAGENTS FOR EXTRACTION, PURIFICATION, AND EM OF SAF

1. Solution A (For detergent homogenization): 10 g N-lauroylsarcosine, sodium salt (Sigma [Poole, Dorset, UK] cat. no. L5125) in 100 mL of sterile deionized water. Adjust to pH 7.4 with $1M$ sodium dihydrogen orthophosphate. Store at 4°C.
2. n-Octanol (defoaming agent, BDH cat. no. 29408).
3. Solution B (High salt solution): 10 g Sodium chloride (BDH, Merck Ltd. [Poole, Dorset, UK] cat no. 10241) and 1 g N-lauroylsarcosine in 100 mL of sterile deionized water. Adjust to pH 7.2 with $1M$ sodium dihydrogen orthophosphate or $0.1N$ sodium hydroxide. Store at 4°C.
4. Solution C (For protease digestion): 1 mg Proteinase K (Sigma cat. no. P0390 type XI 10–20 U/mg derived from *Tritirachium album*) in 100 mL of $0.01M$ Tris-HCl (BDH cat. no. 27119). Store in 1.5-mL aliquots at –70°C.
5. Negative stain (for electron microscopy): 2 g Phosphotungstic acid (Aldrich [Gillingham, Dorset, UK] cat. no. 22420-0) in 100 mL of sterile deionized water. Adjust to pH 6.6 with $10N$ and $1N$ potassium hydroxide. Store at 4°C.

2.1.2.2. REAGENTS FOR DETERGENT EXTRACTION AND PROTEINASE DIGESTION OF PrP

1. Brain lysis buffer (BLB): 10 g N-lauroylsarcosine, sodium salt in 100 mL of $0.01M$ sodium phosphate buffer, pH 7.4. Store at 4°C.
2. $1M$ Tris-HCl pH 7.4: 60.57 g Tris in 450 mL distilled water. Adjust pH to 7.4 with concentrated HCl (25–30 mL). Make up to 500 mL with distilled water. Store at 4°C.
3. 100 mM phenylmethylsulfonyl fluoride (PMSF): 0.435 g PMSF in 25 mL propan-l-ol. Store in a dark bottle at 4°C. **Care: Toxic**.
4. 100 mM N-ethylmaleimide (NEM): 0.313 g NEM in 25 mL propan-l-ol. Store in a dark bottle at 4°C. **Care: Toxic**.
5. Proteinase K (PK): 1 mg PK (Sigma cat. no. P-4914) in 1 mL distilled water. Store in 50-μL aliquots at –70°C.
6. Potassium iodide-high salt buffer (KI-HSB): 1.5 g Sodium thiosulfate, 1.0 g N-lauroyl-sarcosine in 1 mL $1M$ Tris-HCl. Add potassium iodide 10 g (for 10%) or 15 g (for 15% solution). Make up to 100 mL with distilled water. Store at 4°C.
7. 20% sucrose: 20 g sucrose in 100 mL of 10% KI HSB.

2.1.2.3. REAGENTS FOR POLYACRYLAMIDE GEL ELECTROPHORESIS AND IMMUNOBLOTTING

1. Solution D: 91.5 g Tris-HCl in 120 mL $1M$ HCl. Adjust pH to 8.9. Make up to 250 mL with distilled water and filter. Store in a glass bottle at 4°C.
2. Solution E: 30% Acrylamide: Bis (37.5:1) (International Biotechnologies Inc., New Haven, CT). Make up according to manufacturer's instructions. Store in a dark glass bottle at 4°C. **Care: Toxic**.
3. Solution F: 12.1 g Tris-HCl plus 4 mL 20% SDS *(see the following)*. Add 80 mL distilled water and adjust pH to 6.8 with $1M$ HCl. Make up to 100 mL with distilled water and filter. Store in a glass bottle at 4°C.

 4. Sodium dodecylsulfate (SDS): 20 g SDS. Make up to 100 mL with distilled water. Store at room temperature. **Care: Toxic.**
 5. 1M Tris-HCl: 60.57 g Tris-HCl in 450 mL distilled water. Adjust pH to 7.4 with concentrated HCl (25–30 mL). Make up to 500 mL with distilled water. Store at 4°C.
 6. Ammonium persulfate solution (APS): 0.14 g APS in 100 mL distilled water. **Make up fresh**.
 7. Immunostaining Kit (Auroprobe BL plus Code RPN460-467, IntenSE BL Code RPN492. Both obtained from CAMBIO, Cambridge, UK).
 8. Disruption buffer: 2 mL 20% SDS, 1 mL 1M Tris-HCl, pH 7.4. 1 mL 2-mercaptoethanol: 0.6 g sucrose, 1–2 drops bromophenol blue, 15 mL distilled water. Adjust pH to 6.8 with 1M HCl. Store at room temperature. **Care: Toxic.**
 9. Electrode buffer: 6 g Tris, 28.8 g glycine, 10 mL 20% SDS. Make up to 2 L with distilled water.
10. Blotting buffer: 9.09 g Tris, 43.2 g glycine, 600 mL methanol (analar). Make up to 3 L with distilled water. (pH should be approx 8.3—do not adjust.)
11. 12% Acrylamide separating gel (enough for one gel plate): 5 mL Solution D, 15.8 mL Solution E, 0.2 mL 20% SDS, 8.5 mL distilled water, 10 mL APS, 11.5 µL N,N,N',N',-Tetramethylethylenediamine (TEMED). **Care: Toxic**.
12. Stacking gel (enough for one gel plate): 1.28 mL Solution E, 1.5 mL Solution F, 3.23 mL distilled water, 6 mL APS, 3.75 µL TEMED.

3. Methods

3.1. Extraction and Purification of SAF for Electron Microscopy

The extraction techniques used to detect SAF are based on providing a crude mitochondrial pellet that is subfractionated to obtain synaptosomal and synaptic plasma membrane fractions. The detergent renders any membrane protein nonsedimentable at low centrifugal g-forces, but sedimentable at higher forces. The sedimentable pellet is treated with a protease enzyme and with extractions from brains affected by a TSE disease the surviving PrP^{Sc} can be visualized as SAF in the electron microscope.

The method described in this chapter consists of extracting the CNS tissue with N-lauroyl sarcosine detergent (Stage 1) and then purifying SAF in the presence of the enzyme Proteinase K (Stage 2). The method is a modification of two previously published techniques *(26,27)*. A drop of the final extract is placed on a formvar/carbon-coated electron microscope grid, negatively stained with 2% phosphotungtic acid and examined in the TEM microscope at a magnification of between 19 and 30 K (Stage 3). Micrographs of fibrils are taken on the microscope, the negatives are developed, and prints made, which aids interpretation of the results.

All centrifugations are carried out using a Beckman (Fullerton, CA) TL100 benchtop ultracentrifuge with a 100.3 fixed angle rotor and thick walled polycarbonate centrifuge tubes (max. speed of rotor 100,000 rpm giving a g max figure of 540,000g). A diagrammatic representation of each of the three stages of the procedure is shown in Figs. 1A–C.

3.1.1. Stage 1—Detergent Extraction for SAF Detection (Fig. 1A)

1. Take 1 g of appropriate CNS tissue, place in a glass tissue grinder, and add 2.5 mL of Solution A.
2. Homogenize with at least 20 strokes of the piston and pour off into polycarbonate centrifuge tube.
3. If the homogenate has too much foam, add two to three drops of *n*-octanol. Leave homogenate at room temperature for 30 min.
4. Centrifuge at 22,000*g* average (20,000 rpm) for 10 min.
5. Gently recover the supernatant by pipeting and discard pellet.

3.1.2. Stage 2—Proteinase K Digestion for SAF Detection (Fig. 1B)

1. Top up the supernatant with Solution A and mix by aspiration with a pipet.
2. Centrifuge at 540,000*g* average (100,000 rpm) for 20 min.
3. Discard the supernatant and resuspend the pellet by aspiration with a pipet in 2.5 mL of solution B.
4. Centrifuge at 540,000*g* average (100,000 rpm) for 25 min.
5. Discard the supernatant and resuspend the pellet by aspiration with a pipet in 1.5 mL Solution C.
6. Stir for 1 h at 37°C using a Teflon-coated magnet.
7. Centrifuge at 22,000*g* average (10,000 rpm) for 10 min using the Beckman micro-centrifuge tubes (Eppendorf type) and the tube adaptors for the TL100.3 rotor.
8. Discard supernatant and resuspend pellet by aspiration with a pipet in 50 μL of sterile, deionized water.
9. Mix for 30 s using a vortex mixer.

3.1.3. Stage 3—Negative Staining for Transmission Electron Microscopy (Fig. 1C)

1. Coat 300 mesh, 3 mm diameter electron microscope grids with 0.4% formvar (made up in chloroform). Stabilize with evaporated carbon and plasma glo treat prior to use.
2. Pipet one drop (25 μL) of the final extract onto a strip of dental wax and using the Dumont EM forceps place one formvar/carbon/plasma glo-treated EM grid onto this drop.
3. Leave for 10 s, then blot, using the edge of the grid, on the fine grade filter paper.
4. Float the grid on 25 μL of 2% phosphotungstic acid at pH 6.6.
5. Leave for 10 s, then blot, using the edge of the grid, on the fine grade filter paper. Cover with a clean Petri dish and allow to dry.
6. Examine the grid in a transmission electron microscope at magnifications between 19 and 30 k with an accelerating voltage of approx 80 Kv.

3.2. Interpretation of Results

An area of at least 20 grid squares is examined for a minimum of 20 min before the sample is declared negative for SAF or BSE fibrils. The morpho-

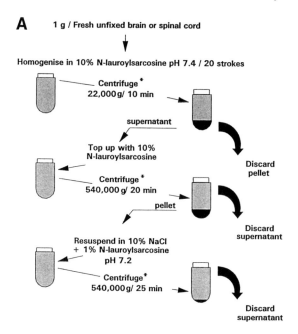

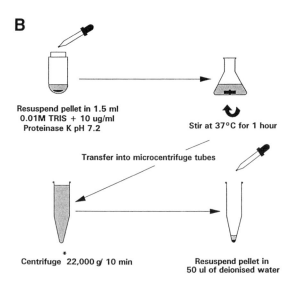

Fig 1. N-lauroylsarcosine and proteinase K extraction technique for the detection of fibrils by negative stain and transmission electron microscopy. **(A)** Stage 1—detergent extraction. **(B)** Stage 2—proteinase K digestion. *Asterisk indicates Beckman TL100 ultracentrifuge with a TL100.3 fixed angle rotor. *(continued on following page)*

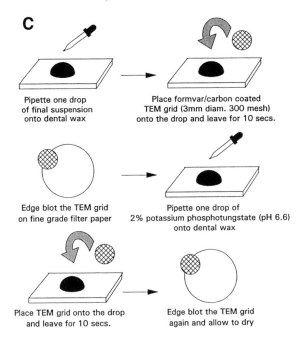

Examine the grid in a TEM at magnifications between 20k and 40k
and at an accelerating voltage of approximately 80Kv

Fig 1. *(continued from previous page)* N-lauroylsarcosine and proteinase K extraction technique for the detection of fibrils by negative stain and transmission electron microscopy. **(C)** Stage 3—negative stain procedure.

logical criteria used for the identification of SAF are those defined in the original report of SAF in which two types of SAF were defined *(2)*. Type I consisted of two filaments measuring 2–4 nm in diameter, forming a fibril 12–16 nm in diameter with a 2–4 nm space between filaments. Fibrils had a marked linearity, and were usually straight but occasionally sharply bent. They were 100–500 nm in length (although some up to 1 μ were observed) and every 40–60 nm the fibril narrowed to 4–6 nm. Type II SAF consisted of four filaments arranged in a parallel fashion forming a fibril 27–34 nm in diameter and sometimes consisted of four filaments at one end and two pairs of filaments forked at the other end. There appeared to be a great deal of variability in the way these individual filaments interacted so intermediate forms were also observed. Typical configurations of ovine SAF and BSE fibrils using the described procedures are shown in Fig. 2A,B. Similar fibrils detected in brain extracts from a domestic cat and a greater kudu are shown in Fig. 2C,D.

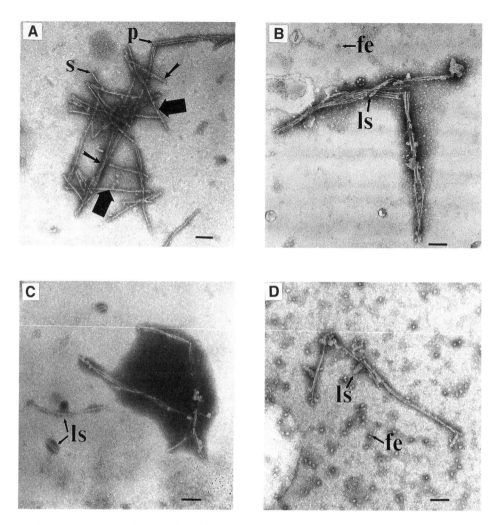

Fig. 2. Electron micrographs of fibrils from four different species obtained using the n-lauroylsarcosine extraction and proteinase K digestion technique followed by staining with phosphotungstic acid to provide negative contrast. **(A)** This micrograph shows the diverse nature of SAF, as found in extracted ovine brain tissue. The Type I structures (large arrows) have a "bendy" appearance but also present are the brittle looking fibrils (small tailed arrows), parallel forms (P), and short forms (S). Barline = 100 nm. **(B)** Micrograph of a cluster of extracted BSE fibrils showing predominately Type I configurations that are closely grouped. Residual n-lauroylsarcosine detergent can be clearly identified by its characteristic striped appearance (ls). Ferritin particles (fe) are often seen in brain tissue preparations after biochemical extraction. Barline = 100 nm. Electron micrographs of fibrils from four different species obtained using the n-lauroylsarcosine extraction and proteinase K digestion technique followed by staining with phosphotungstic acid to provide negative contrast. **(C)** Micrograph showing predominately Type I fibrils extracted from the brain tissue of a domestic cat. *(continued)*

Results are designated as follows:

1. Positive: Identification of undisputed fibrils of any type within the definition of those described in the original publication of SAF *(2)*.
2. Negative: No fibrils fitting the definition of SAF *(2)* found after completing a search of 20 grid squares in a 20-min period.
3. Inconclusive: Where on rare occasions a sample cannot be classified into either positive or negative a further grid or grids should be prepared for examination either from the same extract or an extract from another part of central nervous tissue. Examine as determined and reclassify.

3.3. Extraction and Purification of PrP^{Sc} for Immunoblotting

In outline, the method consists of extracting and purifying the abnormal, fibril form of PrP from brain or cervical spinal cord *(26)*, analyzing the purified protein by polyacrylamide gel electrophoresis (PAGE) *(28)*, followed by Western blotting *(29)*.

All centrifugations are carried out using a Beckman TL100 benchtop ultracentrifuge with a 100.3 fixed angle rotor and thick walled polycarbonate centrifuge tubes (max speed of rotor 100,000 rpm giving a *g* max figure of 540,000*g*), or using a Beckman L8-60M floor standing ultracentrifuge with a 70Ti fixed angle rotor (max speed of rotor 60,000 rpm giving a *g* max figure of 371,000*g*). Diagrammatic representations of each of the two stages of the procedure are shown in Fig. 3A,B.

3.3.1. Stage 1—Detergent Extraction Procedure for PrP (Fig. 3A)

1. Take 4 g of brain or cervical spinal cord.
2. Cut into small pieces and add 5 mL of BLB with 10 µL of 100 m*M* PMSF and 10 µL of 100 m*M* NEM.
3. Homogenize in a sealed unit laboratory mixer (e.g., Silverson) for 30 s.
4. Add another 5 mL of BLB and rehomogenize if necessary.
5. Centrifuge at either:
   a. 20,000*g* average (17,000 rev/min) for 30 min at 10°C in a 70-Ti Beckman ultracentrifuge rotor; or
   b. 22,000*g* average (20,000 rev/min) for 10 min at 10°C in a Beckman TL100 bench ultracentrifuge.
6. Carefully tip the supernatant into clean centrifuge tubes and centrifuge at either:
   a. 177,000*g* average (46,000 rev/min) for 2 h 30 min at 10°C in a 70 Ti Beckman ultracentrifuge rotor; or

---

The detergent residues may appear as oval or in strands (ls). **(D)** Micrograph of fibrils extracted from a greater kudu *(Tragelaphus strepsiceros)*. N-lauroylsarcosine detergent residue is again apparent (ls) and in this micrograph ferritin particles can be seen in abundance (fe). Ferritin particles have been observed in extracts from positive and negative animals from all species.

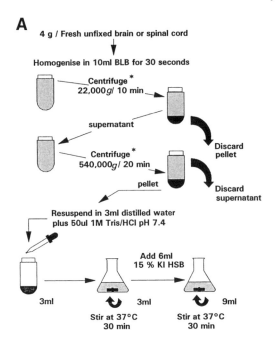

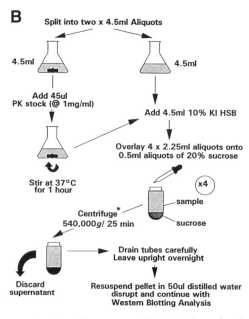

Fig. 3. Extraction method for disease-specific PrP detection by Western blotting. **(A)** Stage 1—detergent extraction. **(B)** Stage 2—proteinase K digestion. *Asterisk indicates Beckman TL100 centrifuge with TL100.3 fixed angle rotor.

    b. 540,000*g* average (100,000 rev/min) for 20 min at 10°C in a Beckman TL100 bench ultracentrifuge.

7. Discard the supernatant and take up the pellet in 3 mL distilled water with 50 μL 1*M* Tris-HCl, pH 7.4 (0.0167*M*), by gentle aspiration with a pipet.
8. Incubate in a waterbath at 37°C for 30 min while stirring.
9. Add 6 mL of 15% KI-HSB and incubate for a further 30 min as in step 8.
10. Divide the mixture above into two aliquots of 4.5 mL.

3.3.2. Stage 2—Proteinase K Digestion and Purification of PrP (Fig. 3B)

1. To one aliquot (from step 10 above) add 45 μL of 1 mg/mL PK and incubate for 1 h as in step 8.
2. To the other 4.5 mL add 4.5 mL of 10% KI-HSB and then either
    a. Overlay the total onto 2 mL of 20% sucrose and centrifuge at 189,000*g* average (51,000 rev/min) for 1 h 30 min at 10°C in a 70 Ti Beckman ultracentrifuge rotor; or
    b. Divide into 4 × 2.25 mL aliquots and overlay each onto 0.5 mL of 20% sucrose and centrifuge at 540,000*g* average (100,000 rev/min) for 25 min at 10°C in a Beckman TL100 bench ultracentrifuge.
13. Carry out step 12 on the PK-treated sample (from step 11).
14. Carefully tip off supernatants and drain the tubes well.
15. Leave tubes standing upright overnight (or 2–3 h) to allow the pellet to dry a little (do not leave inverted because pellet may become detached).
16. Take up the pellet in a small (approx 50 μL) amount of distilled water, by gentle aspiration with a micropipet.

    Samples now can be treated in the appropriate manner and analyzed by polyacrylamide gel electrophoresis and Western blotting.

3.4. Polyacrylamide Gel Electrophoresis (PAGE) and Western Immunoblotting (Fig. 4A,B)

    Methods vary with different electrophoresis systems. The method described here applies to the large format (18 × 15 cm) BRL vertical gel electrophoresis system and the BioRad (Richmond, CA) TransBlot cell. For different equipment, the manufacturer's instruction manuals should be followed. **Note:** Minigel systems that economize on reagents are available, but so far the system described herein has shown better resolution.

3.4.1. Procedure for PAGE and Western Immunoblotting

1. Wash, rinse, and dry the glass electrophoresis plates ensuring that they are free of all grease and detergents.
2. Wipe the inside surfaces of the plates with acetone.
3. Lightly grease the 1.5-mm thick spacers with petroleum jelly (Vaseline) and assemble the plates, clipping firmly together.

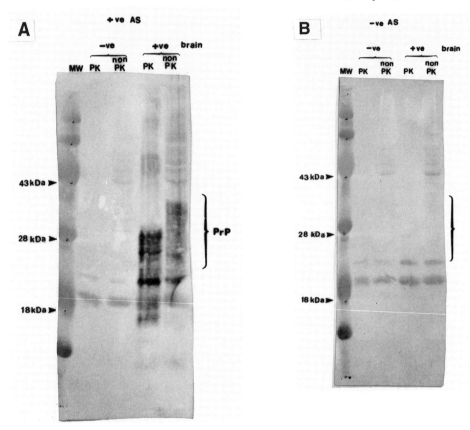

Fig. 4. Immunoblots of brain material from negative and BSE affected animals. **(A)** Positive antiserum (+veAS) tested against negative (-ve) and positive (+ve) BSE brain material, with either proteinase K (PK) or non-proteinase K (non PK) treatment. Immunoblots of brain material from negative and BSE affected animals. **(B)** Negative antiserum (-veAS) tested against negative (-ve) and positive (+ve) BSE brain material, with either proteinase K (PK) or non-proteinase K (non PK) treatment. Far left-hand lane of each immunoblot shows a range of molecular weight markers.

4. Level the plates with a spirit level.
5. Measure down approx 3.5 cm from the top of the smallest glass plate and make a mark. (This will be the top of the separating gel—leaving room for a stacking gel.)
6. Make up the 12% acrylamide separating gel as indicated in Section 2.1.2.3. Volumes are enough for one gel plate.
7. Mix gently and degas using a vacuum pump for 1–2 min. (Acrylamide will not set in the presence of oxygen.)
8. Pour or pipet the gel between the glass plates to approx 2 mm above the marks made earlier (gel will shrink slightly on polymerization).

9. Carefully pipet a layer of distilled water on top of the gel. (Allow the gel to set but not in strong sunlight.) When the gel has set (approx 45 min) a sharp line is visible between the gel/water layer.

10. Make up the stacking gel (*see* Section 2.1.2.3.).

11. Mix and degas as in step 7.

12. Pour the water from the top of the separating gel. Add the stacking gel, with a pipet, to completely fill the unoccupied space.

13. Place the 20-well "comb" into the stacking gel, ensuring that no bubbles are trapped under the teeth. Allow to set. The outside wells should not be used because of edge effects, where "smiling" will be seen. Four wells/sample will be needed, plus two wells for the molecular weight markers. Therefore, four samples can be analyzed on each gel plate. Gels either can be used immediately or covered with damp tissue and kept overnight at room temperature for use the following day.

14. Make up the electrode buffer and pour it carefully into the lower reservoir of the electrophoresis tank (pipet out any froth that may occur).

15. Remove the clamps from around the gel/glass plate assembly and remove the lower spacer.

16. Place the assembly into the lower buffer, ensuring that no air is left in contact with the lower edge of the gel.

17. Clamp the assembly to the top reservoir of the tank and fill with electrode buffer. Check for leaks.

18. Carefully remove the comb from the gel.

19. Mix the pellet in distilled water (from step number 16 of the PrP27-30 extraction and purification procedure) with an equal volume of disruption buffer.

20. Boil for 6 min.

21. Load 10–50 µL of sample into each well as appropriate. PK and non-PK treated samples should be tested against positive and negative antisera so that four wells/sample will be needed (each being loaded with the equivalent of approx 1 g of original brain material).

22. Load a prestained high range protein molecular weight marker (Gibco-BRL [Gaithersburg, MD] cat. no. 560-6041LA), at one end of each batch of samples so that it is easy to cut the membrane in the correct place after Western blotting.

23. Connect the tank to an appropriate power supply and run at a constant current of 30 mA/gel plate until the samples reach the stacking/separating gel interface (approx 1 h).

24. Turn the current to 40 mA/gel and continue electrophoresis until the blue dye front approaches the bottom of the separating gel (approx 4 h). Before the electrophoresis is complete:

25. Make up the blotting buffer and set up 1 blotting cassette/gel as described in steps 26–28.

26. Fill a large tray with blotting buffer and lay the black side of an open cassette into it (ensure the hinges are on the outside of the cassette).

27. Lay one piece of White "Scotchbrite" (BioRad fiber pads) onto each half of the cassette and onto each of those place two pieces of chromatography paper (Whatman [Maidstone, UK] 1 Chr) cut to the size of the "Scotchbrite."

28. Cut a piece of transfer membrane to the size of the gel and "wet" according to the manufacturer's specifications. Immobilon is currently used, which needs wetting with methanol and then soaking in distilled water and blotting buffer. When electrophoresis from step 24 is complete:

29. Turn off the power supply and remove glass plate assembly from the tank.

30. Pry apart the glass plates and cut away the stacking gel.

31. Remove the gel and lay onto the chromatography paper (on the black half of the cassette). Smooth over to ensure that no bubbles are trapped between any of the layers.

32. Lay the "wetted" transfer membrane onto the gel and again smooth over. (Proteins will not transfer where there are air bubbles.)

33. Lay on the other two layers of chromatography paper and "Scotchbrite," smooth over, and close the cassette. There should now be a "sandwich" consisting of the following layers:
    a. Black side of cassette.
    b. Scotchbrite.
    c. 2X chromatography papers.
    d. Gel.
    e. Transfer membrane.
    f. 2X chromatography papers.
    g. Scotchbrite.
    h. Clear side of cassette.

34. Place the cassette into the blotting tank with the black side next to the negative (black) electrode.

35. Top up the blotting tank with blotting buffer, and gently shake the cassette up and down to dislodge any air bubbles.

36. Connect the tank to an appropriate power supply and run at a constant voltage of 30 V overnight.

37. When blotting is complete, remove the cassette and open. The prestained markers should be visible on the transfer membrane.

38. Mark the furthest position traveled of each marker with a pin hole, since the colors may fade during subsequent incubations.

The immunostaining (including blocking) can be carried out according to the instructions supplied with the Auroprobe BL plus and IntenSE BL Kits obtained from CAMBIO.

The primary antibody currently used at CVL is 1B3, a polyclonal rabbit antiserum raised against formic-acid-treated SAF protein extracted from the brains of clinically affected mice infected with the ME7 strain of scrapie, as supplied by The Institute for Animal Health, AFRC/MRC Neuropathogenesis Unit (Edinburgh). The antibody is used at a dilution of 1:1000. (A negative rabbit serum is used, at the same dilution, as a negative control.)

Detection of scrapie PrP can be achieved by silver staining, but this is not as successful for BSE because of the lower levels of PrP in the brain. Kits for silver staining can be obtained commercially (BioRad Laboratories, cat. no. 161-0443).

3.5. Interpretation of Results

3.5.1. Positive Brain—Fig. 4A,B (Lanes Labeled +ve Brain)

1. Treated with positive antiserum (Fig. 4A). For non-PK-treated samples a wide, diffuse region of stained bands is present at sites corresponding to proteins of molecular masses of 33–35 kDa. For PK-treated samples a wide, diffuse region of stained bands is present at sites corresponding to proteins of molecular masses of 27–30 kDa.
2. Treated with negative antiserum (Fig. 4B).

No stained bands can be identified in the discussed regions.

3.5.2. Negative Brain—Fig. 4A,B (Lanes Labeled -ve Brain)

No stained bands can be identified in the described regions with either the positive antiserum (Fig. 4A) or the negative antiserum. **Note:** Bands of approximate molecular masses 19, 22, and 43 kDa occur in blots reacted with positive and negative antisera (Fig. 4A,B); they occur even when no primary antibody has been reacted (picture not shown). These bands are at present considered to be nonspecific.

3.5.3. Inconclusive

A further sample should be prepared from the same or different parts of the frozen central nervous tissue. Examine as determined and reclassify.

References

1. Merz, P. A., Rohwer, R. G., Kascsak, R., Wisniewski, H. M., Somerville, R. A., Gibbs, C. J. J., et al. (1984) Infection-specific particle from the unconventional slow virus diseases. *Science* **225**, 437–440.
2. Merz, P. A., Somerville, R. A., Wisniewski, H. M., and Iqbal, K. (1981) Abnormal fibrils from scrapie-infected brain. *Acta Neuropathol.* **54**, 63–74.
3. Prusiner, S. B., McKinley, M. P., Bowman, K. A., Bolton, D. C., Bendheim, P. E., Groth, D. F., et al. (1983) Scrapie prions aggregate to form amyloid-like birefringent rods. *Cell* **35**, 349–358.
4. Merz, P. A., Somerville, R. A., Wisniewski, H. M., Manuelidis, L., and Manuelidis, E. E. (1983) Scrapie-associated fibrils in Creutzfeldt-Jakob disease. *Nature* **306**, 474–476.
5. Dawson, M., Mansley, L. M., Hunter, A. R., Stack, M. J., and Scott, A. C. (1987) Comparison of scrapie associated fibril detection and histology in the diagnosis of natural sheep scrapie. *Vet. Rec.* **121**, 591.

6. Gibson, P. H., Somerville, R. A., Fraser, H., Foster, J. D., and Kimberlin, R. H. (1987) Scrapie associated fibrils in the diagnosis of scrapie in sheep. *Vet. Rec.* **120,** 125–127.

7. Rubenstein, R., Merz, P. A., Kascsak, R. J., Carp, R. I., Scalici, C. L., Fama, C. L., et al. (1987) Detection of scrapie-associated fibrils (SAF) and SAF proteins from scrapie affected sheep. *J. Infect. Dis.* **156,** 36–46.

8. Scott, A. C., Done, S. H., Venables, C., and Dawson, M. (1987) Detection of scrapie-associated fibrils as an aid to the diagnosis of natural sheep scrapie. *Vet. Rec.* **120,** 280–281.

9. Hope, J., Reekie, L. J. D., Hunter, N., Multhaup, G., Beyreuther, K., White, H., et al. (1988) Fibrils from brains of cows with new cattle disease contain scrapie-associated protein. *Nature* **336,** 390–392.

10. Wells, G. A. H. and Scott, A. C. (1988) Neuronal vacuolation and spongiosus; a novel encephalopathy of adult cattle. *Neuropathol. Appl. Neurobiol.* **14,** 247.

11. Wells, G. A. H., Scott, A. C., Johnson, C. T., Gunning, R. F., Hancock, R. D., Jeffrey, M., et al. (1987) A novel progressive spongiform encephalopathy in cattle. *Vet. Rec.* **121,** 419–420.

12. Wells, G. A. H., Scott, A. C., Wilesmith, J. W., Simmons, M. M., and Matthews, D. (1994) Correlation between the results of a histopathological examination and the detection of abnormal brain fibrils in the diagnosis of bovine spongiform encephalopathy. *Res. Vet. Sci.* **56,** 346–351.

13. Pearson, G. R., Wyatt, J. M., Gryffed-Jones, T. J., Hope, J., Chong, A., Higgins, R. J., et al. (1992) Feline spongiform encephalopathy: fibril and PrP studies. *Vet. Rec.* **131,** 307–310.

14. Peet, R. L. and Curran, J. M. (1992) Spongiform encephalopathy in an imported cheetah *(Actinonyx jubatus). Aus. Vet. J.* **69,** 117.

15. Kirkwood, J. K., Wells, G. A. H., Cunningham, A. A., Jackson, S. I., Scott, A. C., Dawson, M., et al. (1992) Scrapie-like encephalopathy in a greater kudu *(Tragelaphus strepsiceros)* which had not been fed ruminant-derived protein. *Vet. Rec.* **130,** 365–367.

16. Guiroy, D. C., Williams, E. S., Song, K. J., Yanagihara, R., and Gajdusek, D. C. (1993) Fibrils in brains of Rocky Mountain elk with chronic wasting disease contain scrapie amyloid. *Acta Neuropathol.* **86,** 77–80.

17. Liberski, P. P. and Brown, P. (1993) Scrapie-associated fibrils, in *Light and Electron Microscopic Neuropathology of Slow Virus Disorders* (Liberski, P. P., ed.), CRC, Boca Raton, FL.

18. McKinley, M. P., Bolton, D. C., and Prusiner, S. B. (1983) A protease-resistant protein is a structural component of the scrapie prion. *Cell* **35,** 57–62.

19. Bolton, D. C., McKinley, M. P., and Prusiner, S. B. (1982) Identification of a protein that purifies with the scrapie prion. *Science* **218,** 1309–1311.

20. Advisory Committee on Dangerous Pathogens (1994) *Precautions for Work with Human and Animal TSEs.* HMSO ISBN 0-11-321805-2.

21. European Commission for Agriculture (1994) *Transmissible Spongiform Encephalopathies: Protocols Used at CVL for the Laboratory Diagnosis and Confirmation of Bovine Spongiform Encephalopathy and Scrapie. A Report from the*

Scientific Veterinary Committee. European Commission, Directorate General for Agriculture, Unit for Veterinary Legislation and Zootechnics.

22. Scott, A. C., Wells, G. A. H., Chaplin, M. J., and Dawson, M. (1992) Bovine spongiform encephalopathy: detection of fibrils in the central nervous system is not affected by autolysis. *Res. Vet. Sci.* **32,** 332–336.

23. Stack, M. J., Scott, A. C., Done, S. H., and Dawson, M. (1993) Scrapie-associated fibril detection on decomposed and fixed ovine brain material. *Res. Vet. Sci.* **55,** 173–178.

24. Stack, M. J., Scott, A. C., Done, S. H., and Dawson, M. (1991) Natural scrapie: detection of fibrils in extracts from the central nervous system of sheep. *Vet. Rec.* **128,** 539–540.

25. Scott, A. C., Wells, G. A. H., Chaplin, M. J., and Dawson, M. (1990) Bovine spongiform encephalopathy: detection and quantitation of fibrils, fibril protein (PrP) and vacuolation in brain. *Vet. Microbiol.* **23,** 295–304.

26. Himert, H. and Diringer, H. (1984) A rapid and efficient method to enrich SAF-protein from scrapie brains of hamsters. *Biosci. Rep.* **4,** 165–170.

27. Hope, J., Reekie, L. J. D., and Gibson, P. H. (1990) On the pathogenesis of SAF, in *Unconventional Virus Diseases of the Central Nervous System, Paris 1986* (Court, L. A., Dormont, D., Brown, P., and Kingsbury, D. T., eds.), Commissariat à l'Énergie Atomique, Département de Protection Sanitaire, Service de Documentation. Fontenay-aux-Roses Cedex, pp. 536–546.

28. Laemmli, U. K. (1970) Cleavage of structural proteins during the assembly of the head of bacteriophage T4. *Nature* **277,** 680–685.

29. Towbin, H., Staehlin, T., and Gordon, J. (1979) Electrophoretic transfer of proteins from polyacrylamide gels to nitrocellulose sheets. *Proc. Natl. Acad. Sci. USA* **76,** 4350–4354.

6

Exposure to, and Inactivation of, the Unconventional Agents that Cause Transmissible Degenerative Encephalopathies

David M. Taylor

1. Introduction

Some fatal degenerative encephalopathies of mammals form a distinct group because they are caused by unconventional but uncharacterized transmissible agents that evoke no classical immune response in the affected host. The animal diseases are bovine spongiform encephalopathy (BSE), chronic wasting disease of elk and mule-deer, feline spongiform encephalopathy (FSE) of the domestic cat, scrapie in sheep and goats, and transmissible mink encephalopathy (TME). A number of exotic ruminant and felid species maintained in, or originating from, zoological collections in the United Kingdom have also developed fatal encephalopathies that, like FSE, are considered to have been caused by BSE agent; the affected species are cheetah, eland, gemsbok, kudu, nyala, ocelot, oryx, and puma. The human diseases are Creutzfeldt-Jakob disease (CJD), fatal familial insomnia (FFI), Gerstmann-Sträussler-Scheinker syndrome (GSS), and kuru.

Although other tissues can be infected, the highest titers of infectivity are always found within the central nervous system (CNS) during the terminal stage of disease. It is only within such tissues that any histopathological changes can be observed, usually as an intraneuronal spongiform degeneration accompanied by astrocytic hypertrophy with or without discrete extracellular amyloid plaques. Although spongiform change is common, it is not a consistent feature of these diseases, and the term transmissible degenerative encephalopathies (TDE) has been considered to be a more appropriate alternative to descriptions such as "transmissible spongiform encephalopathies" *(1)*. A dominant feature of the TDE is the uncertainty and controversy regarding

From: *Methods in Molecular Medicine: Prion Diseases*
Edited by: H. Baker and R. M. Ridley Humana Press Inc., Totowa, NJ

the nature of their unconventional causal agents, which will be addressed elsewhere in this volume.

A notable characteristic of TDE agents is their relative resistance to inactivation by chemical and physical procedures that are effective with conventional microorganisms *(2,3)*, and they are capable of prolonged survival in the general environment owing to their resistance to desiccation *(4)*, freezing *(5)*, and ultraviolet irradiation *(6)*. Scrapie-infected brain has been shown to retain substantial infectivity after burial for 3 yr *(7)*. Scrapie infectivity is considered to be capable of survival for several years on grazing pastures *(8)*, and CJD agent has been shown to retain infectivity after a 28-mo holding period at room temperature *(9)*.

2. Accidental Transmission of TDE

Among the known instances of accidental transmission of TDE there are some that are suspected or known to be attributable to failure of decontamination procedures. Thus, CJD has been transmitted from human to human *(10,11)*, scrapie from sheep to sheep *(12)* and probably from sheep to cows, giving rise to BSE *(13)*. In addition, a significant number of human patients have developed CJD after injection of growth hormone or gonadotrophin derived from the pituitary glands of human cadavers, or surgical repair with cadaveric human dura mater *(14)*. There had been no perceived risk of such occurrences when these products were first used, but from later studies on the inactivation potential of the manufacturing processes when challenged with scrapie agent it was concluded that infectivity might survive *(15,16)* or crosscontaminate the manufacturing processes *(17)*.

The incidence of scrapie in sheep in some countries is sufficiently high to conclude that the human consumption of infected tissues has been commonplace, given that scrapie agent is relatively resistant to the temperatures achieved during cooking by boiling *(18)* or in the oven *(19)*. Concern that this might be a causal factor for CJD in humans was not supported by a number of epidemiological studies that also dismissed occupational exposure to scrapie agent as a risk factor for CJD (e.g., *20*). Although there are no known examples of transmission of animal TDE to humans, concern regarding this possibility has been rekindled because of the occurrence of BSE that may represent the novel transmission of scrapie to cattle. The concern is that, having breached the cattle species barrier, the cattle-passaged agent might have acquired an enhanced capacity to penetrate the human species barrier but the possibility of this is considered to be low *(21,22)*. Although not statistically significant, the occurrence in recent years of CJD in two farmers who tended herds of cattle affected by BSE has heightened this concern *(23,24)*. As a precautionary measure, specified bovine offals have been prohibited from use as human food on the basis of what is known about the levels of infectivity in these organs in scrapie-infected sheep *(25)* or in calves infected experimentally with BSE agent *(26)*. However, recent studies have shown that apart from the CNS none of the many bovine tissues tested from natural BSE cases have any demonstrable infectivity by mouse bioassay *(27)*. This contrasts sharply with the situation in scrapie-affected sheep in which lymphoreticular and other nonneural tissues can be shown by the same bioassay system to harbor infectivity *(28)*.

Although medical products have been manufactured for many years from sheep tissues sourced from countries where scrapie is endemic without any apparent problems, the emergence of BSE has resulted in stricter guidelines for the use of ruminant tissues in the manufacture of such products; the principal recommendation is that raw materials should be obtained from countries where both BSE and scrapie are known reliably to be absent *(29)*.

3. Occupational Exposure to TDE Agents

As far as occupational exposure to either animal or human TDE agents is concerned, the principal groups at risk appear to be research scientists, healthcare workers involved in human or veterinary medicine, undertakers, embalmers, farmers, butchers, abattoir workers, and those involved in the rendering industry. The degree of actual exposure to TDE agents for these various groups will vary considerably given that these agents have not been found generally to be present in the bloodstream, feces, urine, or secretions from the host. Consequently, those dealing with intact living hosts are at minimal risk of exposure, whereas the theoretical risk is increased for those who have to deal with infected tissues through surgery, postmortem procedures, or as part of an industrial process, especially if the tissues are derived from the central nervous or lymphoreticular systems. Much useful advice for such workers has been published recently *(30)*, but it is important to re-emphasize that the only known examples of humans acquiring CJD accidentally are related to the use of medical procedures involving the use of CJD-contaminated surgical instruments, transplant material, or injectable therapeutic products. This chapter will include information on the currently available practical measures for inactivation of, or reduction of exposure to, TDE agents with particular reference to the laboratory situation but that can be extrapolated to other occupational situations where the level of exposure to TDE agents is perceived to be a theoretical occupational risk.

4. Inactivation Studies on TDE Agents
4.1. Background

Until the 1980s the information on appropriate methods for decontamination of scrapie-like agents was piecemeal and sometimes inappropriate because bioassays were often terminated prematurely, given the delayed dose-response curves that are known to occur frequently after chemical or physical treatment of TDE agents compared with comparable titers of untreated infectivity *(3,16,31–37)*.

Although earlier experiments had shown that acetylethyleneimine *(18)*, ethanol *(19)*, and ethylene oxide *(38)* were ineffective, it was during the 1980s that comprehensive studies were conducted in the United Kingdom and the United States in an effort to establish decontamination standards that could be used reliably for the agents of CJD and other TDE. During these studies further substances were found to be ineffective; these included chlorine dioxide *(36)*, hydrogen peroxide *(39)*, peracetic acid *(40)*, phenolic disinfectants *(36,41)*, potassium permanganate *(41)*, and urea *(36)*. However, in the United States, it was concluded that inactivation could be achieved by exposure for 1 h to 1M sodium hydroxide or gravity-displacement autoclaving at 132°C for 1 h *(42)*. In the United Kingdom, the preferred options have been the use of sodium hypochlorite solution containing 20,000 ppm available chlorine for 1 h *(41)* or porous-load autoclaving at 134–138°C for 18 min *(43)*. These regimens have been adopted largely as international standards but, as will be discussed, more recent studies have cast doubt on the efficacy of all but the hypochlorite decontamination procedure.

4.2. Sodium Hydroxide

The adoption in the United States of a 1-h exposure to 1M sodium hydroxide as a recommended procedure for inactivating CJD agent resulted from the report that 10% homogenates of guinea pig or hamster brain infected with CJD agent or the 263 K strain of scrapie agent, respectively, were inactivated by this procedure *(36)*. However, in the same report it was acknowledged that pre-existing data had demonstrated the survival of scrapie infectivity after exposure to sodium hydroxide. More recently, survival of the 263 K and ME7 strains of scrapie agent has been recorded following exposure to even 2M sodium hydroxide for up to 2 h, the residual titer of 263 K being $10^{4.2}$ ID$_{50}$/g brain *(3)*. These data appear superficially to be at gross variance with those reported in the previous study *(36)* but they were obtained from samples that were undiluted prior to injection through careful adjustment to neutral pH. In contrast, the samples in the earlier study required to be diluted before injection to eliminate problems of acute toxicity to the recipient animals, resulting in a reduction in the sensitivity of the bioassay. The general conclusion regarding inactivation of TDE agents by sodium hydroxide is that although substantial reductions in infectivity titers can be achieved, the procedure cannot be relied on for complete inactivation.

4.3. Sodium Hypochlorite and Sodium Dichloroisocyanurate

From a study involving the use of the 22A and 139A strains of mouse-passaged scrapie agent, it was concluded that a 1-h exposure to a sodium hypochlorite solution containing 20,000 ppm available chlorine is an effective decontamination procedure for TDE agents *(41)*, which was later confirmed for BSE agent *(3)*. However, the expectation that solutions of sodium dichloroisocyanurate containing the same levels of available chlorine would be equally effective was not realized; under the same experimental conditions used to test the effectiveness of sodium hypochlorite solutions, the sodium dichloroisocyanurate solutions failed to release their available chlorine content freely and were thus less effective *(3)*.

4.4. Gravity-Displacement Autoclaving

The endorsement in the United States of gravity-displacement autoclaving at 132°C for 1 h as an effective method for inactivating CJD agent resulted from the report that 10% homogenates of guinea pig or hamster brain infected, respectively, with the K Fu isolate of CJD agent and the 263 K strain of scrapie agent were inactivated by this procedure *(36)*. Unfortunately, subsequent data obtained using the 263 K strain of scrapie agent indicate that this procedure is not entirely reliable *(44,45)*. One proposed solution to the problem of questionable autoclaving standards, based on studies with CJD agent, is to treat contaminated materials for 1 h with $1M$ sodium hydroxide before gravity-displacement autoclaving at 121°C for 30 min *(46)*. However, there may be a problem if autoclaves are used routinely for this purpose because of possible progressive degradative effects on the autoclave chamber.

4.5. Porous-Load Autoclaving

In the United Kingdom, porous-load autoclaving at 134–138°C for 18 min has been recommended for inactivation of CJD agent and, by inference, other TDE agents. This was based on data that showed that 50 mg macerates of mouse brain infected with the 22A or 139A strains of scrapie agent were inactivated by porous-load autoclaving at 136°C for 4 min *(41)*. The security of this procedure appeared to be endorsed by the finding that intact mouse brain infected with 22A, weighing approx 375 mg, was inactivated by porous-load autoclaving at 134°C for 18 min *(47)*, especially since 22A is more thermostable than other strains of scrapie agent *(19,41)*. However, more recent studies using comparable weights of cow brain infected with BSE agent, hamster brain infected with the 263 K strain of scrapie agent, and mouse brain infected with the ME7 strain of scrapie agent have produced conflicting results that suggest that this standard may not be secure *(3)*. The most obvious difference between the most recent study and previous ones is that the brain samples in the most recent study were autoclaved as relatively large aliquots of brain tissue macerates rather than intact tissue that resulted in some smearing (and probably drying) of tissue on the surfaces of the glass containers. It has been reported previously that scrapie-infected brain tissue dried onto glass surfaces is more difficult to inactivate than saline homogenates *(48)*, which may be the explanation for these apparently anomalous findings that currently are being investigated further. However, tissue smearing and dehydration on glass surfaces occurs commonly with laboratory equipment that requires TDE decontamination by autoclaving, and it may be more realistic to ensure that such conditions prevail in future autoclave inactivation studies. Despite the theoretical implications of these findings, we have not observed any anomalous experimental data that might be attributable to cross-contamination; such occurrences would be evident because of the phenotypic characteristics of BSE and different scrapie agents that are observed in mice of particular genotypes. Nevertheless, we have increased our autoclaving sterilization standard for scrapie-like agents to 136°C for 1 h, pending the outcome of further studies. The risk of iatrogenic transmission of human TDE through survival of infectivity on surgical instruments that have been autoclaved has been largely excluded for many years in the United Kingdom. Because the Department of Health recognized the difficulty of establishing secure autoclaving standards, it has recommended that instruments used in neurosurgery or ophthalmology on patients with known or suspected CJD should be destroyed rather than try to decontaminate them *(49)*; more recently, this recommendation has been extended to include newer categories of patients at risk of developing CJD, such as recipients of cadaveric human growth hormone or dura mater *(50)*.

4.6. Rendering

Epidemiological studies have indicated that the emergence of BSE in the United Kingdom during the mid-1980s was associated with the practice of feeding cattle with diets containing ruminant-derived protein that is manufactured mainly from abattoir waste by the rendering industry *(13)*. The cooking procedures used by the rendering industry enable the fat content of animal tissues to be fluidized and collected as commercial tallow; the residual solids are milled to produce meat and bone meal that is the source of the ruminant-derived protein that had been fed to cattle until the UK ban in 1988. It has been hypothesized that scrapie infectivity in ovine tissues survived rendering procedures at sufficiently high titer to represent an effective oral challenge for cattle, with a likely initial exposure at the beginning of the 1980s *(13)*. Although the feeding of meat and bone meal has been commonplace for a much longer period, it was recognized that there had been a rapid reduction in the use of solvent extraction as an adjunct to rendering procedures at a time that coincided with the first putative enhancement of exposure of cattle to scrapie agent *(13)*. Traditionally, solvent extraction had been applied to the solid endproducts yielded by rendering procedures to enhance the yield of tallow and to produce low-fat meat and bone meal that at one time attracted premium prices. Apart from the exposure of the raw materials to heated solvents, such as hexane, benzene, petroleum, and trichloroethane, the solvent extraction process also involved exposure of the processed solids to dry heat and steam at approx 100°C to evaporate residual solvent. It has been suggested that these procedures may have produced sufficient additional inactivation of scrapie agent to produce meat and bone meal with infectivity levels that were insufficient to represent an effective dietary challenge for cattle *(13)*. There is, however, an absence of any information for the inactivation potential of any of the solvents used, and the scrapie agent is known to be relatively thermostable *(51)*. Experiments are in progress to determine the effectiveness of such procedures but results are not yet available.

It has also been recognized that none of the experimental conditions under which the thermostability of TDE agents has been studied previously are directly relevant to rendering procedures *(51)*. Consequently, industrial pilot-scale facsimiles of rendering practices used within the European Community have been spiked with BSE or scrapie agent *(52)*. The scrapie-spiked experiments are still in progress but the BSE-spiked study is now complete. This has demonstrated that, even though the spike level was low, BSE infectivity was detectable in meat and bone meal produced by two types of process. These were: a procedure that involved exposure of the raw materials to heating for 50 min at atmospheric pressure with the final temperature reaching 112 or 122°C; and processing raw materials under vacuum with added preheated tallow for 10 or 40 min, and the final temperatures reaching 120 or 121°C, respectively. In view of these findings, the minimal acceptable temperatures for rendering have been revised within the European Community *(53)*.

Because these types of rendering systems had been introduced increasingly in the United Kingdom during the 1970s, this in itself does not offer an explanation for the emergence of BSE in the 1980s, given that the average age of cattle developing clinical symptoms is 4–5 yr. However, if the solvent extraction process had provided additional inactivation as has been postulated, the rapid decline in the use of this process in the United Kingdom during the late 1970s and early 1980s may have been the key additive factor that permitted sheep scrapie to be transmitted to cattle. The experiments that are currently in progress relating to the inactivation potential of solvent extraction procedures should clarify this situation.

5. Working Practice in the Laboratory

Until 1993 CJD had been observed in 24 individuals who had been health-care workers of various types, including a pathologist and two technicians who had worked in neurohistopathology laboratories, but there was no evident association between their development of CJD and any occupational exposure to

CJD agent *(54)*. Furthermore, there was historically an interval of 40 yr between the recognition of CJD as a clinical entity and the suspicion that the disease might be transmissible. Even though CJD is a rare disease, brain tissue from CJD-infected individuals must have been handled worldwide without significant precautions during this period by pathologists and laboratory personnel but without any apparent increased incidence of the disease in such individuals. Nevertheless, the accidental transmission of CJD to human recipients of CJD-contaminated human growth hormone by intramuscular injection demonstrates that occupationally acquired disease through trauma is a possibility.

Survival of infectivity in brain tissue after exposure to formalin has been described for natural scrapie in sheep *(12)* and for experimental scrapie in hamsters *(55)* and mice *(2)*. The same is true for the agents of BSE *(56)*, CJD *(57)*, and TME, which is known to survive for at least 6 yr *(58)*. When hamster brain containing $10^{10.2}$ ID_{50}/g of the 263 K strain of scrapie agent was fixed in formol saline for 48 h, only 1.5 logs of infectivity were lost *(55)*; even after full histological processing the titer loss was only 2.8 logs *(59)*. Glutaraldehyde treatment is also known to permit survival of CJD *(60)* and scrapie infectivity *(19)*. Consequently, the handling of fixed CJD-infected tissues in the histopathology laboratory has been viewed as a potentially risky activity, and a number of procedures have been recommended to reduce this risk. One suggestion has been to fix such tissues in formol saline containing sodium hypochlorite *(61)*; although, as discussed above, high concentrations of sodium hypochlorite inactivate TDE agents, there has been no validation of its effectiveness when combined with formalin. The addition of phenol to formol saline has also been suggested *(62–64)* but the basis of this proposal was flawed *(65)*, and phenolized formalin subsequently was shown to be not only ineffective *(66)* but also to produce poor fixation *(66,67)*. Sections stained with hematoxylin and eosin, prepared from scrapie-infected formol-fixed brain tissue that was autoclaved at 134°C for 18 min, retained sufficient integrity to permit quantitative scoring of spongiform encephalopathy *(68)*; it has been suggested that autoclaving at 126°C for 30 min *(69)* or 132°C for 60 min *(70)* could be used to inactivate CJD infectivity in formol-fixed brain. However, mouse or hamster-passaged scrapie agent in formol-fixed brain has been shown to survive porous-load autoclaving at 134°C for 18 min *(47)* or gravity-displacement autoclaving at 134°C for 30 min *(55)*, with titer losses of <2 logs. The only procedure that has been shown to result in significant losses of infectivity titer in formol-fixed tissues, without loss of microscopic morphology, is a 1-h exposure to concentrated formic acid *(66)*. In that study the level of infectivity in hamster brain infected with the 263 K strain of scrapie agent was reduced from $10^{10.2}$ ID_{50}/g to $10^{1.3}$ ID_{50}/g; with human brain infected with CJD agent the original titer of $10^{8.5}$ ID_{50}/g was reduced to $10^{2.3}$ ID_{50}/g. However, in another study where mouse brain infected with the 301V strain of BSE agent was fixed using paraformaldehyde-lysine-periodate, a necessary prerequisite for the subsequent immunocytochemical investigation that is an important aspect of TDE investigation, the degree of inactivation by formic acid was calculated to be two logs less than that achieved with formol-fixed 263 K-infected hamster brain, despite the equivalent levels of infectivity of the two agents *(71)*. This suggests either that infected tissues fixed with paraformaldehyde-lysine-periodate are less amenable to the inactivating effect of formic acid than those fixed with formalin or there is a fundamental difference in the susceptibility of the 263 K agent compared with 301V; alternatively, both factors may contribute to this observation. Although further studies are in progress to clarify this situation it is evident that there is no known decontamination procedure that can guarantee the complete absence of infectivity in TDE-infected tissues that have been processed by histopathological procedures. Clearly, autoclaving of histological waste is inappropriate for inactivating scrapie-like agents, and reliance must be placed on incineration.

Precautions in the handling of TDE agents in other types of laboratory are somewhat different from those in the histopathology laboratory. For example, disruption of neural tissues by homogenization has the potential to release many more infectious airborne particles than from section-cutting in the histopathology laboratory, especially if the latter tissues were treated with formic acid. In the biochemistry laboratory there is also the capability, through partial purification procedures, to produce samples that contain infectivity titers higher than those found in naturally infected tissues. Apart from general good laboratory practice, the principal recommendation when handling TDE agents under such conditions is to use microbiological safety cabinets. However, what must be borne in mind is the resistance of TDE agents to inactivation by formalin, which is the customary fumigant for routine decontamination of safety cabinets. The main objective, therefore, is to adopt working procedures that minimize the potential for contamination of the cabinet; these include measures such as the use of disposable covering materials on the work surface, and the prevention of aerosol dispersion, e.g., by retaining cotton-wool plugs in glass tissue homogenizers during sample disruption (and for some time thereafter, if possible). Regardless of these types of precautions, it would be naive to consider that they would guarantee complete freedom from contamination of the internal surfaces of safety cabinets. Although contamination at this sort of level is unlikely to represent any significant risk to the operator, given that such work should always involve the wearing of disposable gloves and laboratory coats, the potential for crosscontamination from different TDE sources in laboratory experiments has to be considered. This can be addressed by adopting a routine of washing the internal surfaces with a solution of sodium hypochlorite containing 20,000 ppm available chlorine; however, a compromise has to be struck between the perceived necessary frequency of such a decontamination procedure and its potential progressive degradative effect on exposed surfaces. Class II safety cabinets are suitable for this type of work, and are popular generally because they combine satisfactory degrees of product and personnel protection under conditions that are not too restrictive for the operator. However, the classical design of such cabinets has been such that contamination of the internal plenum and air-propelling units is likely. Although this is not problematic for conventional microorganisms that can be inactivated by formalin fumigation that penetrates these areas, there is a problem obviously with TDE agents. Although such TDE contamination does not represent any significant risk to the operator or the work activity, there is the problem of how to achieve decontamination before engineers are permitted to carry out repairs or servicing. This is because of the inaccessibility of the plenum to manual hypochlorite decontamination, and additionally the potential degradation effects of hypochlorite on the air-propelling units. An improvement in this situation has been achieved by the manufacture of class II safety cabinets with filters positioned immediately below the working surface, which means that contamination of the plenum and plant are avoided unless there is damage to these filters. Because the main filters are readily accessible in such cabinets, it is an easy matter to prevent particle dispersion during their removal, by prior treatment of the filter surface, e.g., with hairspray or latex solution.

References

1. Taylor, D. M. (1991) Spongiform encephalopathies. *Neuropathol. Appl. Neurobiol.* **17,** 237.
2. Taylor, D. M. (1992) Inactivation of unconventional agents of the transmissible degenerative encephalopathies, in *Principles and Practice of Disinfection, Preservation and Sterilization* (Russell, A. D., Hugo, W. B., and Ayliffe, G. A. J., eds.), Blackwell Scientific Publications, Oxford, pp. 171–179.
3. Taylor, D. M., Fraser, H., McConnell, I., Brown, D. A., Brown, K. A., Lamza, K. A., et al. (1994) Decontamination studies on the agents of bovine spongiform encephalopathy and scrapie. *Arch. Virol.* **139,** 313–326.
4. Wilson, D. R. (1955) Unpublished work cited by Dickinson, A. G. (1976) Scrapie in sheep and goats, in *Slow Virus Diseases of Animals and Man* (Kimberlin, R. H., ed.), North-Holland, Amsterdam, pp. 209–241.
5. Stamp, J. T. (1967) Scrapie and its wider implications. *Br. Med. Bull.* **23,** 133–137.
6. Latarjet, R. (1979) Inactivation of the agents of scrapie, Creutzfeldt-Jakob disease, and kuru by radiations, in *Slow Transmissible Diseases of the Nervous System*, vol. 2 (Prusiner, S. B. and Hadlow, W. J., eds.), Academic, London, pp. 387–407.
7. Brown, P. and Gajdusek, D. C. (1991) Survival of scrapie virus after 3 years' interment. *Lancet* **337,** 269,270.

8. Palsson, P. A. (1979) Rida (scrapie) in Iceland and its epidemiology, in *Slow Transmissible Diseases of the Nervous System*, vol. 1 (Prusiner, S. B. and Hadlow, W. J., eds.), Academic, London, pp. 357–366.

9. Tateishi, J., Hikita, K., Kitamoto, T., and Nagara, H. (1987) Experimental Creutzfeldt-Jakob disease: induction of amyloid plaques in rodents, in *Prions: Novel Infectious Pathogens Causing Scrapie and Creutzfeldt-Jakob Disease* (Prusiner, S. B. and McKinley, M. P., eds.), Academic, New York, pp. 415–426.

10. Bernoulli, C., Siegfried, J., Baumgartner, G., Regli, F., Rabinowicz, T., Gajdusek, D. C., et al. (1977) Danger of accidental person-to-person transmission of Creutzfeldt-Jakob disease by surgery. *Lancet* **i,** 478,479.

11. Foncin, J. F., Gaches, J., Cathala, F., El Sherif, E., and Le Beau, J. (1980) Transmission iatrogene interhumaine possible de maladie de Creutzfeldt-Jakob avec alteinte des grains du cervulet. *Rev. Neurol.* **136,** 280.

12. Gordon, W. S., Brownlee, A., and Wilson, D. R. (1940) Studies in louping-ill, tick-borne fever and scrapie. *Report of the Proceedings of the Third International Congress for Microbiology.* Waverley, Baltimore, pp. 362,363.

13. Wilesmith, J. W., Wells, G. A. H., Cranwell, M. P., and Ryan, J. (1988) Bovine spongiform encephalopathy: epidemiological studies. *Vet. Record* **123,** 638–644.

14. Brown, P., Preece, M. A., and Will, R. G. (1992) 'Friendly fire' in medicine: hormones, homografts, and Creutzfeldt-Jakob disease. *Lancet* **ii,** 24–27.

15. Diringer, H. and Braig, H. R. (1989) Infectivity of unconventional viruses in dura mater. *Lancet* **i,** 439,440.

16. Pocchiari, M., Peano, S., Conz, A., Eshkol, A., Maillard, F., Brown, P., et al. (1991) Combination ultrafiltration and $6M$ urea treatment of human growth hormone effectively minimizes risk from potential Creutzfeldt-Jakob disease virus contamination. *Hormone Res.* **35,** 161–166.

17. Taylor, D. M., Dickinson, A. G., Fraser, H., Robertson, P. A., Salacinski, P. R., and Lowry, R. J. (1985) Preparation of growth hormone free from contamination with unconventional slow viruses. *Lancet* **ii,** 260–263.

18. Stamp, J. T., Brotherston, J. C., Zlotnik, I., McKay, J. M. K., and Smith, W. (1959) Further studies on scrapie. *J. Comp. Pathol.* **69,** 268–280.

19. Dickinson, A. G. and Taylor, D. M. (1978) Resistance of scrapie agent to decontamination. *N. Engl. J. Med.* **229,** 1413,1414.

20. Brown, P., Cathala, F., and Gajdusek, D. C. (1979) Creutzfeldt-Jakob disease in France: III. Epidemiological study of 170 patients dying during the decade 1968–1977. *Ann. Neurol.* **6,** 438–446.

21. Department of Health: Ministry of Agriculture Fisheries and Food (1989) *Report of the Working Party on Bovine Spongiform Encephalopathy.* MAFF, London.

22. Taylor, D. M. (1989) Bovine spongiform encephalopathy and human health. *Vet. Record* **125,** 413–415.

23. Sawcer, S. J., Yuill, G. M., Esmonde, T. G. F., Estibeiro, P., Ironside, J. W., Bell, J. E., et al. (1993) Creutzfeldt-Jakob disease in an individual occupationally exposed to BSE. *Lancet* **341,** 642.

24. Davies, P. T. G., Jahfar, S., Ferguson, I. T., and Windl, O. (1993) Creutzfeldt-Jakob disease in individual occupationally exposed to BSE. *Lancet* **342**, 680.
25. The bovine offals (Prohibition) regulations (1989) Statutory Instrument 1989 No. 2061. HMSO, London.
26. Bovine offal (Prohibition) (Amendment) Regulations (1994) Statutory Instrument 1994 No. 2628. HMSO, London.
27. Spongiform Encephalopathy Advisory Committee (1994) Transmissible Spongiform Encephalopathies: A Summary of Present Knowledge and Research. HMSO, London.
28. Hadlow, W. J., Race, R. E., Kennedy, R. C., and Eklund, C. M. (1979) Natural infection of sheep with scrapie virus, in *Slow Transmissible Diseases of the Nervous System*, vol. 2 (Prusiner, S. B. and Hadlow, W. J., eds.), Academic, London, pp. 3–12.
29. Committee for Proprietary Medicinal Products: Ad Hoc Working Party on Biotechnology/Pharmacy and Working Party on Safety Medicines (1992) Guidelines for minimizing the risk of transmitting agents causing spongiform encephalopathy via medicinal products. *Biologicals* **20**, 155–158.
30. Advisory Committee on Dangerous Pathogens (1994) Precautions for work with human and animal transmissible spongiform encephalopathies. HMSO, London.
31. Dickinson, A. G. and Fraser, H. (1969) Modification of the pathogenesis of scrapie in mice by treatment of the agent. *Nature* **222**, 892,893.
32. Kimberlin, R. H. (1977) Biochemical approaches to scrapie research. *TIBS* **2**, 220–223.
33. Prusiner, S. B., Cochran, S. P., Groth, D. F., Downey, D. E., Bowman, K. A., and Martinez, H. M. (1982) Measurement of the scrapie agent using an incubation time interval assay. *Ann. Neurol.* **11**, 353–358.
34. Lax, A. J., Millson, G. C., and Manning, E. J. (1983) Can scrapie titres be calculated accurately from incubation periods? *J. Gen. Virol.* **64**, 971–973.
35. Somerville, R. A. and Carp, R. I. (1983) Altered scrapie infectivity estimates by titration and incubation period in the presence of detergents. *J. Gen. Virol.* **64**, 2045–2050.
36. Brown, P., Rohwer, R. G., and Gajdusek, D. C. (1986) Newer data on the inactivation of scrapie virus or Creutzfeldt-Jakob disease virus in brain tissue. *J. Infect. Dis.* **153**, 1145–1148.
37. Prusiner, S. B., Groth, D., Serban, A., Stahl, N., and Gabizon, R. (1993) Attempts to restore scrapie prion infectivity after exposure to protein denaturants. *Proc. Natl. Acad. Sci. USA* **90**, 2793–2797.
38. Dickinson, A. G. (1976) Scrapie in sheep and goats, in *Slow Virus Diseases of Animals and Man* (Kimberlin, R. H., ed.), North-Holland, Amsterdam, pp. 209–241.
39. Brown, P., Gibbs, C. J., Amyx, H. L., Kingsbury, D. T., Rohwer, R. G., Sulima, M. P., et al. (1982) Chemical disinfection of Creutzfeldt-Jakob disease virus. *N. Engl. J. Med.* **306**, 1279–1282.
40. Taylor, D. M. (1991) Resistance of the ME7 scrapie agent to peracetic acid. *Vet. Microbiol.* **27**, 19–24.

41. Kimberlin, R. H., Walker, C. A., Millson, G. C., Taylor, D. M., Robertson, P. A., Tomlinson, A. H., et al. (1983) Disinfection studies with two strains of mouse-passaged scrapie agent. *J. Neurol. Sci.* **59**, 355–369.

42. Rosenberg, R. N., White, C. L., Brown, P., Gajdusek, D. C., Volpe, J. J., and Dyck, P. J. (1986) Precautions in handling tissues, fluids, and other contaminated materials from patients with documented or suspected Creutzfeldt-Jakob disease. *Ann. Neurol.* **19**, 75–77.

43. Department of Health and Social Security (1984) Management of patients with spongiform encephalopathy (Creutzfeldt-Jakob disease [CJD]). DHSS Circular DA (84) 16.

44. Ernst, D. R. and Race, R. E. (1993) Comparative analysis of scrapie agent inactivation. *J. Virol. Methods* **41**, 193–202.

45. Pocchiari, M. Unpublished data cited by Horaud, F. (1993) Safety of medicinal products: summary. *Dev. Biol. Stand.* **80**, 207,208.

46. Taguchi, F., Tamai, Y., Uchida, K., Kitajima, R., Kojima, H., Kawaguchi, T., et al. (1991) Proposal for a procedure for complete inactivation of the Creutzfeldt-Jakob disease agent. *Arch. Virol.* **119**, 297–301.

47. Taylor, D. M. and McConnell, I. (1988) Autoclaving does not decontaminate formol-fixed scrapie tissues. *Lancet* **i**, 1463,1464.

48. Asher, D. C., Pomeroy, K. L., Murphy, L., Rohwer, R. G., Gibbs, C. J., and Gajdusek, D. C. (1986) Practical inactivation of scrapie agent on surfaces. *Abstracts of the IXth International Congress of Infectious and Parasitic Diseases*, Munich.

49. Advisory Group on the Management of Patients with Spongiform Encephalopathy (Creutzfeldt-Jakob Disease)[CJD] (1981) Report to the Chief Medical Officers of the Department of Health and Social Security, the Scottish Home and Health Department, and the Welsh Office. DA(81)22, HMSO, London.

50. Wight, A. L. (1993) Prevention of iatrogenic transmission of Creutzfeldt-Jakob disease. *Lancet* **341**, 1543.

51. Taylor, D. M. (1989) Scrapie agent decontamination; implications for bovine spongiform encephalopathy. *Vet. Record* **124**, 291,292.

52. Taylor, D. M. (1994) Deactivation of BSE and scrapie agents, in *Transmissible Spongiform Encephalopathies; Proceedings of a Consultation on BSE with the Scientific Veterinary Committee of the Commission of the European Communities* (Bradley, R. and Marchant, B., eds.), September 14–15, 1993, Brussels, EC document VI/4131/94-EN.

53. Commission Decision (27 June 1994) 94/382/EC on the approval of alternative heat treatment systems for processing animal waste of ruminant origin, with a view to the inactivation of spongiform encephalopathy agents.

54. Berger, J. R. and David, N. J. (1993) Creutzfeldt-Jakob disease in a physician: a review of the disorder in health care workers. *Neurology* **43**, 205,206.

55. Brown, P., Liberski, P. P., Wolff, A., and Gajdusek, D. C. (1990) Resistance of scrapie infectivity to steam autoclaving after formaldehyde fixation and limited survival after ashing at 360°C: practical and theoretical implications. *J. Infect. Dis.* **161**, 467–472.

56. Fraser, H., Bruce, M. E., Chree, A., McConnell, I., and Wells, G. A. H. (1992) Transmission of bovine spongiform encephalopathy and scrapie to mice. *J. Gen. Virol.* **73,** 1891–1897.

57. Tateishi, J., Koga, M., Sato, Y., and Mori, R. (1980) Properties of the transmissible agent derived from chronic spongiform encephalopathy. *Ann. Neurol.* **7,** 390,391.

58. Burger, D. and Gorham, J. R. (1977) Observation on the remarkable stability of transmissible mink encephalopathy virus. *Res. Vet. Sci.* **22,** 131,132.

59. Brown, P., Rohwer, R. G., Green, E. M., and Gajdusek, D. C. (1982) Effect of chemicals, heat, and histopathologic processing on high-infectivity hamster-adapted scrapie virus. *J. Infect. Dis.* **145,** 683–687.

60. Amyx, H. L., Gibbs, C. J., Kingsbury, D. T., and Gajdusek, D. C. (1981) Some physical and chemical characteristics of a strain of Creutzfeldt-Jakob disease virus in mice. *Abstracts of the Twelfth World Congress of Neurology*, Kyoto, September 20–25, p. 5.

61. Armbrustmacher, V. W. Personal communication cited by Titford, M. and Bastian, F. O. (1989) Handling Creutzfeldt-Jakob disease tissues in the laboratory. *J. Histotech.* **12,** 214–217.

62. Kleinman, G. M. (1980) Case records of the Massachusetts General Hospital (case 45-1980) *N. Engl. J. Med.* **303,** 1162–1171.

63. Brumback, R. A. (1988) Routine use of phenolised formalin in fixation of autopsy brain tissue to reduce risk of inadvertent transmission of Creutzfeldt-Jakob disease. *N. Engl. J. Med.* **319,** 654.

64. Esiri, M. M. (1989) *Diagnostic Neuropathology.* Blackwell, Oxford.

65. Taylor, D. M. (1989) Phenolized formalin may not inactivate Creutzfeldt-Jakob disease infectivity. *Neuropathol. Appl. Neurobiol.* **15,** 585,586.

66. Brown, P., Wolff, A., and Gajdusek, D. C. (1990) A simple and effective method for inactivating virus infectivity in formalin-fixed tissue samples from patients with Creutzfeldt-Jakob disease. *Neurology* **40,** 887–890.

67. Mackenzie, J. M. and Fellowes, W. (1990) Phenolized formalin may obscure early histological changes of Creutzfeldt-Jakob disease. *Neuropathol. Appl. Neurobiol.* **16,** 255.

68. Taylor, D. M. and McBride, P. A. (1987) Autoclaved, formol-fixed scrapie mouse brain is suitable for histopathological examination, but may still be infective. *Acta Neuropathol.* **74,** 194–196.

69. Masters, C. L., Jacobsen, C., and Kakulas, B. A. (1985) Letter to the editor. *J. Neuropathol. Exp. Neurol.* **44,** 304–307.

70. Masters, C. L., Jacobsen, P. F., and Kakulas, B. A. (1986) Letter to the editor. *J. Neuropathol. Exp. Neurol.* **45,** 760,761.

71. Taylor, D. M. (1994) Survival of mouse-passaged bovine spongiform encephalopathy agent after exposure to paraformaldehyde-lysine-periodate and formic acid. *Vet. Microbiol.* **44,** 111,112.

7

Surveillance of Prion Diseases in Humans

Robert G. Will

1. Introduction

In contrast to animal prion diseases, the human forms of prion disease are rare and include Creutzfeldt-Jakob disease (CJD), kuru, Gerstmann-Sträussler-Scheinker syndrome (GSS), and fatal familial insomnia (FFI). Meticulous epidemiological research has demonstrated that kuru was caused by cross-contamination in the course of ritual cannibalism (1), whereas advances in molecular biology have provided powerful evidence that familial forms of human prion disease, including GSS and FFI, are linked to mutations within the prion protein (PrP) gene (2–4). The cause of sporadic CJD remains unknown despite both extensive epidemiological research and the remarkable advances in the understanding of pathogenic mechanisms in prion diseases.

Sporadic CJD accounts for about 85% of all cases of human prion disease and has been the subject of extensive media and public interest following the advent of bovine spongiform encephalopathy (BSE) and attendant concerns regarding the possibility of a risk to public health. In the absence of an in vivo diagnostic marker, currently the only mechanism of determining whether BSE has caused disease in the human population is by epidemiological surveillance. In view of the manifest implications of demonstrating such a link between animal and human prion diseases, it is appropriate to critically review the current methodologies for epidemiological research in CJD and to discuss specific difficulties in relation to case definition, bias, and interpretation of results.

2. Case Ascertainment

Epidemiological surveillance of CJD is complicated by the rarity of the disease, the absence of any specific diagnostic test, and the apparently random occurrence of cases. A high level of case ascertainment is nonetheless essential if meaningful results are to be obtained, which is dependent on the systematic

From: *Methods in Molecular Medicine: Prion Diseases*
Edited by: H. Baker and R. M. Ridley Humana Press Inc., Totowa, NJ

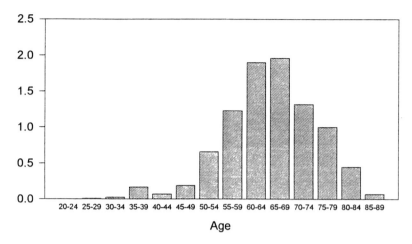

Fig. 1. Age-specific incidence rates in sporadic cases of CJD 1970–1994 (25 yr): $N = 573$.

identification of incident cases. In the United Kingdom data on CJD has been accumulated systematically since 1970 *(5,6)* and has depended on two major mechanisms of case identification:

1. Direct referral of cases from targeted professional groups: neurologists, neuro-pathologists, and neurophysiologists.
2. Death certificates coded under the specific rubrics 046.1 and 331.9 in the 9th ICD revision.

2.1. Direct Referral

CJD is usually a dramatic illness. Patients present with rapidly progressive multifocal neurological dysfunction. The majority develop involuntary myoclonic movements of the limbs and the terminal stage of akinetic mutism supervenes within a few weeks, with median survival from first symptom to death of approx 6 mo. The underlying assumption in the identification of such cases is that this type of clinical presentation, occurring mainly in patients of middle age (Fig. 1), will result in referral for a medical opinion, hospital admission, and referral to a neurologist and/or the recording of an electroencephalogram (EEG) and/or eventual neuropathological examination.

In countries with a developed medical system, it is extremely unlikely that patients who become totally dependent will not be admitted to the hospital and in France it was established that all such patients are subsequently referred to a neurologist *(7)*. In the United Kingdom the 10% of patients with CJD initially admitted to the hospital under the care of a psychiatrist were all later referred to a neurologist. Between 1990 and 1994, 84% of patients with CJD underwent

Table 1
The Electroencephalogram in CJD (UK 1990–1994)

	Typical	Atypical	No EEG	Untraced
Sporadic	99	68	17	4
Familial	5[a]	3	2	0
Iatrogenic	2[b]	7	0	5

[a]3 codon 200 mutations, 2 insert mutations.
[b]2 dura mater recipients.

an EEG and 70% had a postmortem. The referral of CJD cases to a national register depends on the cooperation of neurologists and other professional groups. In the United Kingdom neurologists, neurophysiologists, and neuropathologists are regularly circularized with information on the CJD Surveillance Project and asked to refer any <u>suspect</u> case of CJD.

CJD has been referred to as the archetype exotic neurological disorder *(8)*, perhaps because it is both untreatable and in many cases recognizable from the end of the bed. Although there was almost certainly some improvement in diagnostic efficiency from the 1960s to the 1970s in the United Kingdom, typical cases are now almost certainly promptly recognized. In order to identify a high proportion of cases, including atypical variants, an important component of surveillance is the referral of "suspect" as well as "typical" cases and approx 50% of all referred cases (from all sources, including death certificates) are eventually classified as "possible" CJD or "not CJD" *(6)*.

Another source of cases is from neurophysiology departments. Between 1990 and 1994, 90% of cases of CJD underwent an EEG and 60% of definite or probable cases had a characteristic tracing (Table 1). Although the great majority of cases of CJD were identified from other sources, a small proportion of cases were identified from EEG departments alone.

Neuropathological confirmation of the diagnosis of CJD is an important component of surveillance. The accuracy of clinical diagnosis is over 95% in cases with a typical clinical presentation and a characteristic EEG, but a significant proportion of cases do not undergo EEG or have a single atypical tracing. Similarly, in a small proportion of cases there may be no clinical suspicion of CJD, for example cases referred to nonneurologists. Thirdly, atypical forms of prion disease may be identified. This is an important consideration in relation to the putative risk of BSE because atypical forms of human prion disease may have a very atypical clinical course from classical CJD. The first case of CJD in a human growth hormone recipient in the United Kingdom was certified as dying of encephalitis and it was only after the identification of the typical pathological changes that the diagnosis was confirmed.

Table 2
Accuracy of Death Certification of CJD in the UK[a]

Decade	1960s	1970s	1980s	1990s
Diagnostic classification[b]				
CJD	39	73	66	67
Other	61	27	34	33

[a]Percentage of certified cases classified as CJD after examination of records.
[b]Definite, probable or possible CJD.

Table 3
Case Ascertainment of CJD
from Death Certificates in the United Kingdom (1970–1992)

Period	1970s	1980–1984	1990–1992
Percentage of total cases[a] certified as CJD	62	78	72
Percentage of total cases of CJD[a]			
identified from death certificates alone	42	13	6

[a]Definite and probable cases.

2.2. Death Certificates

It is inevitable that a proportion of cases of CJD are either not referred to a neurologist or not notified to the surveillance center. As a safety net, all death certificates coded under the rubrics 046.1 and 331.9 are obtained and clinical and pathological details sought in every case.

It has been claimed that death certificates alone may be an efficient source of case ascertainment in CJD *(9)*, but this contrasts with experience in the United Kingdom. The accuracy of death certification of CJD in the United Kingdom can be measured by the percentage of cases classified as CJD after examination of case notes (Table 2). Although there has been an improvement in diagnostic accuracy since the 1960s, currently only about two-thirds of certified cases of CJD fulfill diagnostic criteria for the condition, underlining the need to review case records/pathology reports in order to verify the diagnosis in certified cases.

An important question in relation to case ascertainment is whether death certificates, assuming diagnostic verification, can be regarded as an adequate single source of cases for epidemiological studies. This is an important issue because current epidemiological surveillance of CJD in some countries is based primarily on this methodology *(10)*. In the United Kingdom, cases of CJD are identified from a range of sources in order to achieve as high a level of case ascertainment as possible. The proportion of cases of CJD that would have been identified from death certificates is shown in Table 3. In the United Kingdom reliance on death certificates alone for case identification would have resulted in missing 22–38% of cases of CJD. Retrospective surveillance of CJD may have to rely primarily on death certificates as the source of cases. In England and Wales between 1970 and 1979, the overall incidence of CJD was 0.3 cases/ million *(11)* and 42% of these cases were identified from death certificates. Prospective surveillance is more efficient and in both the periods 1980–1984 *(5)* and 1990–1992, the incidence of CJD was higher (0.49 and 0.67, respectively) and the proportion of cases identified from death certificates was significantly lower.

In conclusion, death certificates are an important safety net for the identification of cases of CJD. However, it is essential to verify the diagnosis in certified cases and the use of death certificates alone as a means of case ascertainment is unreliable if a high degree of case ascertainment is to be achieved.

2.3. Electrophysiology

The EEG was first recognized as an important aid to the diagnosis of CJD in 1954 *(12)* and was included as a component of the first published diagnostic criteria for CJD in 1979 *(13)*. In systematic surveys approx 70% of cases of CJD exhibit the typical EEG pattern *(14)*, which consists of generalized triphasic periodic complexes occurring at a frequency of approx 1/s (Fig. 2). In some cases of CJD, no EEG is carried out and in others only a single tracing is obtained early in the course of the illness, and the chances of obtaining a characteristic record are enhanced if serial recordings are carried out. Rarely, the typical EEG does not develop at any stage of the clinical course and there are a number of conditions in which the EEG changes mimic CJD (Table 4). In the majority of these conditions, the differentiation from CJD is usually evident on clinical grounds. The occurrence of a typical EEG tracing in Alzheimer's disease has been described *(15)*, but this is clearly an exceptionally rare occurrence because we have identified only one such case in 25 yr of CJD surveillance in the United Kingdom. It is likely, but unproven, that the occurrence of similar EEG appearances in Lewy body dementia is also exceptional.

The relative specificity of the EEG changes in CJD and the high percentage of cases with such a tracing indicate that EEG departments are a potentially important source for case identification in CJD. In the United Kingdom, EEG departments are circularized and visited when possible, which has led to the identification of the small number of cases not identified from other sources. The EEG is also a crucial component in the diagnostic classification of cases in which there is no pathological confirmation of the diagnosis and, as with death certificates, it is essential to review EEG records.

A major problem with the use of the EEG in the diagnosis of CJD is that there are no established parameters for categorizing the EEG, other than the

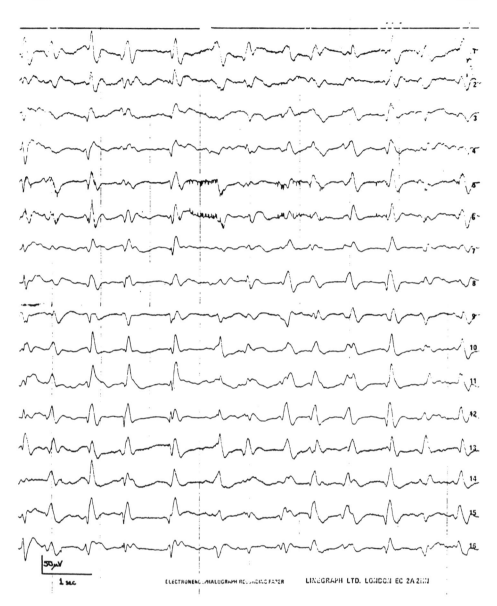

Fig. 2. The typical EEG in CJD.

general description of the typical changes, and there is variation in the accuracy of the reporting of the EEG in CJD. An important element of surveillance is to review the EEGs in all suspect cases in order to build up experience and to allow independent classification of tracings, preferably blind to the clinical or

Table 4
Conditions in Which
a CJD-Like EEG Appearance May Occur

Hyponatremia
Hypernatremia
Hypoglycemia
Hepatic encephalopathy
Hyperammonemia
Lithium toxicity
Metrizamide encephalopathy
Alzheimer's disease
Lewy body dementia

pathological features. Inevitably there are occasional EEG tracings that are difficult to classify and in such cases in the United Kingdom, an independent opinion is sought from an individual with extensive previous experience of the EEG in CJD. The misclassification of EEGs can lead to apparent epidemiological anomalies. For example, CJD was diagnosed in 16 patients in two neighboring hospitals over 2 yr on the basis of the EEG, although review of the tracings and clinical features did not allow the diagnosis of even possible CJD in any case.

The EEG is nonetheless an important component of diagnosis and case classification in CJD. There is, however, variation in the duration of the periodic complexes and the proportion of any record with such suggestive appearances, indicating the need for established EEG criteria for the diagnosis of CJD.

2.4. Neuropathology

The clinical diagnosis of CJD is remarkably accurate in typical cases. Review of the clinical features in two large series of pathologically confirmed CJD has demonstrated that only 10% of cases present atypically *(16)* and in some of these atypical cases the diagnosis was nonetheless suspected. However, the diagnostic classification of the 30% of cases without a typical EEG is dependent on neuropathological verification and clinically atypical cases may only be diagnosed after histological examination of the brain. A high degree of case ascertainment is crucially dependent on obtaining a high postmortem rate in systematic surveys of CJD and the identification of rare or unusual forms of human prion disease may only be possible through neuropathological examination. In the UK study, neurohistological examination has been important in the correct diagnosis of the 5% of cases with a stroke-like presentation and in the 10% of cases with a duration of illness of greater than a year. Neuropathological verification is an important component of case ascertainment and may be crucial to the recognition of atypical cases. The current postmortem rate in the United Kingdom is approx 70% of all suspect cases.

Practical difficulties may be encountered in obtaining postmortem in CJD because of concerns about the potential risks to personnel in pathology departments. Guidelines to minimize any potential risks during postmortems in CJD have been published *(17)* and the epidemiological evidence does not suggest that there is a significant risk to postmortem personnel despite the absence of any precautions in relation to CJD prior to the 1970s. Scientific evidence cannot, however, preclude the possibility of a risk and anxieties have been heightened by the tone of some recent media coverage. A pragmatic approach to the problem of postmortem in CJD is to establish dedicated neuropathology laboratories for the condition and to identify a network of neuropathology departments where postmortem, limited if necessary to examination of the brain, can be carried out.

Brain biopsy is fraught with practical and ethical problems. Most importantly, such a procedure may be difficult to justify if it is unlikely to be of any benefit to the patient. The biopsy may be obtained from a region of the brain unaffected by the pathological process and it may be difficult to reach any definite conclusion from examination of small portions of tissue. The current guidelines in the United Kingdom state that any neurosurgical instruments used in a potential case of CJD must be destroyed postoperatively.

3. Case Definition

The consistent application of diagnostic criteria is a *sine qua non* of epidemiological research. Criteria for the diagnosis of CJD were first proposed in 1979 *(13)* but have had to be adapted to take account of scientific developments. Amyotrophic "CJD", in which there is a combination of dementia and a motor neuron disease-like illness, is no longer regarded as a prion disease in view of the almost uniform failure to demonstrate laboratory transmissibility *(18)* and the absence of detectable PrP on Western blotting in this condition *(19)*. Criteria for a diagnosis of familial CJD have always been problematic. The strict requirement of confirmed or probable CJD in another family member may result in an underestimation of familial cases, whereas allowing any form of dementia in other family members as the criterion for familial CJD may result in an overestimation of familial cases. Advances in the molecular biology of prion diseases now allows an accurate classification of familial cases, provided DNA is obtained for analysis from incident cases.

In the United Kingdom the systematic use of DNA analysis in CJD has demonstrated that about 12% of cases are associated with PrP gene mutations, doubling the estimate of the frequency of familial cases from previous surveys *(11)*. Approximately one-third of these cases have no evident family history and the identification of two cases of FFI in the United Kingdom was dependent on the molecular biological data. On the other hand, three cases of CJD have been identified with a clear family history of dementia and no mutation of the PrP gene.

The systematic analysis of DNA in incident cases of CJD has led to major ethical dilemmas and a marked variation in approach to this problem from country to country. In the United Kingdom informed consent from a relative of the patient is an essential prerequisite before proceeding to DNA analysis. Although the genetic issues in CJD are similar to those in other dominantly inherited diseases, such as Huntington's chorea, there are specific problems with systematic screening carried out in the course of CJD surveillance. Of particular importance are the small proportion of cases in which a mutation of the PrP gene is identified despite the absence of any suggestive family history. In such cases the failure to obtain informed consent leads to extraordinarily difficult ethical problems and the alternative approach, in which no results of DNA testing in CJD are divulged, becomes increasingly difficult to justify in view of the possibility of providing genetic advice to other family members, including prenatal testing *(20,21)*.

Another important development since the formulation of the original diagnostic criteria has been the occurrence of iatrogenic CJD *(22)*, particularly in relation to human pituitary-derived hormones and human dura mater grafts. In human-growth-hormone recipients, the clinical diagnostic criteria for CJD would not allow such cases to be classified as variants of CJD, which has been considered in the formulation of new diagnostic criteria. Through the European Community (EC) a grant was awarded in 1993 for the coordination of CJD surveillance in a number of countries in Europe: France, Germany, Italy, the Netherlands, Slovakia, and the United Kingdom. One of the initial tasks in this collaboration was to review the diagnostic criteria for CJD. The amended criteria are listed in Table 5.

3.1. Atypical Cases of CJD

It is possible that the application of strict diagnostic criteria for CJD will be self-fulfilling and that atypical cases will be missed. There is remarkable consistency in the incidence of sporadic CJD in most systematic surveys and there is similar consistency in the sex distribution, age distribution, and clinical features. Any error in case identification must, therefore, be systematic and uniform.

A crucial assumption in CJD surveillance is that the dramatic nature of the clinical illness will result in hospitalization and an accurate diagnosis of CJD. Although current evidence suggests that there is a high degree of case ascertainment in the young and middle aged, there is a justifiable concern that CJD may be missed in the elderly. The identification of previously undiag-

Table 5
Diagnostic Criteria for CJD

1. Sporadic
   a. Definite:
       i. Neuropathologically confirmed; and/or
      ii. Immunocytochemically confirmed PrP positive (Western blot); and/or
     iii. SAF.
   b. Probable:
       i. Progressive dementia.
      ii. Typical EEG.
     iii. At least two out of the following four clinical features: myoclonus; visual
          or cerebellar; pyramidal/extrapyramidal; or akinetic mutism.
   c. Possible:
       i. Progressive dementia.
      ii. Two out of four clinical features listed above.
     iii. No EEG or atypical EEG.
      iv. Duration <2 yr.
2. Accidental transmission:
   a. Progressive cerebellar syndrome in a pituitary hormone recipient.
   b. Sporadic CJD with a recognized exposure risk.
3. Familial:
   a. Definite or probable CJD *plus* definite or probable CJD in a first degree relative.
   b. Neuropsychiatric disorder *plus* disease-specific PrP mutation.

nosed CJD is exceptional in postmortem series in the elderly demented and it is
unlikely that large numbers of such cases are missed *(23)*. In the UK study
there has been a significant increase in the numbers of elderly patients with
CJD identified in recent years (Fig. 3) but it is of note that these elderly patients
have been identified from throughout the country and that the clinical features
in these cases were typical of sporadic CJD.

Recent studies of genetic forms of prion disease have established marked
phenotypic variation in the clinical and pathological features of CJD *(24)* (and
GSS), and a significant proportion of these cases would not fulfil current diagnostic criteria for CJD. An important question is whether the failure to identify
this type of case significantly prejudices the findings of epidemiological surveillance, and in turn this depends on an assessment of the incidence of atypical forms of CJD.

Review of the clinical features in pathologically confirmed or transmitted
cases of CJD indicates that approx 10% of cases of CJD are clinically atypical
(25). Research in atypical genetic forms of CJD has depended on the identification of highly unusual pedigrees, some of which have been repeatedly stud-

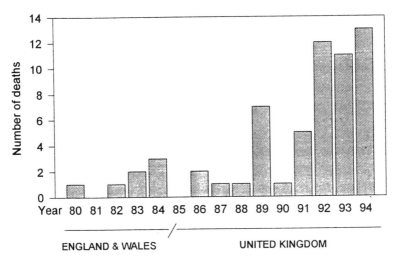

Fig. 3. Sporadic CJD. Number of deaths in patients aged over 75 yr (UK 1980–1994).

ied (and published) over decades. The phenotype in genetic CJD varies with the majority of cases readily identifiable as CJD, e.g., cases associated with a codon 200 mutation *(26)* and some of the cases associated with the codon 178 mutation. Only nine pedigrees of FFI have been identified worldwide *(27)* and the estimated incidence of GSS is 1/10 million/yr, a figure consistent with current findings in the United Kingdom.

Mutations of the PrP gene have been sought in a range of neurodegenerative disorders, almost uniformly without success *(19,28–30)*. Recent evidence suggests that progressive subcortical gliosis may be a prion disease *(31,32)* but again this is an exceedingly rare condition. It has been suggested, on the basis of isolated case reports, that prion disease may occur without characteristic pathology *(33)* with the implication that all cases of dementia without diagnostic pathology may be prion diseases *(34)*. This is not supported by the failure of transmission in laboratory experiments in which brain material from a variety of cases of atypical dementia, including some with minimal pathology, were inoculated into primates *(19)*.

In conclusion, 100% case ascertainment cannot be achieved in CJD surveillance. It is inevitable that some cases are missed, particularly in the elderly or in relation to genetic forms of prion disease. However, current evidence suggests that only small numbers of such cases are likely to be missed in view of the rarity of genetic forms of CJD and that the epidemiological data on CJD is therefore meaningful. The identification of novel human prion diseases, for example in relation to BSE, may depend, by analogy with human growth hormone recipients and genetic forms of CJD, on systematic review of the clinicopathological phenotype as well as systematic study of the descriptive epidemiology of CJD.

4. Case-Control Studies

The aim of the case-control study in CJD is to identify characteristics in the patient group that are distinct from an age- and sex-matched control population, thereby identifying a factor or factors that appear to increase the risk for the development of CJD. Such studies have been carried out in the United States, Israel, Japan, the United Kingdom and currently collaboratively in the EC.

There are many problems with this type of study in CJD. The cause of sporadic CJD is unknown despite advances in basic science, and descriptive epidemiological studies have failed to provide any convincing evidence in relation to the source of infectivity. Indeed, the apparently random distribution of cases in space and time within individual countries and the worldwide and consistent incidence of CJD may be interpreted as indicating that an environmental risk factor is unlikely.

Case-control studies in CJD are therefore not targeted but depend on an assessment of a range of putative risk factors aimed at identifying case-to-case transmission or cross-species transmission from animal prion diseases. The incubation period in human prion diseases may be extremely prolonged with a mean incubation period of 13 yr in human growth hormone recipients *(22)* and a range of incubation periods from 5 to over 30 yr in kuru *(1)*, indicating that examination of risk factors must include analysis of potential exposure decades in the past. This problem is compounded by the necessity of obtaining information from relatives of the patients. A major practical problem in case-control studies is the need to study CJD on a nationwide basis because of its rarity and random occurrence.

The first task in a case-control study is the formulation of a standard questionnaire addressing various potential areas of risk. In the United Kingdom a questionnaire has been constructed including sections on:

1. Past medical history (including previous ocular or neurosurgery and blood donation/ reception);
2. Exposure to medicinal products;
3. Residential history;
4. Occupational history;
5. Family history;
6. Dietary history; and
7. Exposure to animals.

In order to avoid bias the questionnaire must be applied in a standard manner, preferably by an individual experienced in utilizing the questionnaire. Cur-

Table 6
Significant Risk Factors for CJD in Controlled Studies

Author	Method	Risk factors
Bobowick et al. *(36)*	38 "selected" cases; healthy controls	None
Kondo and Kuroiwa *(37)*	Population study: 60 cases; healthy controls	Trauma in males
Kondo *(38)*	88 autopsied cases; autopsied controls	Organ resection
Davanipour et al. *(39)*	26 cases; 40 controls	Trauma or surgery to head or neck Other trauma Surgery needing sutures Tonometry
Davanipour et al. *(40)*	As above	Roast pork, ham, underdone meat, hot dogs
Davanipour et al. *(41)*	As above	Contact with fish, rabbits, squirrels
Harries-Jones et al. *(42)*	92 cases; 184 controls	Herpes zoster Keeping cats Contact with pets other than cats/dogs Dementia in family

rently in the United Kingdom and in the collaborative European study, a research registrar visits each patient in order to obtain clinical information and to interview a relative of the patient. A control case is selected using the following criteria:

1. Age match ± 4 yr.
2. Sex matched.
3. Inpatient in the same hospital as the index case.
4. A relative of the same degree as the index case available for interview.
5. Patient has a condition clearly distinguishable from CJD.

The first available control case fulfilling these criteria is selected for interview, although in practice it is unusual for more than one control case satisfying the criteria to be identified and, indeed, it is not uncommon for no suitable control to be available, requiring a further visit at a later date. The potential difficulties of carrying out a case-control study in a rare disease on a nationwide basis should not be underestimated.

Alternative mechanisms for control selection have been applied in CJD, including the use of healthy relatives and random-digit dialling for community controls, but regardless of methodology the results of these studies have been largely negative and no consistent risk factor for CJD has been identified (Table 6).

There are two major caveats in the interpretation of the results of case-control studies in CJD. The rarity of the disease inevitably results in wide confidence intervals for any identified risk factor, particularly if this is an uncommon exposure, which results in marked fragility of data from year to year. In the United Kingdom a number of statistically significant dietary risk factors for CJD have been identified since 1990 but these vary from year to year, indicating that any apparent positive result should be treated with great caution. Second, the relatives of patients may be aware of the hypotheses being tested, a potential source of bias that has become potentially more important in view of the recent extensive media coverage in relation to BSE and CJD. In the UK study, regular veal consumption was identified as an apparently significant risk factor for CJD (odds ratio: 13:32) in 1994. Analysis of the frequency of veal consumption in suspect cases subsequently classified as "not CJD" demonstrated an almost identical excessive exposure to veal, providing powerful evidence of recall bias *(6)*.

Although there are clear limitations to the case-control methodology in CJD, it is reasonable to conclude that currently there is no consistent evidence of a specific environmental exposure that leads to increased risk of CJD. Implicit in this conclusion is that a number of potential risk factors, including blood transfusion, previous surgery (including eye surgery), and exposure to animals cannot be major risk factors for CJD. The accumulation of evidence in relation to risk factors of CJD may, however, allow the identification of a significant change in relation to novel exposures.

5. Conclusions

The systematic study of the epidemiology of CJD is labor-intensive, time-consuming, and costly. There are major practical difficulties in relation to case ascertainment, visiting hospitals on a nationwide basis, and case-control methodology. The basic premise is that cases of CJD are diagnosed and referred to the surveillance center, which is dependent on a high level of cooperation from the neuroscience community and others. In the United Kingdom the geographical distribution of cases (Fig. 4) indicates that cases have been consistently identified in all regions of the United Kingdom and one measure of the level of case ascertainment is the incidence of CJD (Fig. 5). One interpretation of the rise in incidence is that case ascertainment is improving *(35)* and there is clear evidence in the United Kingdom of an improvement in the identification of CJD in the elderly. Co-operation with this type of study is dependent on maintaining good relations with referring physicians, a task that is not made easier by intensive media scrutiny of individual cases.

The sea change in public perception of prion diseases has occurred largely because of the occurrence of BSE and despite the caveats regarding the surveillance of CJD, it may only be by systematic epidemiological study that any

Fig. 4. Geographical distribution of CJD in the United Kingdom. Definite and probable cases (May 1, 1990–April 30, 1995).

risk to the human population from BSE may be identified. Should a significant change in the pattern of CJD occur, it will be essential to consider whether this is related to the occurrence of BSE or some other factor. The incidence of CJD has increased significantly from study period to study period in each country in which serial surveillance of CJD has been carried out and DNA analysis has resulted in an improvement in the identification of familial cases. It may only be through comparisons of the epidemiology of CJD in different countries that any significant change in one country can be identified.

The cause of CJD in any individual patient cannot be determined from epidemiological evidence. The occurrence of CJD in an adolescent in the United Kingdom and the identification of three dairy farmers in the United Kingdom

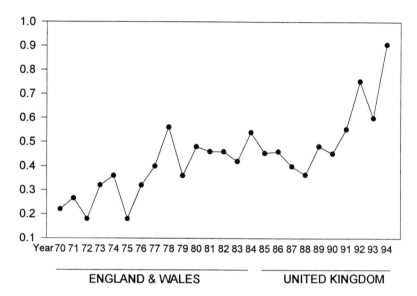

Fig. 5. Incidence (per million) of CJD. Definite and sporadic cases (1970–1994).

with CJD has understandably led to concerns about potential links with BSE. The availability of background epidemiological evidence in other countries may allow such atypical cases to be put into perspective. For example, three adolescents have been identified in countries free of BSE. However, in any individual case, a putative link with BSE can only be determined through other methods, for example, laboratory transmission studies.

The evident difficulties in studying a rare disease with a prolonged incubation period would be resolved if there were an in vivo diagnostic marker for infection. There have already been remarkable advances in the understanding of basic pathogenic mechanisms in prion diseases and it may be that the cause of sporadic CJD will be identified from further basic scientific research rather than through epidemiology.

References

1. Gajdusek, D. C. (1990) Subacute spongiform encephalopathies: transmissible cerebral amyloidoses caused by unconventional viruses, in *Field's Virology* (Fields, B. N. and Knipe, D. M., eds.), Raven, New York, pp. 2289–2324.
2. Hsiao, K., Baker, H. F., Crow, T. J., Poulter, M., Owen, F., Terwilliger, J. D., Westaway, D., Ott, J., and Prusiner, S. B. (1989) Linkage of a prion protein missense variant to Gerstmann Straussler syndrome. *Nature* **338**, 342–345.
3. Hsiao, K., Meiner, Z., Kahana, E., Cass, C., Kahana, I., Avrahami, D., Scarlatto, G., Abramsky, O., Prusiner, S. B., and Gabizon, R. (1991) Mutation of the prion

protein in Libyan Jews with Creutzfeldt-Jakob disease. *New Engl. J. Med.* **324,** 1091–1097.

4. Medori, R., Tritschler, H. J., LeBlanc, A., Villare, F., Manetto, V., Ying Chen, H., Xuf, R., Leal, S., Montagna, P., Cortelli, P., Tinuper, P., Avoni, P., Mochi, M., Baruzzi, A., Hauw, J. J., Ott, J., Lugaresi E., Autilio-Gambetti, L., and Gambetti, P. (1992) Fatal familial insomnia, a prion disease with a mutation at codon 178 of the prion protein gene. *New Engl. J. Med.* **326,** 444–449.

5. Cousens, S. N., Harries-Jones, R., Knight, R., Will, R. G., Smith, P. G., and Matthews, W. B. (1990) Geographical distribution of cases of Creutzfeldt-Jakob disease in England and Wales 1970–84. *J. Neurol. Neurosurg. Psychiat.* **53,** 459–465.

6. CJD Surveillance Unit, Edinburgh (1994) Creutzfeldt-Jakob Disease Surveillance in the United Kingdom—Third Annual Report.

7. Brown, P. and Cathala, F. (1979) Creutzfeldt-Jakob disease in France, in *Slow Transmissible Diseases of the Nervous System,* vol. 1 (Prusiner, S. B. and Hadlow, W. J., eds.), Academic, New York, pp. 213–227.

8. Rudge, P. (1989) Clinical neurology: a review. *J. Roy. Coll. Phys. Lond.* **23,** 83–86.

9. Davanipour, Z., Smoak, C., Bohr, I., Sobel, E., Liwnicz, B., and Chang, S. (1995) Death certificates: an efficient source for ascertainment of Creutzfeldt-Jakob disease cases. *Neuroepidemiology* **14,** 1–6.

10. Holman, R. C., Khan, A. S., Kent, J., Strine, T. W., and Schonberger, L. B. (1995) Epidemiology of Creutzfeldt-Jakob disease in the United States, 1979–1990: analysis of national mortality data. *Neuroepidemiology* **14,** 174–181.

11. Will, R. G., Matthews, W. B., Smith, P. G., and Hudson, C. (1986) A retrospective study of Creutzfeldt-Jakob disease in England and Wales 1970–1979 II: epidemiology. *J. Neurol. Neurosurg. Psychiat.* **49,** 749–755.

12. Jones, D. P. and Nevin, S. (1954) Rapidly progressive cerebral degeneration (subacute vascular encephalopathy) with mental disorder, focal disturbance, and myoclonic epilepsy. *J. Neurol. Neurosurg. Psychiat.* **17,** 148–159.

13. Masters, C. L., Harris, J. O., Gajdusek, D. C., Gibbs, C. J., Jr., Bernoulli, C., and Asher, D. M. (1979) Creutzfeldt-Jakob disease: patterns of worldwide occurrence and the significance of familial and sporadic clustering. *Ann. Neurol.* **5,** 177–188.

14. Will, R. G. and Matthews, W. B. (1984) A retrospective study of Creutzfeldt-Jakob disease in England and Wales 1970–79 I: clinical features. *J. Neurol. Neurosurg. Psychiat.* **47,** 134–140.

15. Miyanaga, K., Takahashi, S., and Fukuda, M. (1982) An autopsy case of Alzheimer's disease accompanied with periodic synchronous discharge and myoclonus. *Clin. Neurol.* **22,** 128–134.

16. Will, R. G. and Matthews, W. B. (1992) Creutzfeldt-Jakob disease epidemiology, in *Prion Diseases of Humans and Animals* (Prusiner, S. B., Collinge, J., Powell, J., and Anderton, B., eds.), Ellis Horwood, Chichester, pp. 188–199.

17. Bell, J. E. and Ironside, J. W. (1993) How to tackle a possible Creutzfeldt-Jakob disease necropsy. *J. Clin. Pathol.* **46,** 193–197.

18. Salazar, A. M., Masters, C. L., Gajdusek, D. C., and Gibbs, C. J., Jr. (1983) Syndromes of amyotrophic lateral sclerosis and dementia: relation to transmissible Creutzfeldt-Jakob disease. *Ann. Neurol.* **14,** 17–26.

19. Brown, P., Kaur, P., Sulima, M. P., Goldfarb, L., Gibbs, C. J., Jr., and Gajdusek, D. C. (1993) Real and imagined clinicopathological limits of "prion dementia". *Lancet* **341,** 127–129.

20. Collinge, J., Poulter, M., Davis, M. B., Baraitser, M., Owen, F., Crow, T. J., and Harding, A. E. (1991) Presymptomatic detection or exclusion of prion protein gene defects in families with inherited prion diseases. *Am. J. Hum. Genet.* **49,** 1351–1354.

21. Brown, P., Cervenakova, L., Goldfarb, L. G., Gajdusek, D. C., Haverkamp, A., Haverkamp, C., Horwitz, J., Creacy, S. D., Bever, R. A., Wexler, P., Sujansky, E., and Bjork, R. J. (1994) Molecular genetic testing of a fetus at risk of Gerstmann-Sträussler-Scheinker syndrome. *Lancet* **343,** 181–182.

22. Brown, P., Preece, M. A., and Will, R. G. (1992) 'Friendly fire' in medicine: hormones, homografts, and Creutzfeldt-Jakob disease. *Lancet* **340,** 24–27.

23. Will, R. G. (1992) BSE and the spongiform encephalopathies, in *Recent Advances in Clinical Neurology* (Kennard, C., ed.), Churchill Livingstone, London, pp. 115–127.

24. Collinge, J., Brown, J., Hardy, J., Mullan, M., Rossor, M. N., Baker, H., Crow, T. J., Lofthouse, R., Poulter, M., Ridley, R. M., Owen, F., Bennett, C., Dunn, G., Harding, A. E., Quinn, N., Doshi, B., Roberts, G. W., Honavar, M., Janota, I., and Lantos, P. L. (1992) Inherited prion disease with 144 base pair gene insertion. 2. Clinical and pathological features. *Brain* **115,** 687–710.

25. Brown, P., Rodgers-Johnson, P., Cathala, F., Gibbs, C. J., Jr., and Gajdusek, D. C. (1984) Creutzfeldt-Jakob disease of long duration: clinicopathological characteristics, transmissibility, and differential diagnosis. *Ann. Neurol.* **16,** 295–304.

26. Brown, P., Goldfarb, L., Gibbs, C. J., and Gajdusek, D. C. (1991) The phenotypic expression of different mutations in transmissible familial Creutzfeldt-Jakob disease. *Eur. J. Epidemiol.* **7,** 469–476.

27. Gambetti, P., Parchi P., Petersen, R. B., Chen, S. G., and Lugaresi, E. (1995) Fatal familial insomnia and familial Creutzfeldt-Jakob disease: clinical, pathological and molecular features. *Brain Pathol.* **5,** 43–51.

28. Schellenberg, G. D., Anderson, L., O'Dahl, S., Wisjman, E. M., Sadovnick, A. D., Ball, M. J., Larson, E. B., Kukull, W. A., Martin, G. M., Roses, A. D., and Bird, T. D. (1991) APP-717 APP-693 and PRIP gene mutations are rare in Alzheimer's disease. *Am. J. Hum. Genet.* **49,** 511–517.

29. Collinge, J., Palmer, M., Sidle, K. C. L., Mahal, S. P., Campbell, T., Brown, J., Hardy, J., Brun, A. E., Gustafson, L., Bakker, E., Roos, R., and Groen, J. J. (1994) Familial Pick's disease and dementia in frontal lobe degeneration of non-Alzheimer type are not variants of prion disease. *J. Neurol. Neurosurg. Psychiat.* **57,** 762–768.

30. Jendroska, K., Hoffmann, O., Schelosky, L., Lees, A. J., Poewe, W., and Daniel, S. E. (1994) Absence of disease related prion protein in neurodegenerative disorders presenting with Parkinson's syndrome. *J. Neurol. Neurosurg. Psychiat.* **57,** 1249–1251.

31. Revesz, T., Daniel, S. E., Lees, A. J., and Will, R. G. (1995) A case of progressive subcortical gliosis associated with deposition of abnormal prion protein (PrP). *J. Neurol. Neurosurg. Psychiat.* **58,** 759–760.

32. Petersen, R. B., Tabaton, M., Chen, S. G., Monari, L., Richardson, S. L., Lynches, T., Manetto, V., Lanska, D. J., Markesbery, W. R., Currier, R. D., Autilio-Gambetti, L., Wilhelmsen, K. C., and Gambetti, G. (1995) Familial progressive subcortical gliosis: presence of prions and linkage to chromosome 17. *Neurology* **45,** 1062–1067.

33. Masters, C. L., Gajdusek, D. C., and Gibbs, C. J., Jr. (1981) The familial occurrence of Creutzfeldt-Jakob disease and Alzheimer's disease. *Brain* **104,** 535–558.

34. Editorial (1990) Prion disease: spongiform encephalopathies unveiled. *Lancet* **336,** 21–22.

35. Editorial (1993) Do epidemiologists cause epidemics? *Lancet* **341,** 993–994.

36. Bobowick, A. R., Brody, J. A., Matthews, M. R., Roos, R., and Gajdusek, D. C. (1973) Creutzfeldt-Jakob disease: a case-control study. *Am J. Epidemiol.* **98,** 381–394.

37. Kondo, K. and Kuroiwa, Y. (1982) A case-control study of Creutzfeldt-Jakob disease: association with physical injuries. *Ann. Neurol.* **11,** 377–381.

38. Kondo, K. (1985) Epidemiology of Creutzfeldt-Jakob disease in Japan, in *Creutzfeldt-Jakob Disease* (Mizutani, T. and Shiraki, H., eds.), Elsevier/Nishimura, Amsterda/Niigate, pp. 17–30.

39. Davanipour, Z., Alter, M., Sobel, E., Asher, D., and Gajdusek, D. C. (1985) Creutzfeldt-Jakob disease: possible medical risk factors. *Neurology* **35,** 1483–1486.

40. Davanipour, Z., Alter, M., Sobel, E., Asher, D. M., and Gajdusek, D. C. (1985) A case-control study of Creutzfeldt-Jakob disease: dietary risk factors. *Am J. Epidemiol.* **122,** 443–451.

41. Davanipour, Z., Alter, M., Sobel, E., Asher, D. M., and Gajdusek, D. C. (1986) Transmissible virus dementia: evaluation of a zoonotic hypothesis. *Neuroepidemiology* **5,** 194–206.

42. Harries-Jones, R., Knight, R., Will, R. G., Cousens, S., Smith, P. G., and Matthews, W. B. (1988) Creutzfeldt-Jakob disease in England and Wales, 1980–1984: a case-control study of potential risk factors. *J. Neurol. Neurosurg. Psychiat.* **51,** 1113–1119.

8

Environmental Causes
of Human Spongiform Encephalopathy

Paul Brown

This chapter reviews all proven or highly probable cases of environmentally acquired human spongiform encephalopathy (cannibalism, neurosurgical procedures, corneal and dura mater homografts, and native pituitary hormone therapy), and evaluates potential but as yet unverified environmental sources of disease, such as peripheral tissue homografts, organ transplants, administration of blood, blood products, and other biologicals, occupational exposures, and zoonotic infections. Implicit in this breviary of known and putative origins of environmentally acquired disease are laboratory and epidemiologic methods for recognizing and if possible preventing subsequent cases both from present and future sources of infection.

Like so much else in the field of human spongiform encephalopathy, it all began with kuru. Long before Carleton Gajdusek introduced this exotic New Guinea Highlands neurologic disease to the medical community in 1957 *(1)*, local missionaries, district officers, bush pilots, and bartenders were already speculating that kuru was being "caught" as a result of ritual cannibalism. Indeed, despite the absence of any clinical or neuropathological signs of an inflammatory process, Gajdusek's first priority was to explore its contagious (and presumably viral) character, for which he initiated an extensive program of inoculation studies involving countless rodent species and tissue culture cell lines. All failed to reveal an infectious agent, because no precedent existed in human virology for experiments to be continued beyond the 1–2 mo periods when most of these studies were terminated.

Genetic, endocrine, and toxic causes of kuru were also being explored, but they too failed to yield the answer, and attention turned to the possibility of a noninflammatory parasitic or fungal infection, with a new set of inoculated

From: *Methods in Molecular Medicine: Prion Diseases*
Edited by: H. Baker and R. M. Ridley Humana Press Inc., Totowa, NJ

animals (including monkeys) held for longer-term observation. After the pathologic similarity of kuru to scrapie was appreciated *(2)*, it was decided to add chimpanzees to the list of experimental animals, and to extend even further the period of surveillance. By the time the first animals showed signs of disease, 2 yr after their inoculation in 1963 *(3)*, kuru was well on its way to becoming extinct as a result of the rapid decline of cannibalism in the late 1950s. The obvious best explanation for the epidemic of kuru, so quickly evident to medical and nonmedical observers alike, and later authenticated by formal epidemiologic study and experimental transmission, turned out to be correct, but required almost 10 yr to prove. Its burden of mortality still far outnumbers deaths from all other sources of disease.

Creutzfeldt-Jakob disease (CJD) had in the meantime been languishing in the backwaters of neurology since the early 1920s as a rare, progressive neurodegenerative process of unknown etiology, but its neuropathologic resemblance to kuru caught the eye of Igor Klatzo *(4)*. Experimental primate inoculation studies demonstrated its transmissibility in 1968 *(5)*, and epidemiologic investigations characterized both its rarity and sporadic character, although a small number of familial cases had also been described. Apart from these few families, in which infection could theoretically pass from affected to unaffected members, it did not seem possible that a randomly occurring disease at an annual frequency of <1 case/million people could be transmitted horizontally; surely not by case-to-case contact, unless a clinically silent carrier state were postulated, and unlikely by any other environmental mode, such as exposure to analogous animal diseases (e.g., scrapie), or to some ubiquitous pathogen that infected only the rare (and presumably genetically susceptible) individual.

It was not until 1974 that a case of CJD was suspected to have resulted from an environmental source. The donor of a corneal graft was found at autopsy to have died from CJD, and, after a latency of 18 mo, the graft recipient also developed CJD *(6)*. This remained the only recognized instance of horizontally transmitted disease until 1977, when a second report described the occurrence of two cases of CJD 16 and 20 mo after stereotactic electroencephalography procedures employing the same (conventionally sterilized) needles that earlier had been used on a patient with CJD *(7)*. To this day, these two cases represent the only fully proven instances of iatrogenic CJD, because the brains of both cases as well as the electrodes used for the EEG procedures later were shown to transmit disease to experimental primates *(8)*.

Given the fact that surgically induced CJD could occur, retrospective studies revealed a high probability that at least three earlier episodes of iatrogenic CJD had occurred as a consequence of neurosurgical procedures *(9,10)*. Among cases of CJD reported by Nevin et al. in 1960 *(11)* were two patients (cases 1 and 3) whose illnesses began 15 and 19 mo after operations in 1952 in which the same instruments had been used 2 wk before on a patient with CJD; another patient (case 7) became ill 18 mo after an operation in 1956 in which the same instruments had been used several hours earlier on a patient with CJD. The third episode occurred in 1965 in France, where a patient developed CJD 28 mo after an operation in which the same instruments probably had been used 3 d before on a patient with CJD *(9)*.

Nothing more was heard about iatrogenic disease for the next several years, but the sanguine outlook encouraged by this lull was to change dramatically in 1985, when the pediatric endocrinologist Raymond Hintz notified the US National Institutes of Health of a case of CJD in a young hypopituitary patient who many years earlier had been treated with native human growth hormone *(12)*, and suggested that pituitary glands from CJD patients inadvertently might have found their way into cadaver pituitary pools from which the growth hormone had been extracted.

With a speed that is almost never encountered in government reactions to potential problems, Mortimer Lipsett, Director of the NIH Institute responsible for overseeing the pituitary treatment program, held advisory meetings and within 2 wk notified pediatric endocrinologists around the country to be on the lookout for unexplained neurologic deaths in their patient population. When, 2 mo later, two more cases surfaced in quick succession *(13,14)*, native growth hormone was withdrawn from circulation and replaced by recombinant hormone, but only time would tell whether we would witness a mere handful of further cases or a full-blown epidemic of iatrogenic CJD *(15)*. As it turned out, the damage was somewhere between the two extremes: The current total stands at 76 cases, with new cases occurring at the rate of about two to three each year, principally in the United States *(16–18;* and unpublished data), the United Kingdom *(19–22;* and unpublished data), and France *(23,24;* and unpublished data) (each using their own sources of pituitary glands), although cases also have been identified in Brazil *(25)* and New Zealand *(26)* in patients given hormone processed in the United States, and in Australia from locally produced hormone (unpublished data). Causality was unequivocally established when a sample of one US lot that was inoculated into experimental primates transmitted disease after an incubation period of 5 yr *(27)*.

The potential for human pituitary gland infectivity had actually been appreciated several years earlier, and a study was carried out in 1979 in which a normal human pituitary gland was mixed with scrapie-infected mouse brain tissue, processed for growth hormone (Lowry method), and the residual infectivity measured at several successive steps *(28)*. Because infectivity was undetectable in the final product, the identically purified hormone distributed for human use was thought to be without risk. Unfortunately, the final sample was only sampled for infectivity (the customary method) rather than inoculated *in*

Table 1
Summary of All Proven
or Highly Probable Cases of Iatrogenic Creutzfeldt-Jakob Disease

Mode of infection	Number of patients	Agent entry into brain	Mean incubation period, range	Clinical presentation
Stereotactic EEG	2	Intracerebral	18 mo (16, 20)	Dem/cereb[b]
Neurosurgery	4	Intracerebral	20 mo (15–28)	Vis/dem/cereb
Corneal transplant	2	Optic nerve	17 mo (16, 18)	Dem/cereb
Dura mater graft	25	Cerebral surface	5.5 yr (1.5–12)	Cereb (vis/dem)
Gonadotrophin	4	Hematogenous	13 yr[a] (12–16)	Cerebellar
Growth hormone	76	Hematogenous	12 yr[a] (5–30)	Cerebellar

[a]Calculated from the midpoint of hormone therapy to the onset of CJD symptoms.
[b]Dem, demential; Cereb, cerebellar; *see text.*

toto (to demonstrate the complete absence of the infectious agent) *(29)*, and, when this more rigorous type of validation protocol was finally carried out several years later (1991), scrapie-infected pituitaries did in fact transmit disease to a few of several hundred inoculated hamsters *(30)*.

The risk of dying from CJD among the treated populations of the United States, United Kingdom, and France show some interesting differences (Table 1). In the United States and United Kingdom, crosscheck analysis of patients and hormone lots revealed that occasional random lots produced during the 1960s and 1970s must have been contaminated in both countries. In contrast, all of the patients in France, where the risk was greatest, had overlapping treatments during 1984 and 1985, suggesting a rather more important degree of contamination confined to lots that were distributed during this period.

Cases of CJD have also occurred in four women treated with cadaver-derived pituitary gonadotrophin, among a total of 1450 treated individuals in Australia *(31,32)* (and unpublished data). Most other pituitary hormones have little or no use in clinical practice, or if used are derived from animal or synthetic sources. The only exception is pituitary-derived prolactin, which was about to be released for clinical trials, then quickly abandoned when the growth hormone story unfolded.

The average incubation period for these 78 cases of CJD from contaminated pituitary hormones (calculated from the midpoint of hormone therapy to the onset of symptomatic CJD), was approx 12 yr; minimum incubation periods (calculated from the end of therapy to the onset of symptoms) ranged from 5–25 yr (Table 2). These latencies are in excellent accord with estimated incubation periods for kuru, which also ranged from approx 4–30 yr, and support the presumption that both diseases resulted from peripheral (as opposed to CNS) body entry routes by small amounts of the infectious agent. Data from

Table 2
Comparison of Risks of Contracting Creutzfeldt-Jakob Disease and Mean Incubation Periods After Treatment with Native Human Growth Hormone in the United States, United Kingdom, and France

Country of origin	Number of patients	Treated population	Risk of CJD, %	Mean incubation period, yr[a]
United States[b]	17	8300	0.2	18 (±6)
United Kingdom	17	1750	1.0	12 (±3)
France	41	1700	2.4	8 (±3)

[a]Calculated from the midpoint of hGH therapy to the onset of CJD symptoms.
[b]Includes two patients from New Zealand and one from Brazil who received hormone prepared in the United States.

experimental primate transmission studies also had shown that small amounts of the agent inoculated peripherally (e.g., intradermal or subcutaneous injection) may be followed by prolonged incubation times, or even failure to transmit disease altogether *(33)*.

Although our attention was riveted on the increasing number of iatrogenic pituitary hormone cases, bad news came from yet another direction: surgical operations employing cadaveric dura mater graft patches. In the United States in 1987, a case of CJD was identified in a young woman who 19 mo earlier had a surgical procedure that included placement of a cadaveric dural homograft *(34)*. Additional cases have since been identified in the United States *(35)*, the United Kingdom *(36,37)* (and unpublished), Canada (unpublished), Italy *(38,39)*, Spain *(40)*, Germany *(41)*, Australia (unpublished), New Zealand *(42)*, and Japan *(43,44)*. The total number of cases currently stands at 25, and like growth hormone cases, is slowly increasing every year.

Unlike the hormone situation, however, in which multiple random lots independently produced in several different countries were contaminated over a period of many years, all but two of the dura mater cases had received grafts distributed between the years 1982 and 1986 by a single producer ("Lyodura" from B. Braun Melsungen AG). The commercial processing protocol, which includes exposure to H_2O_2, acetone, and ionizing radiation (25 kGy), later was found to have very little effect on infectivity present in dura mater from scrapie-infected hamsters; but exposure to $1N$ NaOH (a procedural step that was added to commercial protocols in 1987) resulted in almost total inactivation of the infectious agent *(45)*.

The average incubation period for these dura mater cases is approx 5 yr, ranging from 16 mo to 12 yr after implantation of the graft, and thus intermediate in latency between cases in which the infectious agent was introduced directly into the brain by contaminated instruments, and cases in which the agent was introduced from peripheral inoculations of contaminated pituitary hormone (Table 2). The clinical syndromes typically shown by each of these groups of patients also differed. After intracerebral or ocular nerve infection (surgical instrument, EEG electrode, and corneal graft cases), clinical presentations and evolutions mimicked those seen in sporadic CJD, with many patients showing an important early demential component. After cerebral surface contact infection (dura mater cases), most patients presented with cerebellar abnormalities, alone or in combination with visual signs or mental deterioration; and after peripheral infection (pituitary hormone cases), the disease took on an almost stereotyped evolution of progressive cerebellar signs, with little or no mental deterioration until late in the course of illness.

The first four cases of surgically transmitted CJD that occurred during the 1950s and 1960s were not recognized for the simple reason that the disease had not yet been shown to be infectious, let alone have an incubation period that could extend over a period of years or even decades. The corneal graft and stereotactic EEG cases that occurred in the 1970s were recognized because by that time the infectious character and long incubation period of CJD had been experimentally documented, and just as important, the knowledge had been widely disseminated to the medical community; because the donors as well as the recipients were neuropathologically confirmed cases of CJD; and because the interval between donor and recipient deaths was not so long as to have obscured their relationship. This was also true for the dura mater cases, among which the first case had (providentially) the shortest incubation period (19 mo), and thus was more readily evaluated with respect to antecedent neurosurgery. In contrast, the first growth hormone-related case was suspected by only one of the many physicians who saw him, and more than anything else reflects a remarkable exercise of intuition.

Although the root cause of iatrogenic disease is the awesome resistance of the infectious agent to procedures designed to inactivate conventional pathogens, including exposure to ethanol, H_2O_2, permanganate, iodine, ethylene oxide vapor, detergents, organic solvents, formaldehyde, UV or gamma irradiation, and even standard autoclaving (46–51), genetics has been found to play an important subsidiary role. A gene on chromosome 20 encoding the protein that in patients with spongiform encephalopathy is transformed into an amyloidogenic isoform contains a polymorphism at codon 129 that normally can specify either of two amino acids (methionine or valine). In the general population, heterozygotes and homozygotes occur at about the same frequency; however, in the iatrogenic CJD population, whatever the origin or route of infection, this equilibrium is heavily tilted toward homozygosity (in about a 10:1 proportion) (Table 3). Thus, codon 129 heterozygotes are, for reasons that are not yet understood, significantly less vulnerable to iatrogenic infection than are homozygotes (52–54). It is also interesting that of the four patients who

Table 3
Codon 129 Genotypes in Patients
with Iatrogenic Creutzfeldt-Jakob Disease

Tested groups[a]	Codon 129 genotype			
	Met/Met	Met/Val	Val/Val	Homozygous
CNS route of infection				
Stereotactic-EEG electrode	1	1	0	1/2
Neurosurgery	1	1	0	1/2
Corneal transplant	1	0	1	2/2
Dura mater graft	13	0	1	14/14
CNS subtotal	16 (80%)	2 (10%)	2 (10%)	18/20 (90%)[b]
Peripheral route of infection				
Human gonadotrophin	2	1	1	3/4
Human growth hormone	20	4	15	35/39
Peripheral subtotal	22 (51%)	5 (12%)	16 (37%)	38/43 (88%)[b]
All iatrogenic cases	38 (60%)	7 (11%)	18 (29%)	56/63 (89%)[b]
Normal controls	97 (37%)	135 (52%)	29 (11%)	126/261 (48%)

[a]Pooled information from the United Kingdom *(52)*, United States *(53)*, France *(54)*, and unpublished data.
[b]$P < 0.001$ compared to normal controls (Chi-square test).

were "inoculated" intracerebrally (by neurosurgical instruments and stereotac-tic EEG electrodes), two were heterozygous and two were homozygous, sug-gesting that heterotypic resistance can be overcome by direct introduction of the infectious agent into the brain.

To date, all known cases of iatrogenic CJD have resulted from exposure to infectious brain, pituitary, or ocular tissue, almost certainly owing to the high levels of infectious agent in the central nervous system (CNS) (and organs to which it is directly connected). However, from tissue distribution studies in both humans and experimental animals (Table 4), it has been well established that the infectious agent is widespread in the body, albeit in much lower con-centration than in the brain, and with an irregular and unpredictable occur-rence; thus, there is reason to suppose that at some point in the future we will learn of a case for which a peripheral tissue is implicated as the source of con-tamination. We already are aware of a few cases of CJD in patients who 2–5 yr earlier had a pericardial homograft for tympanic membrane closure *(55)*, or bone, kidney, or liver transplants (unpublished data). In none of these instances, however, was the donor identified as having had CJD, so their iatrogenic ori-gin still must be considered conjectural.

The question of CJD associated with the administration of blood or blood products merits special consideration because blood repeatedly has been shown

Table 4
Distribution of Comparative Frequency of Infectivity in Organs
of Humans or Animals with Spongiform Encephalopathy[a]

Host tissue	Human CJD/kuru[b]	Sheep/goat scrapie	Cattle BSS
Brain	+++	+++	+++
Spinal cord	++	+++	(++)
Cerebrospinal fluid	++	+++	(0)
Eyeball	+++	+++	(0)
Peripheral nerve	(0)	+++	(0)
Pituitary gland	NT[c]	+++	NT
Spleen	+	+++	(0)
Lymph nodes	+	+++	(0)
Leukocytes	+	NT	(0)
Serum	(0)	0	(0)
Whole or clotted blood	0	±	(0)
Bone marrow	(0)	0	(0)
Lung	+	±	(0)
Liver	+	+	(0)
Kidney	+	0	(0)
Pancreas	NT	0	(0)
Thymus	NT	±	NT
Intestine	(0)	+++	(0)
Heart	0	0	(0)
Skeletal muscle	0	0	0
Fat	(0)	NT	(0)
Testis	(0)	0	(0)
Semen	(0)	0	0
Ovary	NT	+	(0)
Uterus	NT	+	(0)
Placenta	(+)	(++)	0
Amniotic fluid	(0)	(±)	(0)
Cord blood	(+)	NT	(0)
Colostrum	(+)	0	NT
Milk	(0)	(0)	0

[a]Based on isolations of the infectious agent from the natural hosts of each disease.

[b]Infectivity: +++ almost always present, ++ frequent, + irregular, ± rare, 0 undetectable. Parentheses indicate very few tested specimens.

[c]NT: not tested.

to be infectious during the incubation period and clinical phase of experimentally infected animals, and occasionally has been found infectious in humans with CJD (56–59), and because iatrogenic disease from this source would dwarf in importance all other sources by virtue of the sheer number of people who theoretically have been or could be at risk. Although a small proportion of individuals among the CJD patient population is known either to have donated or received blood, epidemiologic case control studies covering the years 1980–1984 and 1990–1992 in the United Kingdom found no increase in the incidence of CJD in areas where blood from CJD donors had been distributed, nor any difference between CJD patients and the general population in the frequency of having received blood *(60)*.

This report generated two echoes. The first reverberated from Australia under the alarming title "Transmission of Creutzfeldt-Jakob disease by blood transfusion" *(32)*. Four patients dying of CJD were found to have received transfusions 5 yr before the onset of nondemential cerebellar clinical presentations. However, unlike the British studies, no information was provided about the comparable frequency of transfusions in a non-CJD control population, nor was an effort made to identify CJD patients among the blood donors. The uniformly cerebellar presentation of the four patients is troubling, because cerebellar onsets are the hallmark of iatrogenic disease from peripheral routes of infection, but because nearly one-third of sporadic cases of CJD may also present with cerebellar signs, this fact alone cannot be relied on to prove iatrogenic causality.

The second report echoed from Germany, where an effort was made to trace all recipients of blood from a regular donor who later died of CJD *(61)*. Nearly 70% (35/55) of the recipients were identified, of whom 21 had died from non-CJD illnesses up to 22 yr after having received transfusions, and 14 were still alive without evidence of neurologic disease from 2–21 yr later (mean survival, 12 yr). Similar efforts are presently being pursued for several more such incidents by the American Red Cross and the European BioMed-1 CJD surveillance study.

Finally, it is worth recalling that no case of CJD has been identified among patients whose diseases require (or required before recombinant technology) repeated administration of whole blood, blood components (e.g., plasma or leukocytes), or blood derivatives (e.g., albumin, immune globulin, interferon, α-1 antitrypsin, clotting factors VIII and IX). Such potentially high risk groups include patients with congenital anemias and clotting deficiencies (e.g., sickle cell disease, thalassemia major, hemophilia), immune deficiency/suppression syndromes, multiple sclerosis, α-1 antitrypsin deficiency, bone marrow transplants, and multiple surgical procedures.

None of these epidemiologic observations should be considered definitive, because population-based studies can never prove or disprove the possibility that one or two cases do indeed result from contaminated blood products but are insufficient to break through the barrier of statistical significance, and single incident studies, even if all recipients are traced (which is almost never possible), may be criticized because a proportion of recipients will have died from other illnesses before CJD has had a chance to declare itself. On balance, the available evidence favors the conclusion that blood-borne infection is not a cause of CJD; however, rather than come down hard with either a guilty or not guilty verdict, for the present we might be better advised to render the Scottish judgment of "not proven."

Biologicals derived from tissues other than blood are also candidates for evaluation as potential sources of iatrogenic infection. Certain products, such as CNS gangliosides (of either human or bovine origin) have an obvious potential to be contaminated, whereas other products that derive from animal species not known to be susceptible to spongiform diseases, or even from nonanimal sources, nevertheless may be exposed to culture media and protein broths containing small amounts of tissue extracts from a variety of animals, and stabilized with human albumin in the course of processing protocols. Each such product must be evaluated on an individual basis, and although an informed judgment usually can be made about the potential risk inherent in its components, no amount of reasoning can take the place of "validation" experiments to determine whether in fact infectious starting material retains any infectivity in the final product.

Such studies have been performed only for human growth hormone and bovine gangliosides, and have shown that residual infectivity is indeed present in the final products. However, it was found that growth hormone remained biologically active after exposure to $6M$ urea and ultrafiltration steps that effectively decontaminated the purified hormone *(30)*, and that gangliosides could withstand exposure to NaOH and steam autoclaving that totally inactivated the infectivity present in the starting brain material *(62)*. Similar validation studies will be required to determine if the harsh physicochemical treatments and ultrafiltration steps that are required to disinfect the agent of CJD are compatible with biological activity of the final products.

Two other potential but as yet unproven environmental sources of disease relate to workplace infections in healthcare professionals, and zoonotic infections from animals with analogous spongiform diseases, such as scrapie or bovine spongiform encephalopathy (BSE). During the last 15 yr, several instances of CJD in health professionals have been recorded, either in systematic epidemiologic CJD surveillance studies *(63–65)*, or as individual case reports *(66–71)*, and include physicians, nurses, nurses aides, dentists, and pathology technicians.

In surveillance studies, comparisons of the observed and expected incidence of CJD in these professions usually has not differed significantly from that of the general population, and even when they have differed must be viewed with caution because statistics tend to lose their power of persuasion when dealing with small differences in the incidence of a rare disease, and because equally "significant" differences have been observed among nuns and vicars, for whom no evident risk factors can be imagined *(64,72)*.

Nevertheless, the occurrence of disease in an individual with a recognized high-risk potential never can be dismissed out of hand, and thus the reports of CJD in a neurosurgeon *(66)*, a pathologist *(69)*, and two neuropathology technicians *(67,68)*, even if no definite infecting event was identifiable, are cause for concern. Another recent report described a case of CJD in an orthopedic surgeon who 20 yr earlier had worked with both sheep and human dura mater (the latter material having been sent to the same firm that inadvertently distributed CJD-contaminated human dura mater) *(70)*. Eventually, it may happen that the accumulated weight of such anecdotal reports will become great enough to be persuasive even without the buttress of statistical significance. Fortunately (for the author), no case of CJD has yet occurred in a research laboratory worker.

As concerns zoonotic infection, scientists and the general public have been much exercised over the possibility that the current UK epidemic of BSE could pose a serious risk to human health, from eating beef or other tissues, or drinking milk from animals incubating the disease that might find their way to the marketplace. The question at the moment is unanswerable, because information about the tissue distribution of infectivity and its ability to transmit disease by ingestion (a very inefficient route of infection) will take some years to acquire. To date (and we are now approx 10 yr down the road from the onset of the epidemic) the accumulated epidemiological evidence indicates that BSE (like scrapie, which has been prevalent in the United Kingdom for at least 250 yr) does not pose a significant risk to human health. Although CJD has occurred in four farmers with BSE-infected herds *(72* and unpublished data), more convincing data come from the ongoing active surveillance of CJD in several European countries, in which the annual mortality rates of CJD do not differ between countries that have BSE and those that do not *(73)*.

In conclusion, we know enough today to suspect a wide variety of sources and modes of exposure as potential risks of environmentally acquired CJD; to evaluate in a logical manner the likelihood of these sources to contain the infectious agent and transmit disease by routes of infection with differing efficiencies; to know that we may never encounter incriminating case "clusters" because of the very small amounts of infectious agent to which the victims may have been exposed, and the consequent length and variability of incubation periods that can extend up to 30 yr; and to be alert to the fact that the predominant disease symptoms may be cerebellar rather than mental. In patients that come to autopsy, we also may be able to distinguish environmentally acquired from sporadic cases of CJD by different regional patterns of amyloid protein distribution in the CNS *(59,74)*.

All of this knowledge, gained both from planned experiments and natural misfortunes, should provide us enough information to reason, predict, and, if possible, prevent future occurrences. But surprises may still lie ahead, perhaps in the form of an isolated case of CJD years or even decades after a marrow transplant for leukemia, dura mater grafting for pericardial repair, or leukocyte interferon treatment for multiple sclerosis. For these unforeseen events, we shall need the intuitive genius that, although informed both by knowledge and reason, only comes to us unbidden and alone in magical moments, a gift from the gods.

References

1. Gajdusek, D. C. and Zigas, V. (1957) Degenerative disease of the central nervous system in New Guinea: epidemic occurrence of "kuru" in the native population. *N. Engl. J. Med.* **257**, 974–978.
2. Hadlow, W. J. (1959) Scrapie and kuru. *Lancet* **2**, 289,290.
3. Gajdusek, D. C., Gibbs, C. J. J., and Alpers, M. (1966) Experimental transmission of a kuru-like syndrome in chimpanzees. *Nature* **209**, 794–796.
4. Klatzo, I., Gajdusek, D., and Zigas, V. (1959) Pathology of kuru. *Lab. Invest.* **8**, 799–847.
5. Gibbs, C. J., Jr., Gajdusek, D. C., Asher, D. M., Alpers, M. P., Beck, E., Daniel, P. M., et al. (1968) Creutzfeldt-Jakob disease (spongiform encephalopathy): transmission to the chimpanzee. *Science* **161**, 388,389.
6. Duffy, P., Wolf, J., Collins, G., DeVoe, A. G., Streeten, B., and Cowen, D. (1974) Possible person-to-person transmission of Creutzfeldt-Jakob disease. *N. Engl. J. Med.* **290**, 692,693.
7. Bernoulli, C., Siegfried, J., Baumgartner, G., Regli, F., Rabinowics, T., Gajdusek, D. C., et al. (1977) Danger of accidental person-to-person transmission of Creutzfeldt-Jakob disease by surgery. *Lancet* **i**, 478,479.
8. Brown, P. (1994) Transmissible human spongiform encephalopathy (infectious cerebral amyloidosis): Creutzfeldt-Jakob disease, Gerstmann-Sträussler-Scheinker syndrome, and kuru, in *Neurodegenerative Diseases* (Calne, D. B., ed.), Philadelphia, pp. 839–876.
9. Foncin, J., Gaches, J., Cathala, F., El Sherif, E., and Le Beau, J. (1980) Transmission iatrogène interhumaine possible de maladie de Creutzfeldt-Jakob avec atteinte des grains du cervelet. *Rev. Neurol. (Paris)* **136**, 280.
10. Will, R. G. and Matthews, W. B. (1982) Evidence for case-to-case transmission of Creutzfeldt-Jakob disease. *J. Neurol. Neurosurg. Psychiat.* **45**, 235–238.
11. Nevin, S., McMenemey, W. H., Behrman, S., and Jones, D. P. (1960) Subacute spongiform encephalopathy—a subacute form of encephalopathy attributable to vascular dysfunction (spongiform cerebral atrophy). *Brain* **83**, 519–564.
12. Koch, T. K., Berg, B. O., De Armond, S. J., and Gravina, R. F. (1985) Creutzfeldt-Jakob disease in a young adult with idiopathic hypopituitarism. *N. Engl. J. Med.* **313**, 731–733.

13. Gibbs, C. J., Jr., Joy, A., Heffner, R., Franko, M., Miyazaki, M., Asher, D. M., et al. (1985) Clinical and pathological features and laboratory confirmation of Creutzfeldt-Jakob disease in a recipient of pituitary-derived human growth hormone. *N. Engl. J. Med.* **313,** 734–738.

14. Tintner, R., Brown, P., Hedley-Whyte, E. T., Rappaport, E. B., Piccardo, C. P., and Gajdusek, D. C. (1986) Neuropathologic verification of Creutzfeldt-Jakob disease in the exhumed American recipient of human pituitary growth hormone: epidemiologic and pathogenetic implications. *Neurology* **36,** 932–936.

15. Brown, P., Gajdusek, D. C., Gibbs, C. J., Jr., and Asher, D. M. (1985) Potential epidemic of Creutzfeldt-Jakob disease from human growth hormone therapy. *N. Engl. J. Med.* **313,** 728–731.

16. Marzewski, D. J., Towfighi, J., Harrington, M. G., Merril, C. R., and Brown, P. (1988) Creutzfeldt-Jakob disease following pituitary-derived growth hormone therapy: a new American case. *Neurology* **38,** 1131–1133.

17. New, M. I., Brown, P., Temeck, J. W., Owens, C., Hedley-Whyte, E. T., and Richardson, E. P. (1988) Preclinical Creutzfeldt-Jakob disease discovered at autopsy in a human growth hormone recipient. *Neurology* **38,** 1133,1134.

18. Fradkin, J. E., Schonberger, L. B., Mills, J. L., Gunn, W. J., Piper, J. M., Wysowski, D. K., et al. (1991) Creutzfeldt-Jakob disease in pituitary growth hormone recipients in the United States. *JAMA* **265,** 880–884.

19. Anderson, J. R., Allen, C. M. C., and Weller, R. O. (1990) Creutzfeldt-Jakob disease following human pituitary-derived growth hormone administration. *Neuropathol. Appl. Neurobiol.* **16,** 543.

20. Buchanan, C. R., Preece, M. A., and Milner, R. D. G. (1991) Mortality, neoplasia, and Creutzfeldt-Jakob disease in patients treated with human pituitary growth hormone in the United Kingdom. *Br. Med. J.* **302,** 824–828.

21. Markus, H. S., Duchen, L. W., Parkin, E. M., Kurtz, A. B., Jacobs, H. S., Costa, D. C., et al. (1992) Creutzfeldt-Jakob disease in recipients of human growth hormone in the United Kingdom: a clinical and radiographic study. *Q. J. Med.* **297,** 43–51.

22. Powell-Jackson, J., Weller, R. O., Kennedy, P., Preece, M. A., Whitcombe, E. M., and Newsom-Davies, J. (1985) Creutzfeldt-Jakob disease after administration of human growth hormone. *Lancet* **ii,** 244–246.

23. Billette de Villemeur, T., Beauvais, P., Gourmelon, M., and Richardet, J. M. (1991) Creutzfeldt-Jakob disease in children treated with growth hormone. *Lancet* **337,** 864,865.

24. Job, J. C., Maillard, F., and Goujard, J. (1992) Epidemiologic survey of patients treated with growth hormone in France in the period 1959–1990—preliminary results. *Horm. Res.* **38,** 35–43.

25. Macario, M. E., Vaisman, M., Buescu, A., Neto, V. M., Araujo, H. M. M., and Chagas, C. (1991) Pituitary growth hormone and Creutzfeldt-Jakob disease. *Br. Med. J.* **302,** 1149.

26. Croxson, M., Brown, P., Synek, B., Harrington, M. G., Frith, R., Clover, G., et al. (1988) A new case of Creutzfeldt-Jakob disease associated with human growth hormone therapy in New Zealand. *Neurology* **38,** 1128–1130.

27. Gibbs, C. J., Jr., Asher, D. M., Brown, P. W., Fradkin, J. E., and Gajdusek, D. C. (1993) Creutzfeldt-Jakob disease infectivity of growth hormone derived from human pituitary glands. *N. Engl. J. Med.* **328,** 358,359.
28. Taylor, D. M., Dickinson, A. G., Fraser, H., Robertson, P. A., Salacinski, P. R., and Lowry, P. J. (1985) Preparation of growth hormone free from contamination with unconventional slow viruses. *Lancet* **ii,** 260–262.
29. Brown, P. (1985) Virus sterility for human growth hormone. *Lancet* **ii,** 729,730.
30. Pocchiari, M., Peano, S., Conz, A., Eshkol, A., Maillard, F., Brown, P., et al. (1991) Combination ultrafiltration and 6 M urea treatment of human growth hormone effectively minimizes risk from potential Creutzfeldt-Jakob disease virus contamination. *Horm. Res.* **35,** 161–166.
31. Cochius, J. I., Burns, R. J., Blumbergs, P. C., Mack, K., and Alderman, D. P. (1991) Creutzfeldt-Jakob disease in a recipient of human pituitary-derived gonadotrophin. *Aust. NZ. J. Med.* **20,** 592,593.
32. Klein, R. and Dumble, L. J. (1993) Transmission of Creutzfeldt-Jakob disease by blood transfusion. *Lancet* **341,** 768.
33. Brown, P., Gibbs, C. J., Jr., Rodgers-Johnson, P., Asher, D. M., Sulima, M. P., Bacote, A., et al. (1994) Human spongiform encephalopathy: the National Institutes of Health series of 300 cases of experimentally transmitted disease. *Ann. Neurol.* **35,** 513–529.
34. Thadani, V., Penar, P. L., Partington, J., Kalb, R., Jansson, R., Schonberger, L. B., et al. (1988) Creutzfeldt-Jakob disease probably acquired from a cadaveric dura mater graft. *J. Neurosurg.* **69,** 766–769.
35. Lane, K. L., Brown, P., Howell, D. N., Crain, B. J., Hulette, C. M., Burger, P. C., et al. (1994) Creutzfeldt-Jakob disease in a pregnant woman with an implanted dura mater graft. *Neurosurgery* **34,** 737–740.
36. Esmonde, T., Lueck, C. J., Symon, L., Duchen, L. W., and Will, R. G. (1993) Creutzfeldt-Jakob disease and lyophilised dura mater grafts: report of two cases. *J. Neurol. Neurosurg. Psychiat.* **56,** 999,1000.
37. Willison, H. J., Gale, A. N., and Mclaughlin, J. E. (1991) Creutzfeldt-Jakob disease following cadaveric dura mater graft. *J. Neurol. Neurosurg. Psychiat.* **54,** 940.
38. Masullo, C., Pocchiari, M., Macche, G., Alema, G., Piazza, G., and Panzera, M. A. (1989) Transmission of Creutzfeldt-Jakob disease by dural cadaveric graft. *J. Neurosurg.* **71,** 954,955.
39. Pocchiari, M., Masullo, C., Salvatore, M., Genuardi, M., and Galgani, S. (1992) Creutzfeldt-Jakob disease after non-commercial dura mater graft. *Lancet* **340,** 614,615.
40. Martínez-Lage, J. F., Poza, M., Sola, J., Totosa, J. G., Brown, P., Cervenáková, L., et al. (1994) Accidental transmission of Creutzfeldt-Jakob disease by dural cadaveric grafts. *J. Neurol. Neurosurg. Psychiat.* **57,** 1091.
41. Lang, C. J. G., Schüler, P., Engelhardt, A., Spring, A., and Brown, P. (1995) Probable Creutzfeldt-Jakob disease after a cadaveric dural graft. *Eur. J. Epidemiol.* **11,** 1.
42. Nisbet, T. J., MacDonaldson, I., and Bishara, S. N. (1989) Creutzfeldt-Jakob disease in a second patient who received a cadaveric dura mater graft. *JAMA* **261,** 1118.

43. Miyashita, K., Inuzuka, T., Kondo, H., Saito, Y., Fujita, N., Matsubara, N., et al. (1991) Creutzfeldt-Jakob disease in a patient with a cadaveric dural graft. *Neurology* **41**, 940,941.

44. Yamada, S., Aiba, T., Endo, Y., Hara, M., Kitamoto, T., and Tateishi, J. (1994) Creutzfeldt-Jakob disease transmitted by a cadaveric dura mater graft. *Neurosurgery* **34**, 740–744.

45. Diringer, H. and Braig, H. R. (1989) Infectivity of unconventional viruses in dura mater. *Lancet* **i**, 439,440.

46. Dickinson, A. G. and Taylor, D. M. (1978) Resistance of scrapie agent to decontamination. *N. Engl. J. Med.* **299**, 1413,1414.

47. Gibbs, C. J., Jr., Gajdusek, D. C., and Latarjet, R. (1978) Unusual resistance to ionizing radiation of the viruses of kuru, Creutzfeldt-Jakob disease, and scrapie (unconventional viruses). *Proc. Natl. Acad. Sci. USA* **75**, 6268–6270.

48. Brown, P., Rohwer, R. G., Green, E. M., and Gajdusek, D. C. (1982) Effect of chemicals, heat, and histopathologic processing on high infectivity hamster-adapted scrapie virus. *J. Infect. Dis.* **145**, 683–687.

49. Brown, P., Gibbs, C. J., Jr., Amyx, H. L., Kingsbury, D. T., Rohwer, R. G., Sulima, M. P., et al. (1982) Chemical disinfection of Creutzfeldt-Jakob disease virus. *N. Engl. J. Med.* **306**, 1279–1282.

50. Brown, P., Rohwer, R. G., and Gajdusek, D. C. (1986) Newer data on the inactivation of scrapie virus or Creutzfeldt-Jakob disease virus in brain tissue. *J. Infect. Dis.* **153**, 1145–1148.

51. Brown, P., Liberski, P. P., Wolff, A., and Gajdusek, D. C. (1990) Resistance of scrapie infectivity to steam autoclaving after formaldehyde fixation and limited survival after ashing at 360°C: practical and theoretical implications. *J. Infect. Dis.* **161**, 467–472.

52. Collinge, J., Palmer, M. S., and Dryden, A. J. (1991) Genetic predisposition to iatrogenic Creutzfeldt-Jakob disease. *Lancet* **337**, 1441,1442.

53. Brown, P., Cervenáková, L., Goldfarb, L. G., McCombie, W. R., Rubenstein, R., Will, R. G., et al. (1994) Iatrogenic Creutzfeldt-Jakob disease: an example of the interplay between ancient genes and modern medicine. *Neurology* **44**, 291–293.

54. Deslys, J.-P., Marcé, D., and Dormont, D. (1994) Similar genetic susceptibility in iatrogenic and sporadic Creutzfeldt-Jakob disease. *J. Gen. Virol.* **75**, 23–27.

55. Tange, R. A., Troost, D., and Limburg, M. (1990) Progressive fatal dementia (Creutzfeldt-Jakob disease) in a patient who received homograft tissue for tympanic membrane closure. *Eur. Arch. Otorhinolaryngol.* **247**, 199–201.

56. Manuelidis, E. E., Kim, J. H., Mericangas, J. R., and Manuelidis, L. (1985) Transmission to animals of Creutzfeldt-Jakob disease from human blood. *Lancet* **ii**, 896,897.

57. Tateishi, J. (1985) Transmission of Creutzfeldt-Jakob disease from human blood and urine into mice. *Lancet* **ii**, 1074.

58. Tamai, Y., Kojuma, H., Kitajima, R., Taguchi, F., Ohtani, Y., Kawaguchi, T., et al. (1992) Demonstration of the transmissible agent in tissue from a pregnant woman with Creutzfeldt-Jakob disease. *N. Engl. J. Med.* **327**, 649.

59. Deslys, J. P., Lasmézas, C., and Dormont, D. (1994) Selection of specific strains in iatrogenic Creutzfeldt-Jakob disease. *Lancet* **343**, 848,849.
60. Esmonde, T. F. G., Will, R. G., Slattery, J. M., Knight, R., Harries-Jones, R., de Silva, R., et al. (1993) Creutzfeldt-Jakob disease and blood transfusion. *Lancet* **341**, 205–207.
61. Heye, N., Hensen, S., and Müller, N. (1994) Creutzfeldt-Jakob disease and blood transfusion. *Lancet* **343**, 298.
62. Di Martino, A., Safar, J., Ceroni, M., and Gibbs, C. J., Jr. (1992) Purification of non-infectious ganglioside preparations from scrapie-infected brain tissue. *Arch. Virol.* **124**, 111–121.
63. Masters, C. L., Harris, J. O., Gajdusek, D. C., Gibbs, C. J., Jr., Bernoulli, C., and Asher, D. M. (1979) Creutzfeldt-Jakob disease: patterns of world wide occurrence and the significance of familial and sporadic clustering. *Ann. Neurol.* **5**, 177–188.
64. Brown, P., Cathala, F., Raubertas, R. F., Gajdusek, D. C., and Castaigne, P. (1987) The epidemiology of Creutzfeldt-Jakob disease: conclusion of a 15-year investigation in France and review of the world literature. *Neurology* **37**, 895–904.
65. Harries-Jones, R., Knight, R., Will, R. G., Cousens, S., Smith, P. G., and Matthews, W. B. (1988) Creutzfeldt-Jakob disease in England and Wales, 1980–1984: a case control study of potential risk factors. *J. Neurol. Neurosurg. Psychiat.* **51**, 1113–1119.
66. Schoene, W. C., Masters, C. L., Gibbs, C. J., Jr., Gajdusek, D. C., Tyler, H. R., Moore, F. D., et al. (1981) Transmissible spongiform encephalopathy (Creutzfeldt-Jakob disease). Atypical clinical and pathological findings. *Arch. Neurol.* **38**, 473–477.
67. Miller, D. C. (1988) Creutzfeldt-Jakob disease in histopathology technicians. *N. Engl. J. Med.* **318**, 853,854.
68. Sitwell, L., Lach, B., Atack, E., and Atack, D. (1988) Creutzfeldt-Jakob disease in histopathology technicians. *N. Engl. J. Med.* **318**, 854.
69. Gorman, D. G., Benson, D. F., Vogel, D. G., and Vinters, H. V. (1992) Creutzfeldt-Jakob disease in a pathologist. *Neurology* **42**, 463.
70. Weber, T., Hayrettin, T., Holdorff, B., Collinge, J., Palmer, M., Kretzschmar, H. A., et al. (1993) Transmission of Creutzfeldt-Jakob disease by handling of dura mater. *Lancet* **341**, 123,124.
71. Berger, J. R. and David, N. J. (1993) Creutzfeldt-Jakob disease in a physician: a review of the disorder in health care workers. *Neurology* **43**, 205,206.
72. Sawcer, S. J., Yuill, G. M., Esmonde, T. F. G., Estibeiro, P., Ironside, J. W., Bell, J. E., et al. (1993) Creutzfeldt-Jakob disease in an individual occupationally exposed to BSE. *Lancet* **341**, 642.
73. Alperovitch, A., Brown, P., Weber, T., Pocchiari, M., Hofman, A., and Will, R. (1993) Incidence of Creutzfeldt-Jakob disease in Europe in 1993. *Lancet* **343**, 918.
74. Brown, P., Kenney, K., Little, B., Ironside, J., Will, R., Cervenáková, L., et al. (1995) The intracerebral distribution of infectious amyloid protein in patients with spongiform encephalopathy. *Ann. Neurol.*, in press.

9

Bovine Spongiform Encephalopathy

Methods of Analyzing the Epidemic in the United Kingdom

John W. Wilesmith

1. Introduction

It seems unlikely that anyone could have foretold the interest, controversy, and concern that the occurrence of bovine spongiform encephalopathy (BSE) in cattle in Great Britain would cause when the author was commissioned to investigate its epidemiology in the spring of 1987. The epidemic has proved to be the largest food-borne epidemic of a transmissible spongiform encephalopathy (TSE), and has been the subject of detailed scrutiny epidemiologically, by international agencies and governments, and by representatives of all types of media worldwide. During the course of the epidemic an attempt has been made to decipher the epidemiology of BSE with the usual objectives, the most important being to provide a model of causation such that necessary statutory controls could be identified and enacted to protect animal and human health, both in Great Britain and abroad.

In this chapter a brief history of the epidemiological studies of BSE in Great Britain is described. Inherent in these studies are the problems of a protracted incubation period, the absence of a valid diagnostic test in the live animal, the novelty of the disease in cattle, and the usual difficulties of conducting epidemiological research as outlined by Rothman *(1)*. Hopefully, these hurdles are not used as excuses and the following may be of help in investigating diseases with similar characteristics that may well occur in the future.

2. The Identification of BSE in Great Britain

BSE was first identified in the south of England in November 1986 *(2)* as a result of what could be termed the background surveillance of animal disease

From: *Methods in Molecular Medicine: Prion Diseases*
Edited by: H. Baker and R. M. Ridley Humana Press Inc., Totowa, NJ

in Great Britain. Although this is a passive rather than active system it involves a relatively high degree of active communication between animal keepers and their veterinary surgeons, who in turn seek help from the network of Veterinary Investigation (VI) Centres, whose staff may seek specialist advice from the Central Veterinary Laboratory (CVL). More specifically, BSE was identified as a result of the occurrence of multiple cases of unusual neurological disease in adult cattle in large dairy herds and the informal exchange of information between herd owners about unusual cases of disease. An initial difficulty was securing sufficiently well-fixed brain material for histological examination from suspect cases of BSE. This, however, did not delay the identification of the disease unduly and fortunately electron microscopists at CVL were experienced in the detection of scrapie-associated fibrils (SAF) as a result of a then recently completed study of SAFs in sheep scrapie *(3)*. The identification of SAFs in bovine brains with spongiform encephalopathy provided some confidence that the new disease was a TSE *(2)*.

In retrospect, it is clear that the national animal disease surveillance system identified BSE at a relatively early stage of the epidemic. This is probably a result, in part, of the absence of a federal structure and the national infrastructure of the State Veterinary Services VI Centres, which intrinsically has a high degree and rapid rate of communication and the required expertise in neuropathology.

3. Case Definition of BSE

Previous research on TSEs in other species had indicated that it was highly unlikely that there would be a diagnostic test for the presence of the abnormal PrP^{Sc} in the live animal in the foreseeable future. Although the chronic clinical course and ultimate clinical signs presented by cases of BSE appeared to be pathognomonic, it was apparent that the initial clinical signs were not. Two options were therefore available in selecting a valid diagnostic method for the case definition of BSE, and both were pathological. The first was the electron microscopical (EM) examination of fresh brain tissue for SAFs and the second was a histological examination of fixed brain. The initial comparison of these two diagnostic methods was made difficult by the competition for the same anatomical areas of the brain, since the prevalence of SAFs was likely to be greatest at sites where histological changes are most marked. This comparison was made more difficult because the histological findings in sheep scrapie indicated a variation in the distribution of lesions *(3)*.

The outcome of this contest was the decision to conduct a histological examination of 64 neuron groups of the brains of suspect cases of BSE *(4)* during the course of the initial epidemiological study. Eventually the opportunity was taken to examine the possibility of histological examination of a reduced number of brain sections for routine diagnosis, especially because of the apparent

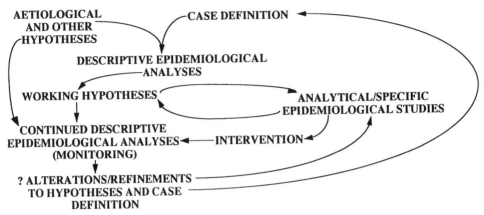

Fig. 1. Dynamics of the epidemiological research plan.

remarkable uniformity of the distribution of brain lesions *(5,6)*. This resulted in the present method of diagnosis based on a single section, which is possible because the uniform distribution of lesions was confirmed *(7)*. Subsequent studies of the validity of EM examinations for the presence of SAF have indicated that histological examination is the preferred method of diagnosis *(8)*. The exception is the examination of autolysed brain tissues, for which EM examination has proved to be superior *(9)*.

The validity of the histological examination has been the subject of a continued reassessment as described in Section 6. This has been a necessary aspect of the research effort, and has been extended to assessments of the more novel methods of diagnosis, such as immuno-cytochemistry. These are in progress, but raise the issue of the detection of infection rather than disease. Studies of the pathogenesis of BSE *(10)*, still in progress, should facilitate the validation of these and other methods of diagnosis. In the meantime, the histological examination of the brain stem remains as the routine method of confirming a clinical diagnosis.

4. The Initial Epidemiological Studies

4.1. The Basic Plan

Much has been written about epidemiological concepts and methods of epidemiological research. These essentially philosophical treatises are important, but a more pragmatic description of the scientific methodology is simply to develop and identify causal hypotheses and test their validity. This naturally results in a dynamic process in conducting epidemiological research. The basic plan, in its simplified form, for this as applied to BSE is shown in Fig. 1. Despite the absence of a complete understanding, it has been necessary throughout this process to make causal inferences in order that control measures to protect public and animal health could be implemented in a timely manner. Therefore, overlaying the basic plan was a systematic means to assist in differentiating causal from noncausal associations. The most attractive set of standard criteria is that proposed by Hill, which are strength, consistency, specificity, temporality, biological gradient, plausibility, coherence, experimental evidence, and analogy *(11)*. These have been discussed in some detail by Rothman *(1)*.

Expertise in veterinary epidemiology and some experience in balancing the need for rapid results with the requirement to carry out rigorous medium- to long-term studies are needed to establish such a plan. An element of sound epidemiological judgment is therefore required along the way.

4.2. Identification of the Causal Hypotheses and the Means of Their Investigation in the Initial Epidemiological Study

Despite the histological findings of a spongiform encephalopathy and the detection of SAFs by electron microscopy, there was a degree of uncertainty about the possible etiology of the disease. The etiological hypotheses were not restricted to BSE being a member of the TSEs. This was important because alternative hypotheses could have abounded and, indeed, have been continually proffered. Also, even if BSE proved to be a TSE, other factors could be component, rather than sufficient, causes. Those investigated can be summarized as a scrapie-like agent, an intoxication, or a purely genetic disease *(12,13)*. For the first of these the possible sources or vehicles of infection were investigated as far as was possible. These were imported cattle, contact with sheep, contact with wildlife, and contaminated biological products, including feedstuffs.

Having established these basics, the next step was to decide the most appropriate design for the initial epidemiological study. There were three obvious options: an abattoir survey, a random survey of farms, and detailed case studies of affected herds and animals. These needed to be considered with the other desirable objectives of the study, which included determining whether BSE was a truly new disease, and if it was, the time when the first cases occurred; securing a sufficient number of brains to establish a case definition; obtaining an improved description of the clinical signs and the clinical course; and collecting the essential descriptive epidemiological data. In the outcome, the evaluation of the three options was not protracted. An abattoir survey has the inherent difficulty of tracing animals back to their last and natal herds. This, together with the fact that a large proportion of cases would not be clinically acceptable for slaughter and the incidence and prevalence of the disease appeared to be low *(13)*, made this an inefficient option. Similarly, for the last reason a national survey of herds would clearly have been an inefficient method. The remaining option, involving detailed case studies, was therefore chosen, but required that the ascertainment of suspect cases be maximized. This was achieved by reversing the operation of the surveillance network and raising awareness of the disease and, most importantly, disseminating information about what was known of the clinical signs. The resulting voluntary notifications of animals suspected of having BSE provided the cases for this study. An evaluation of the availability of the required documented data and information and the development of a questionnaire and lines of questioning was the next step. This was achieved by the author visiting farms that had suspect cases and interviewing herd owners and workers using a proforma questionnaire. Fortunately, herd owners were very cooperative and it became clear that there would be sufficient documented data to determine the necessary life histories of the cases and the parent herd. A standard questionnaire was therefore developed together with a list of potential sources of data from the range of farm records encountered during the feasibility stage.

4.3. Assessment of the Initial Causal Hypotheses and the Formulation of Additional Hypotheses

The target for the initial study was to obtain the detailed epidemiological data from 200 cases. This was achieved by December 1987, having commenced in May of that year. It resulted in a relatively large information base and instigated the development of the BSE computer-based database *(14)*.

The assessment of the causal hypotheses proved to be relatively straightforward *(12)*. Essentially, intoxication either from veterinary therapeutic and prophylactic products or agricultural chemicals could be ruled out since no particular product or generic chemical compound emerged as a common factor. There was also considerable evidence against BSE being merely of genetic origin; for example, a wide range of breeds were affected. In examining the TSE hypothesis, there was no possibility that BSE had been introduced with the importation of cattle. Cases had occurred in totally closed herds with no contact with other cattle. Contact with sheep was also a highly improbable reason for the infection of cattle, because only a small proportion of cases occurred on farms that maintained breeding flocks or, indeed, any type of flock. Similarly, the presence of species of wildlife, such as deer, was not identified as a common factor. Although new bovine vaccines had become available during the period of interest, these and existing vaccines together with other biological products were recorded at very low rates of usage.

This process of elimination left the possibility that animal protein, in the form of meat and bone meal, included in commercial cattle feedstuffs, or some other unidentified medium, was the vehicle of infection for a scrapie-like agent.

The first step was to double-check the feeding histories obtained from the affected farms and then seek the formulation of the commercial feedstuffs consumed during the lifetime of the animals affected from the compounders of these rations. The result of this was that in every case of BSE for which full records were available, meat and bone meal had been incorporated as a protein supplement in at least one product. This, therefore, provided circumstantial evidence for a common exposure factor. It also highlighted the complexity, and therefore difficulties, of investigations involving data on the exposure of cattle to individual ingredients in cattle rations. First, cattle, even in their first 6 mo of life, can be fed a number of different products not necessarily produced by the same company. Second, the details of individual products would not always be available from the manufacturers, either because of the passage of time, with records not being kept for a sufficiently long period, or because companies had been taken over or had simply ceased trading. These factors were certainly important in thinking about the potential analytical epidemiological studies. This identification of a common exposure factor was clearly important, but needed to be examined for its plausibility in the first instance with respect to the descriptive epidemiological picture, especially because meat and bone meal was a commonly used ingredient in commercially produced cattle rations.

The first intriguing aspect in the occurrence of BSE was its appearance more or less coincidentally throughout Great Britain (15). However, this needed to be considered with the evidence of the initial occurrence of BSE. This was identified as April 1985, from clinical histories and examination of archived fixed brain, suggesting that BSE was a truly novel disease (16). Therefore, the widespread occurrence of BSE was consistent with an exposure factor common throughout Great Britain. Another striking aspect of the descriptive epidemiology was the markedly greater incidence of disease in dairy herds compared with beef suckler herds (14). This difference was certainly explicable by the meat and bone meal hypothesis because the feeding of concentrated rations is relatively uncommon in commercial beef herds, in which nutrition during the winter period is largely based on conserved grass and homegrown cereals.

The descriptive epidemiological features that were most difficult to reconcile with the working hypothesis were, first, the geographical variation in risk, with animals in the south of England having a notably greater risk of disease than those in the north of England and Scotland. The second was that meat and bone meal had been used as a protein supplement in commercial foodstuffs for decades; its utilization was not a recent event. In addition, the question of why BSE had only occurred, or been reported, in Great Britain pervaded, at least subconsciously, epidemiological thinking.

Two observations were made tending to countermand these findings that were potentially detrimental to the working hypothesis. The first was the occurrence of BSE on the Channel Islands of the British Isles. In the early years, BSE was confirmed on the island of Guernsey, but not on Jersey. This stimulated an investigation of the suppliers of commercial cattle feedstuffs to these two main Channel Islands, and revealed that the principal manufacturer of cattle rations imported by each island was different, although both were in mainland Great Britain and feed was not imported from any other sources. The main supplier to Guernsey was found to use meat and bone meal relatively frequently, whereas that for Jersey had not considered meat and bone meal as a nutritionally/economically valued ingredient in the formulation of the majority of its cattle rations. The eventual occurrence of BSE on the island of Jersey, but at a much lower incidence, provided the necessary natural experiment, the results of which supported the hypothesis.

The second observation was the occurrence of cases of TSE in a nyala in 1986 and a gemsbok in 1987 in a zoological collection in the south of England *(17)*. These had been identified in the course of the normal surveillance of animal diseases. At the time, thoughts on the etiology centered on a solely genetic origin. However, further investigations following the establishment of the working hypothesis for BSE proved helpful. The nutrition of the collections of these two species had included commercially produced feedstuffs, but meat and bone meal had been specifically excluded from the ingredients at the request of the keepers, for reasons that were not related to any concern about the risk of infection with a scrapie-like agent. Vegetable proteins, such as soya, had been used, but when the price of this commodity increased markedly, meat and bone meal was used for economy for a very short period of time *(13)*. The most probable incubation period was shorter than that which had been observed in cattle, but the coincidence could not be ignored.

In retrospect, these "observations" were only made possible by a good degree of communication and enticing the curiosity of individuals to assist in explaining unusual occurrences that would be a help in understanding the larger problem. However, the basic surveillance was a key to providing supporting epidemiological evidence as advised by Hill *(11)*.

One problem with the meat and bone meal hypothesis was that meat and bone had been used in the formulation of cattle rations for many years; it was not a recent development. At this stage, discussions had obviously taken place with nutritionists in the animal feedstuffs industry, but not with the rendering industry, which produces meat and bone meal. Instead, efforts were concentrated on an attempt to determine the most likely time of the onset of effective exposure of the British cattle population.

The occurrence of the AIDS epidemic has stimulated the development of methods of estimating the time of exposure to infection *(18)*. However, a relatively simple deterministic simulation approach was used for BSE when the number of cases was small *(12,13)*. This involved investigating the probabilities of exposure as calves, yearlings, and adults, estimating the incubation period distribution, and examining the possibilities of exposure at various times. The validity of the outputs from the simulations was assessed by the fit to the age-specific incidences. The best fit indicated that the onset of effective exposure occurred suddenly in 1981–1982 and that the majority of clinical cases observed had been infected in calfhood. More sophisticated statistical techniques that have been used during the course of the epidemic have not altered these basic conclusions, but have provided improved estimates of the incubation period distribution *(19)*.

This finding provided some confidence in addressing the question of why cattle became exposed. This was directed at examining the production processes used to render waste animal tissues to produce meat and bone meal. A prerequisite for this was to discuss the epidemiological evidence with representatives of both the animal feedstuffs and rendering industries, to gain their confidence and cooperation. As a result, a survey of all rendering plants was conducted in the second half of 1988. The main objectives of this were:

1. To obtain details of the changes in rendering processes during the 1970s and 1980s;
2. To determine current time–temperature treatments in the various processes; and
3. To obtain details of the species composition of the waste tissues rendered in each plant over time.

The survey was conducted by three veterinary officers with a knowledge of the rendering industry who had been briefed specifically about the objectives. Data were obtained by a questionnaire and time–temperature measurements were recorded in a uniform manner, having evaluated a number of potential methods.

The results of this survey indicated that the estimated onset of effective exposure was coincident with a marked reduction in the use of hydrocarbon solvents to maximize the yield of tallow (fat) during the rendering process *(20)*. Only limited laboratory-based studies had been carried out on the use of organic solvents for the disinfection of surgical and laboratory instruments and equipment. However, there was some indication that this change was a biologically plausible explanation for the onset of exposure. The survey also provided a further explanation for the geographical variation in the risk of infection of cattle, notably the continued and exclusive use of the solvent extraction process in Scotland, where the majority of meat and bone meal produced had undergone this treatment.

This whole stage of the investigation was obviously important because of the need to assess the validity and strength of the initial hypothesis. If the assessment had failed to provide substantiating and explanatory data and information, second thoughts on investigating the epidemic would have been required. This, fortunately, proved not to be the case and the success, at least in part, could be attributed to the collection of sound descriptive epidemiological and other data based on documented sources. The results, therefore, were used as a basis for formulating the first statutory control measure, introduced in July 1988, to prevent further exposure from the feed-borne source, and for designing further, more analytical, epidemiological studies.

5. Analytical Epidemiological Studies to Test the Working Hypothesis and to Determine the Occurrence of Other Potential Means of Transmission

5.1. Analytical Epidemiological Studies

The most useful epidemiological approach to examine both the validity of the meat and bone meal hypothesis in general and the solvent-extracted meat and bone meal hypothesis in particular would have been a cohort study. This was not possible for either because of the changes in feed supplier that can occur even during calfhood, and because the contents of rations change resulting from the use of the least-cost formulation method and variation in the suppliers of individual ingredients.

A case-control approach *(21)*, therefore, needed to be contemplated in early 1988, following the identification of the most likely source. The usual considerations of selecting an appropriate control population and minimizing all sources of bias were necessary. At this stage sample size was not considered and efforts were concentrated on identifying a suitable control population. Because the disease had not yet been made statutorily notifiable, but became so in June 1988, it was considered important to prevent the misclassification of herds, and, therefore, possibly individuals. Also, the early identification of potential control herds was an attempt to select controls before the meat and bone meal hypothesis had gained general currency. To these ends the use of neighborhood controls was the chosen approach, but with the knowledge that overmatching could ultimately be a problem. Veterinary practices that had reported cases voluntarily were contacted to identify potentially suitable control herds. The criteria for these included:

1. The herds were subject to regular routine visits by the attendant veterinary surgeon;
2. The herd owners or workers generally sought veterinary advice when animals became ill;
3. Record keeping was a routine; and
4. The majority of replacement animals for the adult herd were homebred.

This was successful and the necessary list was compiled.

It was intended to conduct the next phase, obtaining detailed feeding histories, by visiting the farms and interviewing herd owners and workers to secure comparable information to that being obtained routinely for BSE cases, with confirmatory documentation from receipts and other records, but this proved impossible because insufficient field staff resources were available. As a result, a postal questionnaire had to be employed that, although less satisfactory, did produce the information that was essential.

The final decision in completing this study involved three aspects. First, which birth cohort would be used in the analyses? This was a relatively straight-forward decision and was based on the modal age of onset of 4–5 yr. In order to conduct the analyses at the earliest opportunity, sufficient cases must have occurred in the cohort born nearest to the estimated time of the onset of effec-tive exposure. The second aspect, sample size, was necessarily of lesser impor-tance, essentially because the number of neighborhood control herds could not be increased prospectively. However, the remaining control herds, which had not experienced a case of BSE, were considered to be sufficiently numerous to be used in the analysis. The third aspect, which was an overriding one, was the assessment of the completeness of the details available from the manufacturers of the ingredients of the feedstuffs consumed by both the cases and controls. Although it was clearly not possible to examine the potential dose-response relationship because of the lack of records on the inclusion rates of meat and bone meal, the data on the presence or absence of meat and bone meal was considered to be sufficient. The final results of this study provided supporting evidence for the hypothesis (22).

This was, therefore, a difficult study to design and if the disease had been less important the study may not have been pursued, notably because of the possibility of the specific problem of overmatching and the necessity to limit the study to calfhood exposure. These could have conspired to produce a false negative result. However, it could be regarded as an example of having to cope with such difficulties, and, providing the basic design could be made sound, persevering rather than abandoning all hope.

5.2. The Question of Maternal Transmission

It is, of course, good scientific practice to read the appropriate scientific literature when "new" diseases occur. The literature on kuru in the Fore-speak-ing people of Papua New Guinea indicates that maternal transmission does not occur (23). However, a cursory reading of the literature is persuasive that maternal transmission is the means by which sheep scrapie is naturally main-tained. A more critical review of the published studies suggests quite strongly that this could have been a self-perpetuating myth, and that in high-incidence flocks horizontal transmission is of at least equal importance (24).

The initial approach to the question of maternal transmission of BSE was to determine whether or not this means of spread, on its own and irrespective of the mechanism, could maintain the disease in the cattle population. The only method available to examine this question was simulation modeling, which revealed that maternal transmission alone could not sustain the epidemic *(5)*.

This result was of considerable importance, but, returning to the literature, the only excreted material in which the sheep scrapie agent was known to be present was the placenta. Therefore, exposure to this tissue could result in transmission to either offspring (maternal) or to unrelated animals (horizontal). In the quest to determine whether BSE was transmitted naturally, it seemed likely that the occurrence of maternal transmission would be identifiable earlier than the occurrence of horizontal transmission. If evidence of maternal transmission was found, then a worst case scenario assuming horizontal transmission would have to be considered in deciding what animal disease control measures needed to be taken.

As a result, the decision was made to expend a great deal of time and effort on determining whether or not there is maternal transmission of the BSE agent, and a cohort study was instigated in July 1989 *(25)*. This is a long-term study for which the result is not available at the time of writing this chapter, but in the meantime other opportunities have been identified and acted on to obtain the earliest possible valid answer to the question (*see* Section 7.).

It is unfortunate that the reasons for studying the question of maternal transmission in such detail have often been misinterpreted, and that the myth of the occurrence and importance of maternal transmission in animal TSEs, particularly sheep scrapie, has apparently been perpetuated. With hindsight, this is a problem of education to counter preconceived ideas and myths, but even with the benefit of hindsight it is difficult to know how the problem, which in the case of BSE has proved to be important in discussions on the trade in animals and animal products, could have been prevented.

6. Monitoring the Epidemic

The monitoring of the BSE epidemic was facilitated by the disease being statutorily notifiable, particularly because this maximized the ascertainment of cases and therefore removed any bias. The main aims of monitoring the epidemic were to assess the effects of the statutory action, to identify any changes in the disease, to investigate means of transmission other than the feed-borne source, and to provide epidemiological data to construct analytical studies. It required the continued allocation of veterinary field staff resources to visit animals suspected of having BSE, examine them clinically, and complete the standard epidemiological questionnaire, not an inconsiderable undertaking given the size and duration of the epidemic and therefore the fortitude and endurance necessary.

An underlying need was the computing facility to handle the epidemiological data efficiently and conduct the analyses to fulfil, at least, the first aim. The main descriptive epidemiological features used to assess the effects of the statutory control measures that have been taken were age-specific incidences, the weekly reporting rate of suspect cases, the monthly animal and herd incidence, and the within herd incidence. All of these have indicated that the statutory ban on the feeding of ruminant-derived protein to ruminants has had a positive effect *(26–28)*. The fulfilment of the second aim was naturally intermingled with investigating the first, but required additional analyses—notably to detect any change in the incubation period distribution. This was achieved by cohort analyses *(29)*, which fortunately did not reveal any changes that would have made the analysis and interpretation of the epidemiological findings more complex.

Specific analyses have concentrated on two other potential changes: the clinical manifestation of the disease and any change in the distribution of lesions in the brain that could invalidate the routine histological diagnosis. The first was examined by a comparison of the frequency of clinical signs between cohorts, but no change was identified *(30)*. The second change could have occurred because of a change in the BSE agent as a result of its passage in cattle. This was examined by taking a systematic sample of whole brains from suspect cases throughout the epidemic and studying the distribution of lesions. Again, no changes have been observed.

The third aim, to determine whether other means of transmission had occurred, was achieved by a number of analyses in addition to those used to determine the effects of the statutory control measures. The possibility of maternal transmission was assessed by comparing the observed incidence of BSE in the offspring of confirmed cases with that expected as a result of exposure from the food-borne source. The latter was estimated in a number of ways, but essentially employed the annual incidences in successive birth cohorts *(27,28)*. These analyses have been conducted at three monthly intervals and are not regarded as the ultimate means of determining whether or not there was evidence of maternal transmission, but more as a background monitoring in an attempt to identify this means of transmission at the earliest opportunity.

Other analyses to this end included within-herd incidences in affected herds, especially those in which the only cases were in purchased animals and there was no evidence that the recipient herd had itself been exposed to infection. The population of beef suckler herds was of interest here because commercial concentrates were fed much less frequently than in dairy herds, and because a proportion of these cross-bred females are purchased from dairy herds to provide replacements for the adult herd.

The inevitable long-term monitoring of the epidemic was an important component of the epidemiological studies, especially in view of the national and international interest resulting in a constant flow of questions of either a political or scientific nature. The results and findings have also been used to reassess the original thoughts and hypotheses on the origin of the epidemic. Although normal descriptive epidemiological analyses have provided the basis for the assessments and reassessments, it has also been necessary to identify novel means of analysis to address specific aspects, such as the possibility of other means of transmission.

7. Further Analytical Studies

As described in Section 5., the best approach to determining whether or not maternal transmission occurs is a cohort study. This was supplemented by conducting the possible appropriate analyses to examine the incidence of BSE in offspring of confirmed cases compared with that expected from the food-borne source.

The occurrence of cases of BSE in animals born after July 18, 1988, when statutory measures were introduced to prevent further exposure from the food-borne source, provided an opportunity for additional analytical studies. The prerequisite of these was naturally a more detailed study of each case, i.e., to obtain descriptive epidemiological data to formulate hypotheses and, ultimately, design analytical studies. The outcome of these case studies indicated that there was a clear risk of exposure for animals born in 1988, and possibly later, from the feed-borne source, as a result of feedstuffs manufactured before the ban still being in the food chain or on farms. The recommended "shelf life" of these products is of the order of 3 mo, but they can remain fit for consumption for years. Case studies of later-born animals proved to be unrewarding with respect to an assessment of the risk from the food-borne source. Therefore, the feasibility of conducting a case-control study was examined to assess the risks of infection from feedstuffs, maternal and horizontal transmission, and associated population-attributable risks.

It proved possible to design a within herd study to assess all these risks except that from feed. More than 300 herds with homebred cases in cattle born between November 1, 1988 and June 30, 1990 participated, and in addition to the cases, up to four control animals were randomly selected from unaffected homebred animals that had been born in the same herd and calving season as each case. In order to take account of the matched study design, conditional logistic regression was used with herd and calving season as the matching variable.

A simulation model was developed to determine the rates of transmission of BSE in the population that would be consistent with the odds ratios observed and supplement the population-attributable risk. Essentially the results indicated that neither maternal nor horizontal transmission could account for the majority of the cases, leaving the feed-borne source as the only known possibility. This would have been at a reducing rate in successive birth cohorts since July 1988 because the changes in the risk of infection over time have been estimated by calculating what is akin to standardized morbidity ratios for animals born in each month and year since the ban *(31)*. These continued analyses have indicated that the risk of infection was reduced by at least 40% immediately after the introduction of the feed ban and the risk of infection for animals born in December 1990 was 10% of that for animals born in December 1987.

The question of obtaining evidence for a continued risk from the feed-borne source, albeit at a much reduced rate, was therefore raised. The magnitude of the dose exposure in natural infection had been a subject of interest throughout the epidemic and the weight of evidence was that it was a low-dose phenomenon *(32)*. Evidence is currently emerging from the experimental oral exposure of calves with 1, 10, and 100 g of brain from terminal cases, each on a single occasion, and 100 g on three occasions *(33)*, that the epidemiological interpretation was correct. There was, therefore, some biological plausibility for accidental contamination of cattle feedstuffs involving relatively small quantities of ruminant-derived meat and bone meal.

The possibility of such contamination was examined by recourse to the basic descriptive epidemiology, studying the regional proportional distribution of homebred cases by 12-mo birth cohorts before and after the feed ban. This revealed that following the feed ban the proportion of homebred cases in the eastern and northern regions of England increased *(34)*. This was of interest because the national poultry and pig populations are largely concentrated in these regions and the feedstuffs for these species could legally have included ruminant protein derived from potentially high-risk tissues (although not from clinically suspect animals) until September 1990. More importantly, there are a number of mills producing commercial feedstuffs for both ruminants and monogastric animals. The correlation between the incidence of homebred cases of BSE in animals born after October 30, 1988 in each county in England and Wales and the ratios of adult cattle to the number of pigs, the number of poultry, and the combined population sizes of pigs and poultry in each county was examined as a further analysis of the descriptive epidemiological data. A highly statistically significant correlation was found *(34)*. Such an ecological correlation cannot obviously be taken as evidence of a causal association, but it can provide confidence in the further assessment of the hypothesis. A between-herd case-control study was the appropriate means for further study of the problem, and preliminary visits to a small sample of feed mills were made to assist in finalizing the study design. However, these visits provided evidence that accidental cross-contamination could have occurred in mills producing rations for cattle and nonruminant species. The envisaged case-control study has yet to be completed, but efforts have been made to ensure the enforcement of the statutory bans on both the feeding of ruminant-derived protein to ruminants (July 1988) and the specified bovine offals (September 1990) considered likely to contain significant titers of the BSE agent, and the legislation controlling the separation and disposal of specified bovine offals was further strengthened in April and July 1995.

8. The Future Course of the Epidemic

All epidemiologists who become involved in studying diseases of national and/or international concern will become aware of the inherent interest of the general human population in foretelling what the future holds. There is, therefore, a need to recognize the difference between the perception, by the media and their attendant population, of the power of astrology and that of epidemiological methods. On the other hand, there has been a specific need to estimate the number of animals to be slaughtered as suspect cases of BSE in the coming years, since money must be allocated to compensate the owners of animals that are compulsorily slaughtered, and to pay for the incineration of the carcasses of slaughtered cattle. Modeling studies to provide these estimates have, therefore, been necessary. A number of approaches have been used *(12,33,35,36)*, but an age-period-cohort approach has proved to be remarkably successful and accurate, and the current decline in the incidence is in agreement with that predicted. Although publishing estimates of the future number of cases would not be prudent, the present evidence is that the incidence of BSE will continue to decline such that by the turn of the millennium the incidence will be insignificant. This said, there is a need to continue to study the epidemic in some detail in order to detect any changes in the epidemiology that could result in an undue and unexpected prolongation of the epidemic.

9. An Epidemiological Epilogue

As indicated at the beginning of this chapter, the occurrence of BSE was unexpected. Its occurrence presented a number of difficult problems, the apparent answers for some of which naturally raising further questions that increased the level of concern about how cattle had become exposed. However, from a personal point of view, a systematic approach involving an adherence to the basic study plan proved to be of great help in accumulating the necessary evidence and therefore confidence, especially for providing advice for the formulation of control policies. Important in this process was obtaining valid and detailed enough descriptive epidemiological data to conduct relatively straightforward analyses. Although subsequent, more complex, analytical studies provided the epidemiological perspective in terms of population-attributable risks, the descriptive epidemiological analyses provided a potent basis for hypothesis formulation and a general understanding of the epidemiology.

The whole process of studying the epidemiology was facilitated by the availability of a multi-disciplinary team, with pathology and epidemiology as the most important disciplines in the initial stages. An experience in neuropathology was naturally of importance and in the case of epidemiology the availability of a team capable of handling, validating, and analyzing large sets of data was probably crucial, in addition to the general experience in dealing with national veterinary epidemiological problems.

In the veterinary field, epidemiologists have the advantage of utilizing the results of experimental studies in the species of interest, a benefit not available to human epidemiologists. Interestingly, because of the protracted incubation period, laboratory-based studies were not of immediate help in formulating hypotheses of causality. A number of these are now coming to fruition and will be of enormous interest in attempting to piece together some of the more complex parts of this epidemiological jig-saw, but it seems that none will alter the general theses. An exception is perhaps the evidence of an absence of any genetic component in the susceptibility of cattle to the BSE agent. The results of the initial epidemiological studies did not indicate that this was a significant factor, but molecular genetic studies *(37–39)*, the results of the initial parenteral exposure of calves using brain homogenates, and a population approach using statistical models of possible modes of inheritance of susceptibility confirmed that the genotype of cattle was not a risk factor that required consideration in the design of epidemiological studies. This is in contrast to studies of sheep scrapie. If the judgement in the early stages that this was not a factor had been wrong, then a number of the analytical studies would have been flawed. In the outcome, this did not prove to be a problem and, in fact, the study of the epidemiology of BSE has been relatively simple without such confounding factors.

As indicated in Section 5.2., all researchers have resorted to the published literature to formulate their ideas, but there had been few, if any, rigorous studies of the epidemiology of naturally occurring TSEs in animals, notably on sheep scrapie. This statement is perhaps a little harsh because since the heyday of studies of natural sheep scrapie epidemiological methodology has developed, and molecular genetic studies of sheep have provided a stimulating, additional tool for epidemiological studies that can take account of all known risk factors. This said, there has to be some concern in taking account of, and extrapolating from, the results of experimental studies in laboratory animals, notably inbred strains of mice and hamsters, neither of which are naturally susceptible species. This is a somewhat controversial point, but there has to be some doubt, and therefore word of caution, in using the results of such studies, especially with respect to the species-barrier phenomenon. An element of judgment is required in assessing this polemical view, but the translation of results from the studies in laboratory animals to naturally susceptible species will always be difficult and contentious.

Finally, but of some considerable importance, the early identification of BSE and the subsequent investigation of the epidemiology of BSE was only achieved by the presence and use of the infrastructure of the State Veterinary Service in Great Britain. The inherent surveillance network identified the disease at the earliest opportunity, and the national structure of the veterinary field service facilitated both the continued monitoring of the epidemic and assisted in obtaining the necessary data for studies. Epidemiological studies are heavily dependent on the provision of valid data, and the contribution made by field staff of the State Veterinary Service in providing this has been exemplary.

References

1. Rothman, K. J. (1986) *Modern Epidemiology*. Little, Brown and Co. Boston/ Toronto, pp. 7–21.
2. Wells, G. A. H., Scott, A. C., Johnson, C. T., Gunning, R. F., Hancock, R. D., Jeffrey, M., and Bradley, R. (1987) A novel progressive spongiform encephalopathy in cattle. *Vet. Rec.* **121**, 419,420.
3. Dawson, M., Mansley, L. M., Hunter, A. R., Stack, M. J., and Scott, A. C. (1987) Comparison of scrapie associated fibril detection and histology in the diagnosis of natural sheep scrapie. *Vet. Rec.* **121**, 591.
4. Wells, G. A. H., Hawkins, S. A. C., Hadlow, W. J. and Spencer, Y. I. (1992) The discovery of bovine spongiform encephalopathy and observations on the vacuolar changes, in *Prion Diseases in Humans and Animals* (Prusiner, S. B., Collinge, J., Powell, J., and Anderton, B., eds.), Ellis Horwood, London, pp. 256–274.
5. Wilesmith, J. W. and Wells, G. A. H. (1991) Bovine spongiform encephalopathy, in *Transmissible Spongiform Encephalopathies. Current Topics in Microbiology and Immunology*, vol. 172 (Chesebro, B. W., ed.), Springer-Verlag, Berlin/Heidelberg, pp. 21–38.
6. Wells, G. A. H. and Wilesmith, J. W. (1989) The distribution pattern of neuronal vacuolation in bovine spongiform encephalopathy (BSE) is constant. *Neuropathol. Appl. Neurobiol.* **15**, 591.
7. Wells, G. A. H., Hancock, R. D., Cooley, W. A., Richards, M. S., Higgins, R. J., and David, G. P. (1989) Bovine spongiform encephalopathy: diagnostic significance of vacuolar changes in selected nuclei of the medulla oblongata. *Vet. Rec.* **125**, 521–524.
8. Wells, G. A. H., Scott, A. C., Wilesmith, J. W., Simmons, M. M., and Matthews, D. (1994) Correlation between the results of a histological examination and the detection of abnormal brain fibrils in the diagnosis of bovine spongiform encephalopathy. *Res. Vet. Sci.* **56**, 346–351.
9. Scott, A. C., Wells, G. A. H., Chaplin, M. J., and Dawson, M. (1992) Bovine spongiform encephalopathy: detection of fibrils in the central nervous system is not affected by autolysis. *Res. Vet. Sci.* **52**, 332–336.
10. Wells, G. A. H., Dawson, M., Hawkins, S. A. C., Green, R. B., Dexter, I., Francis, M. E., Simmons, M. M., Austin, A. R., and Horigan, M. W. (1994) Infectivity in the ileum of cattle challenged orally with bovine spongiform encephalopathy. *Vet. Rec.* **135**, 40,41.
11. Hill, A. B. (1965) The environment and disease: association or causation? *Proc. R. Soc. Med.* **58**, 295–300.

12. Wilesmith, J. W. (1993) BSE: epidemiological approaches, trials and tribulations. *Prev. Vet. Med.* **18**, 33–42.
13. Wilesmith, J. W., Wells, G. A. H., Cranwell, M. P., and Ryan, J. B. M. (1988) Bovine spongiform encephalopathy: epidemiological studies. *Vet. Rec.* **123**, 638–644.
14. Wilesmith, J. W., Ryan, J. B. M., Hueston, W. D., and Hoinville, L. J. (1991) Bovine spongiform encephalopathy: descriptive epidemiological features 1985–1990. *Vet. Rec.* **130**, 90–94.
15. Wilesmith, J. W. (1991) Epidemiology of bovine spongiform encephalopathy. *Semin. Virol.* **2**, 239–245.
16. Wilesmith, J. W. (1994) Bovine spongiform encephalopathy and related diseases: an epidemiological overview. *N. Z. Vet. J.* **42**, 1–8.
17. Jeffrey, M. and Wells, G. A. H. (1988) Spongiform encephalopathy in a nyala *(Tragelaphus angusi). Vet. Pathol.* **25**, 398,399.
18. Brookmeyer, R. and Gail, M. H. (1988) A method for determining short-term predictions and lower bounds on the size of the AIDS epidemic. *J. Am. Stat. Soc.* **83**, 301–308.
19. Wooldridge, M. J. A. (1995) A study of the incubation period, or age at onset, of the transmissible spongiform encephalopathies/prion diseases. Ph.D. Thesis, University of London.
20. Wilesmith, J. W., Ryan, J. B. M., and Atkinson, M. J. (1991) Bovine spongiform encephalopathy: epidemiological studies on the origin. *Vet. Rec.* **128**, 199–203.
21. Schlesselman, J. J. (1982) *Case-Control Studies: Design, Conduct, Analysis.* Oxford University Press, New York.
22. Hoinville, L. J., Wilesmith, J. W., and Richards, M. S. (1995) An investigation of risk factors for cases of bovine spongiform encephalopathy born after the introduction of the "feed ban". *Vet. Rec.* **136**, 312–318.
23. Brown, P. (1994) Vertical transmission of prion diseases. *Hum. Reprod.* **9**, 1796,1797.
24. Hoinville, L. J. (1995) A review of the epidemiology of scrapie in sheep. *Rev. Sci. Tech. Off. Int. Epiz.* (submitted for publication).
25. Wilesmith, J. W. (1992) Bovine spongiform encephalopathy: a brief epidemiography, 1985–1991, in *Prion Diseases of Humans and Animals* (Prusiner, S. B., Collinge, J., Powell, J., and Anderton, B., eds.), Ellis Horwood, London, pp. 243–255.
26. Wilesmith, J. W. (1994) Update on the epidemiology of bovine spongiform encephalopathy in Great Britain, in *Transmissible Spongiform Encephalopathies* (Bradley, R. and Marchant, B., eds.), Commission of the European Communities, September 14–15, Brussels, pp. 1–12.
27. (1995) Bovine Spongiform Encephalopathy. An Update Report, May 1995. Ministry of Agriculture, Fisheries and Food. London.
28. Wells, G. A. H. and Wilesmith, J. W. (1995) The neuropathology and epidemiology of bovine spongiform encephalopathy. *Brain Pathol.* **5**, 91–103.
29. Hoinville, L. J (1992) MSc Thesis, London University.
30. Wilesmith, J. W., Hoinville, L. J., Ryan, J. B. M., and Sayers, A. R. (1992) Bovine spongiform encephalopathy: aspects of the clinical picture and analyses of possible changes. *Vet. Rec.* **130**, 197–201.

31. Hoinville, L. J. (1994) Decline in the incidence of BSE in cattle born after the introduction of the "feed ban". *Vet. Rec.* **134,** 274,275.
32. Kimberlin, R. H. and Wilesmith, J. W. (1994) Bovine spongiform encephalopathy (BSE): epidemiology, low dose exposure and risks. *Ann. NY Acad. Sci.* **724,** 210–220.
33. Spongiform Encephalopathy Advisory Committee (1995) Transmissible Spongiform Encephalopathies. A summary of present knowledge and research, September 1994. HMSO, London.
34. Wilesmith, J. W. (1995) Recent observations on the epidemiology of bovine spongiform encephalopathy, in *Proceedings of VIth International Workshop on Bovine Spongiform Encephalopathy: The BSE Dilemma* (Gibbs, C. J., ed.), Serono Symposia. Norwell, VA, in press.
35. Richards, M. S., Wilesmith, J. W., Ryan, J. B. M., Mitchell, A. P., Wooldridge, M. J. A., Sayers, A. R., and Hoinville, L. J. (1993) Methods of predicting BSE incidence, in *Proceedings of the Society for Veterinary Epidemiology and Preventive Medicine* (Thrusfield, M. V., ed.), The Society, pp. 70–81.
36. Holford, T. R. (1992) Analysing the temporal effects of age, period and cohort. *Stat. Meth. Med. Res.* **1,** 317–337.
37. Hunter, N., Goldmann, W., Smith, G., and Hope J. (1994) Frequencies of PrP gene variants in healthy cattle and cattle with BSE in Scotland. *Vet. Rec.* **135,** 400–403.
38. Dawson, M., Wells, G. A. H., and Parker, B. N. J. (1990) Preliminary evidence of the transmissibility of bovine spongiform encephalopathy to cattle. *Vet. Rec.* **126,** 112,113.
39. Dawson, M., Wells, G. A. H., Parker, B. N. J., Francis, M. E., Scott, A. C., Hawkins, S. A. C., Martin, T. C., Simmons, M. M., and Austin, A. R. (1994) Transmission studies of BSE in cattle, pigs and domestic fowl, in *Transmissible Spongiform Encephalopathies* (Bradley, R. and Marchant, B., eds.), Commission of the European Communities, September 14–15, Brussels, pp. 161–167.

10

Handling the BSE Epidemic in Great Britain

David A. J. Tyrrell and Kevin C. Taylor

1. Introduction

Henry Ford thought that history was bunk; Hegel is supposed to have said that "people and governments never have learnt anything from history," and Disraeli advised us to "read no history." The authors of this chapter are less dismissive, taking the alternative view that those who ignore history are destined to repeat it, which is the view that underscores this essay on the response to Bovine Spongiform Encephalopathy (BSE); an unforeseen disease, induced by human activity, that had important impacts on government and veterinary and medical affairs. It is unlikely that such an event will occur again, at least in exactly the same way, but it is worth telling the story of how scientific findings contributed in planned and unplanned ways and how the complex nexus of regulations, the legislature, the media, and international relations developed; we learned lessons along the way that we believe could be applied with profit in the face of another such epidemic. The account is not intended to be comprehensive: It is a selective and incomplete history that concentrates on the most important issues. A chronology (*see* Table 1) that relates events to control measures gives further details, and a brief reading list is provided for those who wish to study the subject in greater detail.

2. Before the Storm

It is common to talk of "the calm before the storm," but in fact BSE emerged in a period of many and varied activities, only some of which, and then only recognized in retrospect, were relevant to the epidemic.

During the second half of the 20th century the British farming industry was busy adapting itself to changes in the market and scientific knowledge. The number of farms decreased and the size of herds increased. New breeds were

(text continued on p. 182)

From: *Methods in Molecular Medicine: Prion Diseases*
Edited by: H. Baker and R. M. Ridley Humana Press Inc., Totowa, NJ

Table 1
Chronology of Events in the United Kingdom (to December 31, 1994)

Date	Event	Statutory action
November 1986	Disease identified by Central Veterinary Laboratory following study of affected cow referred to Weybridge for investigation and post mortem. Transmission experiments needed that required fresh material from animals thought to be suffering from the same problem	
April 1987	Initial epidemiological studies started. Objective was to obtain detailed data from a case study of 200 herds	
June 5, 1987	CVO informs Ministers about new disease. Transmission experiment then put under way. Normal time for disease to develop in mice about 10 mo. Results available September 1988 and published October 1988 in *Veterinary Record*	
December 15, 1987	Initial epidemiology studies completed. Concluded ruminant-derived meat and bonemeal was only viable hypothesis for cause of BSE	
January–March 1988	Double checking of feeding histories of affected animals initiated; request sent to compounders for details of inclusion of meat and bonemeal in rations fed	
April–May 1988	Responses from compounders further substantiated hypothesis for cause of BSE. Government indicated that they would legislate to make BSE notifiable and to ban feeding of rations that contained protein derived from ruminants	
April 21, 1988	Southwood Working Party announced	
June 1988	Discussions with major compounders on timing of ruminant feed ban	
June 20, 1988	Southwood Working Party held first meeting and decided to issue interim advice immediately	
June 21, 1988		Disease made notifiable, and isolation of suspects when calving required, by BSE Order 1988

Date		
June 22, 1988	Interim advice received from Southwood—destroy affected cattle: proposed feed ban welcomed	
July 7, 1988		Ruminant feed ban comes into force (not advised by Southwood but later welcomed by Working Party in their report). Included in BSE Order 1988 but implementation delayed until July 18. Ban to apply until December 31, 1988
July 18, 1988	Decision to introduce slaughter policy announced	
August 8, 1988		Slaughter policy introduced. Compensation paid at 50% value for confirmed cases, 100% for negative; both subject to a ceiling
October 1988	Transmission to mice following intracerebral inoculation of BSE brain tissue reported in *Veterinary Record*	
November 15, 1988	Further interim advice received from Southwood—extend feed ban and destroy milk from suspect cattle	
November 30, 1988	Decision announced to prolong feed ban and destroy milk from affected cattle	
December 30, 1988		Legislation came into force to prolong feed ban and to prohibit use of milk from suspect cattle for any purpose other than feeding to cow's own calf.
February 9, 1989	Southwood Report received by Ministers	
February 27, 1989	Southwood Report published and Government response announced (all recommendations have or will be introduced)	
July 28, 1989	EC ban on export of cattle born before July 18, 1988 and offspring of affected or suspect animals (Decision 84/469/EEC)	
November 13, 1989		Offals ban Regulation came into force in England and Wales (following consultation—a legal requirement—and consideration by top experts)

(continued)

Table 1 *(continued)*

Date	Event	Statutory action
January 9, 1990	Publication of Tyrrell Report and Government response (all top and medium priority work recommended under way or would be undertaken). Publication delayed so Government could ensure financing for R&D was in place. Research itself was *not* delayed	
January 30, 1990		Offals ban introduced in Scotland following additional consultation
January 31, 1990	Announcement that five antelopes have succumbed to a spongiform encephalopathy (great kudu, arabian oryx, eland, nyala, and gemsbok). The last two were referred to in Southwood report	
February 3, 1990	Cattle-to-cattle transmission following intracerebral and intravenous inoculation of BSE brain tissue, and into mice via the oral route, reported in *Veterinary Record*, following press briefing on February 2	
February 14, 1990		Full compensation up to a ceiling introduced. There was no sudden upsurge of cases, and pattern of reporting was unaffected
March 30, 1990		Ban on export of specified offal and certain glands and organs (for uses other than human consumption) to other Member States
April 1, 1990	Disease made notifiable to European Commission (Decision 90/134/EEC made March 6)	
April 3, 1990	Announcement about the establishment of permanent advisory group on spongiform encephalopathies under chairmanship of David Tyrrell	
April 9, 1990	EC decision to ban exports of SBO and other tissues (90/200/EEC)	
April 11, 1990	Humberside County Council withdraw British beef from school meals	

Date	Event	
May 10, 1990	Announcement about cat with a spongiform encephalopathy	
May 17, 1990	Announcement that decisions about breeding should be left to individual farmers and their veterinary advisors	
June 8, 1990	Council of Ministers agree on arrangements for trade in beef and calves from UK (Decision 90/261/EEC made June 8)	
July 12, 1990	Publication of Tyrrell Committee's detailed reasoning on why there was no need to give official advice on breeding from offspring of BSE cases	
July 12, 1990	Report of Agriculture Committee published	
July 23, 1990	UK progress report to OIE meeting	
September 25, 1990		Ban on use of specified bovine offals in any animal feed introduced. Exports of such feed also effectively banned to other Member States. (Third country exports banned under DTI legislation on July 10, 1991)
September 28–29, 1990	OIE meeting in Paris; recommendations made regarding trade in live cattle, beef, dairy, and bovine products and coordination of research	
October 2–5, 1990	OIE Conference in Sofia (Bulgaria); recommendations made regarding trade, prevention, control and surveillance of BSE, and the need for further consideration on trade in live animals	
October 15, 1990		New record-keeping arrangements come into force requiring cattle farmers to maintain breeding records. These and movement records to be retained for 10 yr
November 21, 1990	Publication of Government response to Agriculture Committee report	
March 27, 1991	First case of BSE in animal born after feed ban announced	
May 1991	UK Progress Report to the OIE General Assembly	
July 10, 1991		The Export of Goods (Control) (Amendment No. 7) Order 1991 came into force controlling export of SBOs to third countries. (Dept. of Trade and Industry legislation)

(continued)

Table 1 *(continued)*

Date	Event	Statutory action
September 16–20, 1991	Meeting of OIE International Animal Health Code Commission in Paris	
November 6, 1991		Main legislation re-enacted as a single statutory instrument including new provisions to prevent the use of meat and bone meal produced from SBOs as a fertilizer
March 4, 1992	Results of further experiments on the host range of BSE announced. Also that the Tyrrell Committee had considered the latest BSE research and concluded that the measures at present in place provide adequate safeguards for human and animal health	
May 1992	UK Progress Report to the OIE General Assembly	
May 1992	OIE General Assembly in Paris agree on trading conditions for bovine products from countries affected by BSE	
May 14, 1992	EC Commission Decision prohibiting intra-Community trade in bovine embryos derived from BSE suspect or confirmed dams or dams born after 7/18/88 (Decision 92/290/EEC)	
June 30, 1992	Publication of the "Interim Report on Research" by the Spongiform Encephalopathy Advisory Committee	
November 1992	UK Progress Report presented to the EC Standing Veterinary Committee	
November 24, 1992	Announced by PQ that details of the total number of cases (by county) would be placed regularly in the library of the House of Commons	
May 24, 1993	UK Progress Report presented to the OIE General Assembly	
June 10, 1993	UK Progress Report presented to the EC Standing Veterinary Committee (same as OIE Progress Report)	
July 14, 1993	100,000th confirmed case of BSE in Great Britain announced in response to a Parliamentary Question, as an update to the UK Progress Report to the OIE	

November 25, 1993	GB Progress Report placed in the library of the House of Commons
January 21, 1994	Changes to the BSE compensation arrangements introduced
April 26, 1994	GB Progress Report placed in the library of the House of Commons
May 1994	UK Progress Report presented to the OIE General Assembly
June 27, 1994	Commission Decision 94/381 on BSE and feeding of mammalian-derived protein made
June 1994	UK Progress Report updated
June 27, 1994	Commission Decision 94/382 on the approval of alternative heat treatment systems for processing animal waste
June 30, 1994	Results of further BSE experiment announced. Extension of SBO ban implemented voluntarily by industry
July 27, 1994	Commission Decision 94/474 introduced new certification requirements for bone-in beef exported to other Member States of EC
November 2, 1994	Bovine Offal (Prohibition) (Amendment) Regulations 1994 came into force, extending ban on use of SBO in human food. The Spongiform Encephalopathy (Miscellaneous Amendments) Order 1994 came into force, extending the ban on use of SBOs in animal feed, banning the use of mammalian protein in ruminant feedingstuffs, and making notifiable laboratory suspicion of spongiform encephalopathies in species other than cattle, sheep, and goats
December 1994	GB progress report placed in the Library of the House of Commons
December 14, 1994	Commission Decision 94/474 amended by Decision 94/794. Beef from cattle born after January 1, 1992 excluded from certification requirement
February 1995	SEAC report "Transmissible Spongiform Encephalopathies—a summary of present knowledge and research" published

being introduced and semen and embryos from UK stock were traded internationally. Endemic bovine diseases, such as tuberculosis and brucellosis, were controlled, and imported infections, such as foot and mouth disease, were promptly eradicated. Infections, such as salmonellosis and listeriosis, had sensitized the public to the dangers of contaminated poultry and dairy products. The rendering industry, which existed to dispose of animal waste from the slaughtering and knackering industries, refined its production methods to improve the value of its products (fat, and meat and bonemeal) and the use of organic solvents to recover additional fat was discontinued in most plants. At the same time, concern about salmonellosis led to the introduction of legislation requiring the end products to be salmonella-free. Changed conditions implied changed risks, and "new" infections, for example, swine vesicular disease, had turned up from time to time. It was an accepted philosophy that the best way of detecting and understanding both "old" and "new" epidemics was to maintain a network of highly competent clinical and laboratory veterinary experts, in good relationship with the veterinary profession in practice, who could identify and study promptly anything unusual that turned up.

Thus it was that among the 11 million cattle in Great Britain cases of unusual neurological disease were identified in 1986 and the brains of two cattle examined at the Central Veterinary Laboratory (CVL) Weybridge in November of that year were recognized as showing the characteristic changes of a spongiform encephalopathy. Although common in sheep, such a condition had never been seen in cattle before and it was quite unclear what the cause was, and whether the findings were simply an isolated curiosity or signified the start of a major epidemic. As a first step it was important to collect information about the new condition to enable epidemiological studies to be carried out and to facilitate further pathological studies, and to attempt experimental transmission to cattle and to mice. Voluntary reporting of suspect cases to Veterinary Investigation Centres was encouraged, and further cases were reported in all parts of Great Britain. In April 1987 an epidemiological study to identify the risk factors was initiated and when completed early in 1988 the study (of some 200 affected cattle) showed that although some possible risk factors, such as contact with sheep or administration of vaccines or medicinal treatments, were not important, a history of ruminant-derived protein having been fed was. It was concluded that the new disease, by now called Bovine Spongiform Encephalopathy (BSE) by the scientific community, and "Mad Cow Disease" by the media, was a common source epidemic caused by the inclusion in cattle feed of ruminant-derived protein (in the form of meat and bonemeal) containing a scrapie-like infectious agent.

In April 1988 the Government decided it was necessary to have an external review of the state of knowledge and the Minister of Agriculture and Secretary of State for Health jointly appointed a small working party of scientific experts under the chairmanship of Sir Richard Southwood to consider the situation and report on it. As background to their discussions the committee could draw on the findings of years of research on scrapie, which had been studied at Compton and Edinburgh and in the United States so that the peculiar properties of the causative agent, the slow way it spread within the organs of infected animals, the importance of genetics, and the route of transmission in determining susceptibility were all known and published. However, scrapie was not considered to be a subject of great importance in 1988, and research was running down.

The decision to introduce statutory control measures for BSE in Great Britain was taken in May 1988, before the Southwood committee had even met, and the first controls were imposed on June 21 of that year when the disease was made notifiable. The objectives were:

1. To discover the true incidence of the disease (it was recognized that under the voluntary reporting system in use until then many suspect cases would escape notice);
2. To facilitate the collection of clinical and epidemiological information from all cases, rather than only a proportion, and so test on a wider database the epidemiological hypothesis on which the control policy was based; and
3. To prevent cattle that were not already infected becoming infected, and so to eventually eradicate BSE.

The statutory obligation to notify suspect cases of BSE underpins other control measures. Following notification, the suspect animal is examined by a veterinary officer of the Ministry of Agriculture, Fisheries, and Food. If BSE is suspected on the basis of the clinical signs observed, a restriction notice is served to prevent the suspect animal being removed from the premises, and to require isolation if about to calve. The notice applies only to the individual suspect animal: BSE is not believed to be a contagious disease, so there is no justification for restricting the movement of other animals in the herd. From June 21 until August 7, 1988 it was permissible to move a restricted animal, under license, to a slaughterhouse. After slaughter, for human consumption or otherwise, the head was removed and taken—again under license—to a laboratory so that the brain could be removed, fixed, and subjected to a histopathological examination to confirm the clinical diagnosis. Since August 7, 1988, as a result of an interim recommendation from Sir Richard Southwood, the carcasses of all suspect cattle have been destroyed after the animal has been compulsorily slaughtered, in order to protect public health against any risk that BSE may pose. Compensation is paid to owners.

The key measure taken to prevent further infection of cattle is a prohibition on feeding ruminant protein (the definition of which excludes milk and milk products, and dicalcium bone phosphate) to ruminant animals. This was first implemented on July 18, 1988, 1 mo after the disease had been made notifiable: The delay was intended to allow feed that had been manufactured already and that contained ruminant protein to be used up. If rigorously observed the ruminant feed ban prevents scrapie-like agents being transmitted in feed from one ruminant species to another and, probably more important, prevents recycling of infection within the same species. If there is no other significant route of transmission this measure alone will eradicate BSE.

A theoretical alternative approach, which was considered and rejected, would have been to destroy any infectivity in meat and bone meal by heat treatment. This was not possible because the combinations of time and temperature needed to destroy the BSE and scrapie agents had not been determined, and without such information it was impossible to know which, if any, commercial rendering processes could continue to be used. In 1994 the preliminary results of a collaborative study on the effect of different rendering protocols on BSE infectivity showed that at least two systems used in Great Britain were ineffective, but the limited sensitivity of the study means that it is still impossible to identify "safe" systems and the ruminant protein feed ban remains the only practical method of preventing infection through feed. The banning of processes that are known to be ineffective is, however, a useful additional measure to strengthen the effectiveness of the feed ban.

The possibility that BSE would be transmitted via the placenta of the parturient female, as is scrapie in sheep, was also recognized. To counter this, legislation required that a suspect must be isolated in approved accommodation while calving, and for 72 h afterward, and that the placenta, discharges, and bedding must be burned or buried and the isolation premises cleaned and disinfected after use. The purpose of this measure is not to prevent infection of the calf, which would be impractical, but to reduce the opportunity for horizontal spread of infection to other cattle in the herd.

3. Monitoring the Epidemic

It was realized that it would be important to monitor the progress of the epidemic, and that clinical diagnosis without any supporting laboratory evidence would be problematic, at least until farmers and veterinary surgeons became more familiar with a disease that few had seen at that time. Videos showing clinical cases therefore were prepared and widely distributed, and diagnostic capacity was increased by training Investigation Centre staff in the techniques developed at the Central Veterinary Laboratory (CVL). The results of the laboratory investigations were sent to the State Veterinary Service headquarters, where veterinary staff assessed the information and confirmed those cases for which there was appropriate evidence. The field data from all these cases were sent to the CVL epidemiology department, which had to be greatly

Table 2
BSE Statistics as of December 31, 1994[a]

Year	Cases reported	Number slaughtered	Number confirmed	Percent confirmed
1988	2516	2376	2184	91.9
1989	8447	8061	7137	88.5
1990	17,323	16,641	14,181	85.2
1991	30,009	29,025	25,032	86.2
1992	44,846	43,154	36,680	85.0
1993	42,932	41,081	34,370	83.7
1994	30,247[b]	26,443[b]	20,884	82.7[b]

[a]Total cases = 140,910 on 31,747 farms.
[b]These figures are subject to change.

expanded to deal with the problem. About 93% of clinical diagnoses were confirmed in the early stages, but this declined with time: In 1993, 83.7% of suspect cases were confirmed (*see* Table 2). Regular reports were issued on the number of cases being confirmed, and the epidemic curve was plotted and frequently updated (Fig. 1). The likely future course of the epidemic was estimated from mathematical models.

The ability to forecast the size and shape of the epidemic was of more than academic interest. It was necessary to find the funds to pay compensation to farmers whose cattle were killed, to provide sufficient diagnostic capacity, and to encourage the private sector to provide incinerators operating at a hearth temperature of 800°C or higher in which carcasses could be destroyed without having to be dismembered first. In the early stages of the epidemic these were not available and at first some 30% of bodies had to be buried in sites that were chosen carefully to avoid contamination of aquifers; others had to be burned in the open. Since May 1991 virtually all carcasses have been incinerated in purpose-built plants.

Some pessimists argued that suspect cases would not be reported by farmers because of adverse effects on trade and sales, but there was never any evidence to support this view. New cases were only infrequently detected at markets or prior to slaughter, and when in February 1990 the level of compensation was increased from 50–100% of the animal's value there was no effect on the number of notifications received. It seems that notifications were honest and complete. The numbers of reports did vary from one week to another in response to other events on the farm, and always dropped precipitately over the Christmas holidays, but any shortfall in one week was balanced by an increase in the following weeks. Suspect cases were more likely to be observed and reported

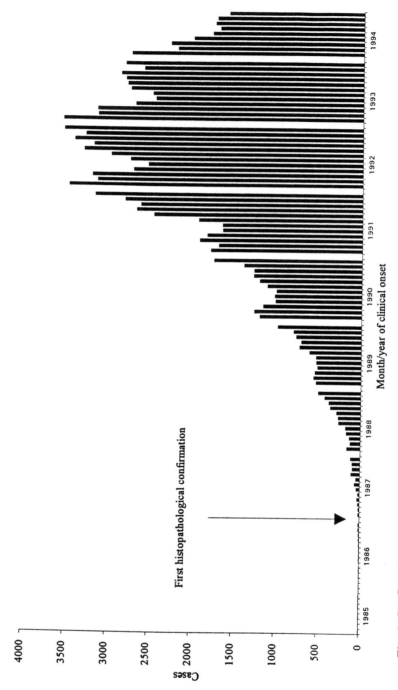

Fig. 1. Confirmed cases of BSE plotted by month and year of clinical onset. (Data to end of September 1994.)

186

in the autumn, winter, and spring than during the summer months, although this may reflect closer observation of housed cattle rather than a true seasonal variation in incidence.

4. The Public Health

There was ample evidence that scrapie did not pose a risk to humans; in particular, that it was not associated epidemiologically with Creutzfeldt-Jakob disease and other spongiform encephalopathies of humans. Individuals had eaten sheep's brain for centuries without apparent harm.

The initial BSE control measures in Great Britain were intended only to protect animal health, it being assumed that if scrapie posed no risk to human health, neither would BSE. The carcasses and organs of suspect cases, other than the brains (which were needed to confirm the clinical diagnosis) therefore could be used for human consumption. The Southwood working party, however, concluded that even though scrapie may have caused BSE it could not be assumed that BSE could not affect humans, even by the oral route, and in an interim recommendation received the day after statutory controls were first introduced advised that although there was no known hazard to human health it would be a sensible precaution to remove suspect cattle from the food chain while the working party considered the evidence. A slaughter policy was introduced on August 8, 1988, the first public health control measure to be implemented.

The change introduced a number of practical problems for which solutions had to be found. The accuracy of diagnosis based on clinical signs alone was of particular concern, since the disease was still unfamiliar to most veterinary surgeons and there was no diagnostic test available that could be used to assist differential diagnosis in the live animal. Slaughter would be compulsory, and any inaccuracy of the clinical diagnosis would be apparent to all when the results of histopathological examination of the brain was received. New legislation therefore distinguished between *suspicion* of disease, justifying movement restrictions, and *conviction* that an animal was affected with BSE, justifying slaughter. Compensation is paid for cattle that are compulsorily slaughtered, and an efficient disposal operation has been developed to deal with their carcasses. After slaughter by injecting barbiturate intravenously, the carcass is moved under license to one of a number of plants in different parts of the country where, after the head has been removed and sent to a Veterinary Investigation Centre for histopathological examination of the brain, the rest of the carcass is destroyed.

The effect of this change was to prevent any part of a clinically suspect animal being used in human food, or fed to animals. It was clear, however, that the measure did not prevent the consumption of tissues from cattle that were infected and incubating the disease but had not yet developed clinical symptoms. The Southwood Working Party report, published in February 1989, identified the systems in which infectivity was most likely to be present, and although it did not recommend statutory action, it did suggest that baby food manufacturers should avoid the use of ruminant offal and thymus. Government, however, concluded that preventing the use of these tissues in *all* human food would provide a valuable additional safeguard.

Work on natural scrapie in sheep had indicated that infectivity could be present in the lymphoreticular and central nervous systems of apparently normal animals and it was prudent to take account of this by assuming that the same could be true of BSE; establishing the facts for the latter disease would require an experiment of several years duration and serious harm might be done if action were to be delayed. A ban on the use of specified bovine offals (SBO) therefore was introduced, first for human use, and later also for animals. The age at which cattle would be subject to SBO controls had to be timed precisely, and the decision was based on the times at which scrapie infectivity was first detectable in different tissues in sheep and goats, as well as careful checks on the distribution of the lymphoid tissue within the alimentary tract of cattle. The objective is to prevent the BSE agent from getting into any food chain, human or animal, and the bovine offals ban, if properly implemented, should ensure that humans do not eat any bovine tissue containing significant infectivity. Subsequent research has suggested that the BSE agent may be less widely distributed in infected cattle than is the scrapie agent in infected sheep, and that the SBO ban is more extensive than necessary. This is no bad thing in a public health protection measure, and much more study is necessary before the extent of the SBO controls can be reassessed with confidence. It is also pertinent to wonder whether a slaughter and destruction policy for suspect cattle is needed *in addition* to SBO controls: In theory, the latter alone would suffice to protect public health (and animal health is protected in other ways), but a cautious view rightly has been taken and both sets of controls continue to be applied.

Despite the effectiveness of present control measures, however, it is clear that humans could have ingested infectivity before the disease was recognized and the ban was introduced. The risk is likely to have been very small, but one cannot not say it did not exist. Public concern about BSE, fanned by media pressure and often inaccurate reporting, has been intense at some periods. Rather curiously, the first report of spongiform encephalopathy in a cat—not a species that is eaten—apparently had a greater effect on beef consumption in Great Britain than the epidemic in cattle, presumably reflecting a fear that if one species barrier could be breached, so could others. A policy of scientific openness, combined with independent advice from the Spongiform Encephalopathy Advisory Committee (SEAC), has helped to provide a more balanced assessment of the situation and of the measures taken to protect human and animal health. In the long term this has probably been more effective than publicity stunts, however well intentioned.

5. The Role of Research

It has been made clear already that BSE was recognized quickly because an effective surveillance system for new and emerging diseases existed in Great Britain. The transmissible spongiform encephalopathies (TSEs), especially scrapie, had been studied in great detail already, and an immediate response was possible because of the general laboratory and epidemiological resources of the MAFF and the specialized research staff and facilities at the Neuropathogenesis Unit (NPU) of the Agriculture and Food Research Council (AFRC). The scientists recognized the pivotal importance of transmission experiments, which were set up before formal funding had been arranged.

The Southwood Committee perceived the importance of research and recommended that a large observational study be set up to establish whether BSE was maternally transmitted from dam to offspring. This was done, and results are expected in 1997. They also recommended that a further committee be set up to review the need for research. Thus, the first "Tyrrell" committee, composed of a small but diverse group of scientists with interests in the area, was constituted in February 1989. It reviewed the questions that needed answering and suggested experimental or other approaches, and graded the priority of each proposal. The committee reported in June 1989, but their report was not published until the following year, when sources for the required funding had been identified. By the time the report was published all high and medium priority research that it identified had been started. Funds were made available to support MAFF/NPU research and also to support a new research initiative by AFRC, which was open to universities and research institutes, to study basic aspects of TSEs. This committee was then converted into the Spongiform Encephalopathy Advisory Committee (SEAC), still under the chairmanship of Dr. Tyrrell, which met every 4 mo, and more often if necessary, to review the development of the epidemic and the results of research as they became available, and to answer questions posed by the Department of Health and the Ministry of Agriculture.

The government's intention was to base its response to the epidemic on independent scientific advice, and to be open about what was being done and why. Previously it had been said that the work of Ministry scientists was not published sufficiently freely and it was intended that there should be no delay in implementing changes recommended as a result of any new findings. These intentions were not always easy to put into practice. The Minister might say in

Parliament, "What the Tyrrell Committee recommends, I will do," but despite their scientific expertise the Committee members often knew little of the detail of farming practices or of the relevant laws and regulations and could only express general views and scientific principles that had to be interpreted and implemented by others with appropriate expertise. Nor could the committee assess how scrupulously the regulations were observed and enforced. As to openness and publication, it was sometimes decided to report experiments that were only partly completed, in spite of the risk that the conclusions drawn and consequent action taken would have to be modified when all the results came in. Work therefore was sometimes published sooner than would have been the case if purely scientific editorial criteria had been used. Although there may be benefits in avoiding delay, such a policy also carries the risk of overreaction to an isolated finding seen out of context. The Committee generally tended to take a "fail safe" view of experiments that had possible public safety implications. For instance, when unexpectedly parenteral injection of BSE caused infection and clinical disease in pigs, they advised that the ban that already prevented the use of specified bovine offal in human food should be extended to prevent feeding to pigs and poultry and all other animals and birds as well, even though there had been no evidence that a TSE had ever occurred naturally in either pigs or poultry, despite the fact that both had consumed large quantities of ruminant protein in feed. On the other hand when, earlier than expected, infectivity was found in the distal ileum of calves that had been given a massive dose of BSE by mouth, their advice was that the epidemic was declining in response to the control measures already in place, so that any risk that tissues from calves under 6 mo old would expose humans to risk was "minuscule or absent." Despite this advice the government decided to be ultracautious and amend the regulations to prevent the use of intestines and thymus from cattle of any age, rather than only applying the ban to these tissues from cattle 6 mo of age and older.

6. Medicinal Products

The Medicines Control Agency, through the Committee on the Safety of Medicines, also had a role in protecting the public health. If BSE was a hazard to humans then the greatest risk would be from products given or used parenterally, such as vaccines, suture materials, and so on. This was a potentially serious problem since healthy animals *might* be infected, and by analogy with scrapie there was no sterilization method that was likely to eliminate infectivity without destroying the product. In addition, any risk from parenteral use would be expected to be greater than from oral consumption. The Agency therefore issued guidelines and sought information from all manufacturers to find out if bovine materials were used in their products. One attractive solution was to source bovine materials from irreproachably BSE-free herds: The ideal would be from well-supervised herds in a country where neither scrapie nor BSE were present. It is impressive that industry took up this challenge very effectively; for instance, intestines for the manufacture of surgical catgut were sourced from Australasia within months.

7. Spongiform Encephalopathies in Other Species

It gradually became clear that the host range of BSE was wider than that of scrapie. Exotic ungulates in zoos had been fed the same ruminant-derived protein as cattle and several species developed SEs. In due course, laboratory studies showed that the agent causing disease in these species behaved like BSE when inoculated into a panel of genotypically susceptible experimental mice. In addition, domestic cats and great cats in zoos developed SEs and strain typing, again suggested that the agent responsible was similar to BSE. Members of the Pet Food Manufacturers Association voluntarily had decided to exclude specified bovine offals from their products when the possibility of banning these tissues for human consumption was first discussed in 1989, but the ban nevertheless was extended to cover all animal species in September 1990 after the experimental transmission of BSE to pigs by simultaneous intracerebral, intravenous, and intraperitoneal inoculation. There are no data to indicate the number of other species that received the same potentially contaminated feeds as cattle, or were fed specified bovine offals before the bans were imposed, but nevertheless did not develop a spongiform encephalopathy. It is probably far greater than the number in which disease has been observed. No specific monitoring of unaffected species has been planned but if spongiform encephalopathy is suspected as a result of laboratory examination of brain tissue from any animal species, the fact must be notified so that further official investigations can be undertaken.

8. CJD Epidemiology

The Spongiform Encephalopathy Advisory Committee considers that all the measures necessary to protect public health are now in place, and that beef produced in the United Kingdom is safe to eat. The Southwood working party considered the possibility that BSE would affect humans to be "remote," even if no preventative measures were taken. However, the possibility of human susceptibility could not be ruled out entirely, and control measures were recommended and implemented. Heeding the historical injunction to "think it possible that ye may be mistaken" the working party also recommended that an earlier survey of the occurrence of CJD in England and Wales should be reactivated nationally. This was done by providing a small unit in Edinburgh with resources and expertise to ascertain and investigate all clinically possible cases and carry out the necessary postmortem and epidemiological studies. The CJD Surveillance Unit produces annual reports and has had to cope with the problems of publishing information about preliminary, and thus unreliable, findings, such as when the incidence of CJD appeared to rise the year after they started work, and when dietary enquiries in 1992 suggested a link between eating "puddings" (a kind of sausage) and developing BSE. Neither finding was repeated in the 1993 study; the number of CJD cases fell and the spurious association with "pudding" consumption disappeared, only to be replaced by other, equally improbable, supposed risks. The problem of recall bias makes such studies particularly difficult to interpret.

9. Relations with the Media

It was not difficult to predict that there would be public concern about a previously unknown and invariably fatal disease of cattle, transmitted by contaminated feed and that resembled a number of known fatal diseases of humans. Furthermore, the public was surprised to be told that there was no test that could identify infected live animals, and that the epidemic would have to be investigated, at least at first, by methods of clinical observation, histopathology, and "shoe leather" epidemiology that had been used for 100 yr or more. The new epidemic also took place against a background of MAFF and DoH involvement in controversy over diseases such as salmonellosis and listeriosis, which originated from contaminated food.

The catchy, but inaccurate, name "mad cow disease" was coined and the sad spectacle of a sick ataxic cow sinking to the ground, taken from the video produced to educate farmers and veterinary surgeons about the new disease, was screened and rescreened. It might, in retrospect, have been better to name the new disease "Cattle Scrapie"—as the French did subsequently—than to coin the scientifically correct but hard to say "bovine spongiform encephalopathy." Cases of CJD in two dairy farmers were claimed to be evidence that BSE had been transmitted to humans, and much publicity was given to a young woman reduced to a chronic vegetative state with a condition that resembled (but since she was still alive could not be confirmed as) CJD, who had eaten hamburgers earlier in her life. Even the lack of any definitive diagnosis in the latter case did not deter critics from claiming that the case was further "proof" that BSE had been transmitted to humans, even though the detailed studies made this very unlikely.

In these and other ways public concern was fuelled by intelligent individuals who questioned the assumptions, and later the basic science, on which control measures (particularly those that protected public health) were based. Rather than using the well developed scientific fora to debate these issues, they made it their business to express their doubts to the population in general through the newspapers, television, and other areas of the media, often in apocalyptic terms and with a highly selective presentation of information. Such confrontational tactics made excellent publicity, but did considerable disservice to attempts to implement a rational control policy firmly based on scientific evidence.

10. Effects on Farming

No epidemic involving the death or compulsory slaughter of more than 166,000 adult cattle over a period of more than 7 yr can fail to have an effect on the farming industry, but this has probably not been as great as might have been expected, and many of the problems have been caused by the international response to the UK epidemic rather than domestic circumstances.

Even at the peak of the epidemic, in 1992 and early 1993, only one in every 100 adult cattle were slaughtered as suspects, and two-thirds of breeding herds have never experienced a case of BSE. Of the herds that have had cases, 37% have had only one, and 72% have had four or fewer. In terms of normal culling and replacement these figures are insignificant, although the onset of disease does deprive the farmer of the ability to choose which animal to cull. Compensation has been generous, and except in the case of pedigree cattle there has been little or no financial loss when a suspect animal has been slaughtered. The costs of diagnosis and disposal also have been borne by the government, and some £157 million (approx $240 million) has been paid already in compensation and disposal costs.

Perhaps the greatest domestic effect has been the temporary drop in beef consumption that occurred in 1990, following the discovery of spongiform encephalopathy in a domestic cat. Beef consumption fell by 30%, but subsequently recovered over a period of about 1 yr. A handful of education authorities continue to ignore scientific advice and prohibit the use of British beef in school meals, but the great majority either took no action in 1990, or have rescinded the decisions they took then. UK agriculture also has been adversely affected by the actions of other countries throughout the world in response to the BSE epidemic.

11. International Repercussions

Relatively few countries have reported cases of BSE, and the disease has occurred at high incidence only in the United Kingdom. Nevertheless, there has been considerable interest in the disease, and the response to isolated cases has often been guided by political rather than scientific considerations. The situation has not been helped by the response in some countries when a single case of BSE has been reported in an imported animal in a country that is otherwise free of the disease. Some countries have slaughtered the entire herd in which a case occurred, and others have attempted to identify and destroy all cattle imported from the United Kingdom, even though most were beef cattle and some more than 10-yr-old. Neither response is scientifically justified.

Both the World Health Organization (WHO) and the Office International des Epizooties (OIE) have considered BSE and issued appropriate advice. The latter organization has adopted a chapter in the International Animal Health Code, together with a supporting scientific document, which advises member countries of the measures necessary to allow safe trade in cattle and cattle products from countries where BSE occurs, or may occur, at high or low incidence. Although adopted unanimously the recommendations have been ignored, in whole or in part, by most member countries, and in some cases restrictions have been placed on the importation even of products that OIE recommends can be imported unconditionally without risk.

A number of countries have assessed the risk of BSE occurring as an indigenous disease, but none so far has identified the same combination of factors as existed in the United Kingdom in the early 1980s. The risk of a major indigenous epidemic arising in any country therefore is considered to be low, and the application of preventative measures based on British experience could remove any risk there is. However, sometimes it is more difficult to do what is necessary than to know what is necessary, and there may be considerable resistance to the introduction of preventative measures against a disease that has never been recorded in a country.

12. Continued Monitoring

The BSE epidemic in Great Britain probably has been studied and monitored more intensively than any other cattle epidemic in history, and the information obtained from the epidemiological study has been vital to the control of the epidemic and in monitoring the effectiveness of control measures. The long incubation of the disease made a quick response to control measures impossible, and until 1991 the number of cases approximately doubled from one year to the next. The effect of the feed ban was to remove, or at worst substantially reduce, the risk of infection for cattle born after July 1988, but initially this effect was more than offset by the increasing amount of infectivity present in feed, as a result of recycling, until July 1988. Despite official confidence in the effectiveness of the measures that had been taken, it was a considerable relief to observe the age distribution of cases beginning to change, and the number of suspect cases being reported (and placed under restriction) reaching a peak and then declining (see Fig. 2).

The accelerating decline in the epidemic does not, however, obviate the need for continued careful monitoring. Although much has been discovered, and the effectiveness of the measures taken to control the epidemic been demonstrated,

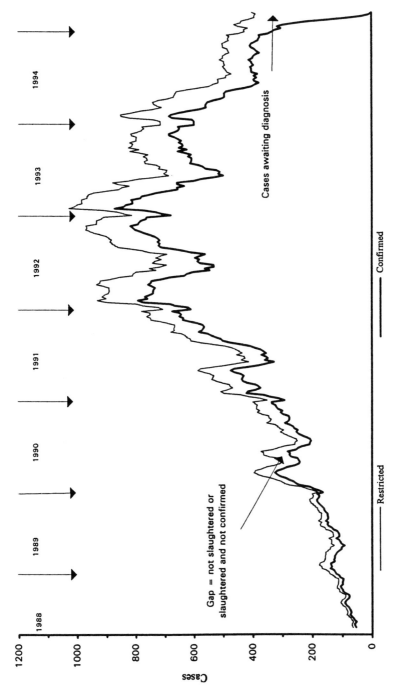

Fig. 2. Four-week rolling mean of restricted and confirmed cases. (Valid to week ending March 24, 1995.)

195

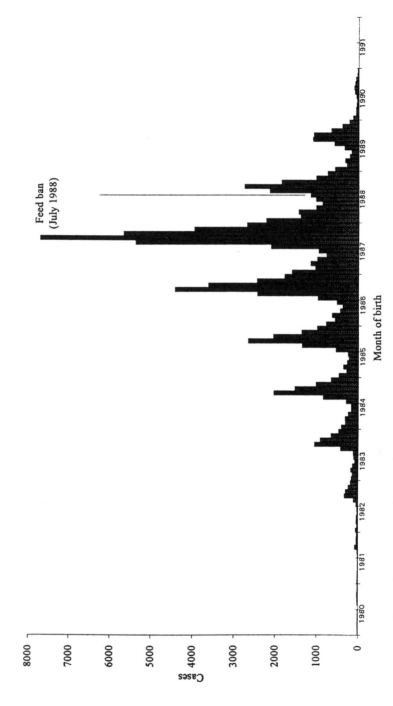

Fig. 3. Confirmed cases of BSE with known dates of birth, plotted by month of birth (data to March 1995).

there is still much to be learned. Continued monitoring is necessary to see whether there is any change in the pattern of the disease or how it is transmitted, and to check that new strains of the agent, perhaps with different characteristics, do not evolve. The continued flow of information obtained from monitoring, and from research, will enable the control program to be adjusted expeditiously if necessary, and its publication will provide the information necessary for any other country to prevent an outbreak occurring, and to control an outbreak that does.

13. Conclusion

The UK response to BSE has provided a model of how to deal effectively with a previously unknown disease, basing the immediate response on sound scientific principles and arranging and funding the research needed to build on that immediate response, while providing detailed information to statutory and scientific authorities throughout the world. Separate but interrelated measures have been taken to protect public health against any risk that BSE may pose, to prevent infection being spread to other animal species, and to eradicate BSE from UK cattle. The costs of doing so—more than £160 million (approx $256 million) on compensation and disposal costs alone at the time of writing—have been enormous, and it should be noted that almost all of these costs have been incurred by the public health protection measures. Animal health is protected in other, less expensive, ways, although the cost of destroying SBOs has been significant. The hypothesis developed after studying the first 200 or so cases, that BSE was basically a food-borne common source epidemic that could be controlled by preventing the use of infected feed, has been vindicated by events. The effect of the feed ban is shown clearly in the epidemic curve (*see* Fig. 2) and in the dramatically reduced number of cases in animals born after its introduction (*see* Fig. 3, *previous page*). Such progress is pleasing, but only partially relevant to public health, which is protected by separate measures designed to be effective whether BSE is controlled and eradicated or not.

References

1. Report of the Working Party on Bovine Spongiform Encephalopathy (The Southwood Report) (1989) DoH, MAFF, HMSO, London.
2. Consultative Committee on Research into Spongiform Encephalopathies. Interim Report ("The Tyrrell Report") (1989) MAFF, DoH; HMSO, London.
3. Spongiform Encephalopathy Advisory Committee. (1992) Interim Report on Research. DoH, MAFF; HMSO, London.
4. Spongiform Encephalopathy Advisory Committee. (1995) Transmissible Spongiform Encephalopathies: A Summary of Present Knowledge and Research. HMSO, London.

5. Animal Health, 1988–1993 inclusive (Reports of the Chief Veterinary Officer) HMSO, London.
6. Transmissible Spongiform Encephalopathies of Animals. OIE Scientific and Technical Review (1992) vol. 11.
7. Bovine Spongiform Encephalopathy in Great Britain. (1994) A Progress Report. MAFF; HMSO, London.
8. Taylor, K. C. (1991) The control of bovine spongiform encephalopathy in Great Britain. *Vet. Rec.* **129,** 522–526.
9. Taylor, K. C. (1994) Bovine spongiform encephalopathy control in Great Britain. *Livestock Prod. Sci.* **38,** 17–21.

11

Special Problems of Genetic Counseling in Adult-Onset Diseases

Huntington's Disease as a Model

Jonathon R. Gray, Jo R. Soldan, and Peter S. Harper

1. Introduction

Neurological disorders impose a severe burden on affected individuals. Inherited neurological disorders of late onset present their own specific difficulties in terms of genetic counseling, management, and patients' psychological adaptation. Huntington's disease (HD) is the archetypal late-onset neurogenetic disorder. It is an autosomal-dominant disorder usually of adult onset in which progressive degeneration of the cerebral cortex and basal ganglia result in choreiform movements and progressive mental deterioration. Closely linked genetic markers have been available since 1983 so that presymptomatic testing protocols have been in development longer for this disorder than for any other similar disorder (for further information on HD specifically *see* ref. *1*). In this chapter we highlight the major issues we consider relevant to achieving optimal diagnostic, predictive, and prenatal testing and to the family management of individuals from HD families. We put forward the protocols we have developed to help clinicians and other allied workers approach these difficult areas.

One of the main issues in late-onset disorders relates to diagnostic and presymptomatic testing. It is important to recognize from the outset that these are separate problems and must be treated as such. Clear distinction is implicit in the term "diagnostic" test, in which the patient is referred for specific mutation analysis because of suspicious clinical features of that disorder. Presymptomatic testing, on the other hand, is performed on healthy individuals in the absence of any clinical features of the disorder.

From: *Methods in Molecular Medicine: Prion Diseases*
Edited by: H. Baker and R. M. Ridley Humana Press Inc., Totowa, NJ

Although the experience and procedures discussed here have been based on HD, they have wide relevance and applicability to other late-onset neurological disorders. In recognition that many issues relating to late-onset disorders are not resolved but still open to debate, we conclude with a consideration of some of the outstanding problems.

2. Diagnostic Testing
2.1. Special Issues

Diagnostic testing in HD aims not only to confirm a diagnosis, but also to rule out other possible underlying causes of the same clinical features, some of which may be treatable. Given the multiple problems presented by HD and the importance of specificity of diagnosis, diagnostic protocols must address problems associated with differential diagnosis, the various referral routes, and associated liaisons with other specialists, for example neurologists and psychiatrists.

Genetic testing for the specific HD mutation now forms an important part of the diagnostic process but should be used in conjunction with other clinical and investigative approaches and not as a substitute for them. Particular care must be taken when investigating relatives of a known case of HD where presence of the mutation may not necessarily be related to the cause of the symptoms. A full diagnostic workup allows the ruling out of the varied alternative diagnoses and if necessary provides the space for full counseling if the tests do suggest HD.

In the majority of cases, previous risk of HD is known, and therefore diagnosis involves a transition from living at risk to having the disease. Often quite subtle changes in symptoms signify a significant change in self-perception, against which a number of defenses may have already developed *(2)*. This transition, therefore, requires careful management. Diagnostic protocols must include an assessment of the timing of diagnosis, which must be carefully considered from the patient's point of view, the result not being given until it is wanted by the patient *(3)*.

2.2. Procedures

2.2.1. Neurological Examination

2.2.1.1. GENERAL INTERVIEW

Be aware of potential diagnostic clues, e.g., minor involuntary movements, altered mental state, abnormal gait or posture.

2.2.1.2. GENERAL EXAMINATION

This is not expected to be abnormal.

2.2.1.3. CRANIAL NERVES

Fundi: normal.
Eye movements: Saccades are a sensitive early sign.
Hearing: normal.
Facial muscles: Early evidence of choreiform movements may be visible.
Tongue: chorea, dyspraxia.
Speech: rarely affected in early disease, but may be severe later.

2.2.1.4. LIMBS

General: choreic movements early in hands and feet. No early wasting.
Power: normal, but may seem to fluctuate because of chorea.
Tone: usually initially normal; increased early in rigid cases.
Reflexes: usually normal, except in juvenile form. Often increased late in some cases, with upgoing plantars. Reflexes impaired or absent in neuro-acanthocytosis.
Coordination: Rarely there may be cerebellar signs.
Sensation: normal.
Gait: normal early but may be bizarre later and affected by chorea/dystonia.

2.2.2. Neuropsychological Assessment

A range of global memory and frontal lobe assessments can ascertain the individual's cognitive performance in HD and hence give an indication of any global or specific problems. In addition to being helpful in the diagnostic profile, the results of objective, standardized tests can be helpful for the patient and family. Ascertaining objective problems can facilitate the understanding and acceptance of what are often bewildering, gradual changes.

2.2.3. Radiological Assessment

1. Computerized tomography (CT). Caudate or generalized cortical atrophy are useful but not diagnostic indicators.
2. Magnetic resonance imaging (MRI) provides similar information but with some improved definition. Its value lies in excluding differential diagnoses, such as tumors and infarcts. Practical limitations include normal scan often seen early in disease and the possibility of poor patient cooperation with scanning in later disease.

2.2.4. Laboratory Investigations

A number of laboratory investigations (many of which will exclude other conditions) may be useful in suspected cases of HD. These are shown in Table 1.

2.3. Outstanding Problems

Diagnostic procedures are normally only instituted with the express consent of the person being tested. This, of course, may not always be possible in a

Table 1
Laboratory Investigations in Suspected Cases of Huntington's Disease

Investigation	Changes in HD	Usefulness
Blood film (fresh thick wet film)	Normal	May suggest neuroacanthocytosis or alcoholic encephalopathy
Creatine kinase	Normal	Raised in neuroacanthocytosis
Red cell indices	Normal	Exclude polycythemia
Copper studies	Normal	Exclude Wilson's disease
Slit lamp	Normal	Exclude Wilson's disease
Thyroid function test	Normal	Exclude thyrotoxicosis
Antinuclear factor	Normal	Exclude systemic lupus erythematosis
Syphilis serology	Normal	Exclude syphilis
Postmortem neuropathology	Usually characteristic but may appear normal in early stages	Distinguishes from other neurodegenerative disorders
DNA analysis	Expanded triplet repeats	Diagnostic for genetic status, but may not be explanation for current status

disorder such as HD when cognitive function may be impaired. Even in the absence of significant cognitive impairment, denial can be a major factor in such patients *(2)* and although a partner and the clinicians involved may recognize a problem, the individual at risk may not see it in the same way. Careful supportive medical care can be helpful in such situations *(3)*, but often these are very difficult areas to address.

Problems of an unresolved diagnosis are less frequent with the advent of molecular testing for the HD mutation, although all patients embarking on both diagnostic and presymptomatic testing should be warned of the very small risk of a result that may not be clearly interpretable. Within our own population a cut-off point of 34 repeats is usually taken as the upper limit of normal, although as evidence is accruing from other centers it is becoming clear that this is not universally applicable.

3. Presymptomatic Testing

3.1. Issues

As already mentioned, presymptomatic testing is normally clearly distinguishable from diagnostic testing in that there are no clinical features of the disorder at the time of the test. One aspect of many presymptomatic testing programs is a neurological examination; if early features of the disease are identified then the patient would be offered the opportunity to switch to a diagnostic protocol. It is important that candidates be made aware and consent to such an approach before they embark on presymptomatic testing *(4)*. An important feature of most presymptomatic testing evaluations has been the finding of abnormal clinical features in a significant proportion of applicants. If the individual is unaware of these, or if they are equivocal in nature, this may give special problems in counseling and may blur the distinction between diagnostic and presymptomatic testing.

The technological simplicity of the laboratory aspects of the presymptomatic test contrasts with the complexity of the decision to take the test and potential psychosocial sequelae of altering one's status from a 50% risk to no risk or to presymptomatic gene carrier. Research on other forms of genetic testing has shown that how a test is offered has a major effect on uptake *(5)*. Developing services need to consider how they offer a presymptomatic test, in order to ensure that individuals know of its availability, without feeling that it is being medically advised. There are a number of ethical and legal dilemmas associated with presymptomatic testing for adult-onset disorders *(6–8)*. The ethical principles of autonomy, beneficience, confidentiality, and justice have been discussed as playing a major role *(9)*, although the practical decisions in any individual situation can usually be resolved by a sensitive and experienced approach.

A major issue surrounding presymptomatic testing is the concept of informed consent *(10)*. In the case of presymptomatic testing, being "informed" requires exploration of such information as potential insurance and employment implications, an understanding of the medical and possible psychiatric implications related to the disease, and the psychological adaptation necessary in order to live with a result. Whether normal or abnormal the result will involve a change in self-perception, and therefore may be perceived as psychologically threatening until the individual achieves a new homeostasis *(11)*.

The aim of presymptomatic counseling is to foster insight and understanding that will help the individual in his or her decision-making process concerning taking the test and subsequent adjustment to the result. It is vital that participation in presymptomatic testing programs is the individual's decision at each stage of the process, and it is important that participants are aware that they can withdraw from the program at any stage. In the knowledge of this the various issues are explored and their possible coping strategies elucidated in the pretest interviews.

Research evidence is suggesting self-selection for presymptomatic testing to be the major factor influencing the perception of psychological variables and perception of ability to cope with the result *(12)*. This may change with time, different external pressures, and in different disease processes, where other variables may play a more major role in determining selection of testing.

A major issue surrounding the whole concept of presymptomatic testing is the concept of who should be accepted into such programs. It could be argued that anyone who can give informed consent should be allowed to enter a presymptomatic testing program, although such issues as the age of consent, the testing of children, those showing definite or equivocal signs, or those with a recent or current psychiatric condition, the testing of pregnant women, and the problems associated with testing people at 25% risk (and thereby generating a result on an intervening parent) all need to be addressed.

An essential part of the HD program is that the counseling about testing and the potential impact of a result allows individuals to consider their decision to be tested and to prepare for the result. This aim is hard to achieve if candidates perceive a need to convince the clinician that they will cope with a result in order to be allowed to proceed, a potential conflict that needs to be addressed in service delivery.

The impact of a presymptomatic result is the subject of much current research *(13–15)*. One result that surprised many was the evidence of adverse reactions to a reduced risk *(16)*, highlighting the adaptation demands previously mentioned contingent on the change in risk status, in whatever direction. The result of a presymptomatic test also has implications for the individual's family, particularly his or her partner, and will have a potential impact on relationships *(13,14)*. For this reason protocols have advocated the inclusion of the partner in the testing program.

3.2. Protocols

Original international and national guidelines *(17,18)* have been revised following the isolation of the HD mutation *(19)*. A number of authors have written concerning their agreed guidelines and clinical programs *(14,20–23)* providing useful information for professionals considering establishing a presymptomatic testing program. The protocol used in our center is based on the recommendations of the UK Predictive Testing Consortium for Huntington's Disease and is outlined here.

3.2.1. Presymptomatic Testing Protocol for HD

3.2.1.1. INCLUSION CRITERIA

1. Confirmed family history of HD.
2. At risk of HD.
3. Aged 18 yr or over.
4. Informed consent freely given.
5. Individuals who have previously received counseling.

3.2.1.2. POSTPONEMENT CRITERIA

1. Clinically affected with HD.
2. Significant risk of suicide.

3. Recent history of significant mental illness.
4. Recent history of drug misuse.
5. Recent major life events/stressors.

3.2.1.3. Preliminary Discussion Interview (Clinical Geneticist and Coworker)

1. Review and update of family and social circumstances.
2. Information booklet explaining the test and research protocol given to applicant.
3. Applicants asked to confirm in writing if they wish to proceed with testing.

The first interview, conducted by a coworker of the clinical geneticist, should consist of the following elements:

1. Collection of basic sociodemographic details. Include items on social and medical history.
2. Confirmation of family and clinical data (discussion of confirmation of mutation in closest affected relative).
3. Assessment of impact (personal, financial, and social) of HD on the applicant and, if appropriate, on the partner, and discussion of existing coping styles and resources.
4. Assessment of knowledge (knowledge questionnaire) of HD and presymptomatic testing. Includes history of learning about HD.
5. Assessment of motivation (motivation questionnaire) for requesting predictive testing and potential impact of results on applicant's life. To include discussion of the possibility that neurological examination may detect early symptoms.
6. Detailed explanation of how the test works, its possible outcomes, and its limitations.
7. Consideration of who the applicant will disclose the result to.
8. Neurological examination and Quantitative Neurological Examination (QNE).
9. First blood sample taken.

A second coworker-conducted interview will consist of these elements:

1. Further counseling and discussion of previously identified problem areas.
2. Explanation of possible results.
3. Review informal social support.
4. Nomination of professional supporter who will be willing to offer support if and when necessary.
5. Rehearsal of disclosure session and travel arrangements.
6. Standardized assessments:
   a. Beck Depression Inventory.
   b. General Health Questionnaire.
   c. State Trait Anxiety Inventory.
7. Signing of consent form. Copy given to applicant.
8. Blood sample.
9. Specific appointment made for third and fourth interview. Applicant reminded that he or she can withdraw from testing at any time.

The third interview, at which the coworker will disclose the test results, should consist of these elements:

1. Applicant to be accompanied by partner / close friend.
2. Disclosure of result by key worker.
3. Applicant is given the opportunity to discuss result with clinical geneticist.
4. Confirm fourth appointment.

The individual should be contacted by telephone 1 wk after disclosure, and clinical appointments should be made 1, 3, and 6 mo and 1 yr after disclosure. There should be a minimum of 2 mo between the first and second appointment, and 2 wk between the second appointment and disclosure.

3.3. Outstanding Problems

As already mentioned, the issues surrounding testing of those at 25% risk and the possible generation of results on intervening parents against their wishes may be difficult to resolve satisfactorily and will inevitably involve careful counseling and consideration in individual situations. The small chance of sample mix-up has been addressed in our laboratory by duplicate samples taken on separate occasions *(24)*. Other problems identified include presymptomatic testing in patients with various psychiatric disorders, in patients who are pregnant, and children *(25,26)*. It is always difficult to define completely the external pressures on individuals to have presymptomatic testing, although identification of such pressures is a major aim of the counseling sessions.

4. Prenatal Testing

4.1. Issues

The major aim of genetic testing in relation to pregnancy is the clarification of the genetic options available to the couple, before pregnancy occurs, if possible. The availability of psychological, medical, and procedural support must be made clear. Close liaison with the obstetric department and the patient's general practitioner is vital. When the first contact is for counseling and possible testing on an already established pregnancy, it may be extremely difficult to resolve the different issues in terms of testing the pregnancy and the parent at risk, as indicated below.

In conditions such as HD where mutation testing is now possible, the potential use of the less specific exclusion testing still needs consideration. In our experience there are situations in pregnancy where exclusion testing is preferable to the pregnant couple *(27)* because it does not alter the risk of the intervening parent. The options can be difficult to understand and the emotional overlay of anxiety about one's own risk, that of the current pregnancy, and time pressures only serve to make understanding and consideration harder.

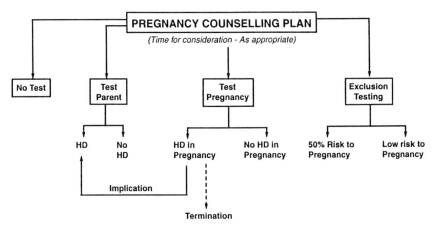

Fig. 1. Pregnancy counseling plan.

4.2. Protocols

Service protocols are harder to establish in pregnancy because of the limited time available in prenatal testing and the different problems of individual situations. The aim is to provide counseling that informs the individual / couple of the options and enables them to make the best decision for themselves in the circumstances. The approach is outlined in Fig. 1.

4.3. Outstanding Problems

A paper from our center has documented the experiences of, and issues raised by exclusion testing *(27)*. Experience in most centers with prenatal diagnosis remains limited and at times traumatic *(28)*. The clinician's view regarding the desirability of termination after a confirmatory diagnosis in the fetus may be at odds with the patient's decision. In such situations, the issues surrounding the clinician's and the patient's views regarding termination of an affected fetus have to be explored fully, and as yet this can still be an area of controversy. Obviously the major ethical implications of testing a fetus for such a late-onset condition remain extremely controversial for many people.

5. Family Management

By definition, services for genetic disorders need to acknowledge the familial context. This is important in relation to clinical management *(29)* and provision of genetic counseling. We have found it important to identify *who is the patient* when providing services and ensuring confidentiality. Our resolution has been to identify the referred patient and ensure that counseling and support systems are provided for that individual initially. It has not been our practice to actively contact other family members directly but to allow the individual to disseminate information among relatives. This practice extends to donation and use of blood samples in testing procedures for family members. We feel that it is important to obtain specific signed consent for research use and for use of blood samples for testing of family members.

It is, however, clinically invaluable, when providing a genetic service, to have access to clinical family information. Further to an ascertainment study *(30,31)*, a confidential register of affected and at-risk family members in Wales has been established. This is regularly updated and has an important role in the long-term management of HD in Wales.

Implicit in identifying and coordinating a support network for affected individuals and family members are close liaisons with other professional groups, for example, those referring patients, and those who may be the patient's local and first line of support and advice. Such individuals obviously include the primary care health team, psychiatrists, and neurologists.

6. Conclusions

It will be apparent that even within the "archetypal" disorder of HD and in an established service many questions remain unresolved. We hope that we have illustrated that despite this a framework can be created for approaching many of the difficult issues. There are many different views and attitudes toward the management and testing in late onset neurological disorders but we have attempted to convey some possible approaches to the clinical issues that we have experienced. Collective thought and discussion with colleagues and patient groups (such as the Huntington's Disease Association) about the areas that we recognize to be contentious is very valuable. As previously discussed, a consortium has been established in the United Kingdom *(17)* in order to effect such discussion concerning presymptomatic testing, and proves a very valuable resource for those providing a service. Not all the varied situations fit with the tentative framework we have suggested; individual situations can be discussed with such a group and broad consensus often reached regarding the optimum course of action.

References

1. Harper, P. S., ed. (1991) *Huntington's Disease.* Saunders, London.
2. Martindale, B. (1987) Huntington's chorea: some psychodynamics seen in those at risk and in the responses of the helping profession. *Br. J. Psychiat.* **150,** 319–332.
3. Bloch, M., Adam, S., Fuller, A., Kremer, B., Welch, J. P., Wiggins, S., Whyte, P., Huggins, M., Theilmann, J., and Hayden, M. R. (1993) Diagnosis of Huntington disease: a model for the stages of psychological response based on experience of a predictive testing program. *Am. J. Med. Genet.* **47,** 368–374.

4. Nance, M. A. and Ludowese, C. J. (1994) Diagnosis of Huntington's disease. Model for a predictive testing programme based on an understanding of the stages of psychological response. *Am. J. Med. Genet.* **52,** 118–119.

5. Bekker, H., Modell, M., Denniss, G., Silver, A., Mathew, C., Bobrow, M., and Marteau, T. (1993) Uptake of CF testing in primary care: supply push or demand pull? *Br. Med. J.* **306,** 1584–1586.

6. Lamport A. (1987) Presymptomatic testing for Huntington's chorea: ethical and legal issues. *Am. J. Med. Genet.* **26,** 307–314.

7. Ball, D. M. and Harper, P. S. (1992) Presymptomatic testing for late onset genetic disorders: lessons from Huntington's disease. *FASEB J.* **6,** 2818–2819.

8. Brandt, J. (1994) Ethical considerations in genetic testing: an empirical study of presymptomatic diagnosis of HD, in *Medicine and Moral Reasoning.* (Fukford, K. W. M., Gillett, G. R., and Soskice, J. M., eds.) Cambridge University Press, Cambridge, pp. 41–59.

9. Huggins, M., Bloch, M., Kanani, S., Quarrel, O. W. J., Theilman, J., Hedrick, A., Dickens, B., Lynch, A., and Hayden, M. (1990) Ethical and legal dilemmas arising during predictive testing for adult-onset disease: the experience of Huntington's disease. *Am. J. Hum. Genet.* **47,** 4–12.

10. Sharpe, N. F. (1994) Informed consent and Huntington's disease: a model for communication. *Am. J. Med. Genet.* **50,** 239–246.

11. Falek, A. (1977) Use of coping process to achieve psychological homeostasis in genetic counselling, in *Genetic Counselling* (Lubs, H. A. and De la Cruz, F., eds.), Raven, New York, pp. 179–192.

12. Codori, A. M., Hanson, R., and Brandt, J. (1994) Self selection in predictive testing for Huntington's disease. *Am. J. Med. Genet.* **54,** 167–173.

13. Codori, A. M. and Brandt, J. (1994) Psychological costs and benefits of predictive testing for Huntington's disease. *Am. J. Med. Genet.* **54,** 174–184.

14. Tibben, A., Vegter van der Vlis, M., Skraastad, M. I., Frets, P. G., Van der Kamp, J. J. P., Niermeijer, M. F., Vannomen, G. J. B., Roos, R. A. C., Rooijmans, H. G. M., Stronks, D., and Verhage, F. (1992) DNA testing for HD in the Netherlands: a retrospective study on psychosocial effects. *Am. J. Med. Genet.* **44,** 94–99.

15. Tibben, A., Duivenvoorden, H. J., Vegter van der Vlis, M., Niermeijer, M. F., Frets, P. G., Van der Kamp, J. J. P., Roos, R. A. C., Rooijmans, H. G. M., and Verhage, F. (1993) Presymptomatic DNA testing for Huntington's disease: identifying the need for psychological intervention. *Am. J. Med. Genet.* **48,** 137–144.

16. Bloch, M., Adams, S., Wiggins, S., Huggins, M., and Hayden, M. R. (1992) Predictive testing for Huntington's disease in Canada: the experience of those receiving an increased risk. *Am. J. Med. Genet.* **42,** 449–507.

17. Craufurd, D. and Tyler, A. (1992) Predictive testing for Huntington's disease: protocol of the U. K. Huntington's Prediction Consortium. *J. Med. Genet.* **29,** 915–918.

18. Went, L. (1990) Ethical issues policy statement on Huntington's disease's molecular genetics predictive test. *J. Med. Genet.* **27,** 34–38.

19. Went, L. (1994) Guidelines for the molecular genetic predictive test in Huntington's disease. *J. Med. Genet.* **31,** 555–559.

20. Bennett, R. L., Bird, T. D., and Teri, L. (1993) Offering predictive testing for Huntington disease in a medical genetics clinic: practical applications. *J. Genet. Counsel.* **2,** 123–137.

21. Quaid, K. A. (1992) Presymptomatic testing for Huntington's disease. Recommendations for counselling. *J. Genet. Counsel.* **1,** 227–301.

22. Benjamin, C. M., Adam, S., Wiggins, S., Theilmann, J. L., Copley, T. T., Bloch, M., Squitieri, F., McKellin, W., Cox, S., Brown, S. A., Kremer, H. P. H., et al. (1994) Proceed with care: direct presymptomatic testing for Huntington disease. *Am. J. Hum. Genet.* **55,** 606–617.

23. Demytteriaene, K., Evers-Kiebooms, G., and Decryenaene, M. (1992) Pitfalls in counselling for presymptomatic testing in Huntington's disease. *Birth Defects: Original Article Series* **28,** 105–111.

24. Lazarou, L. P., Meredith, A. L., Myring, J. M., Tyler, A., Morris, M., Ball, D. M., and Harper P. S. (1993) Huntington's disease: predictive testing and the molecular genetics laboratory *Clin. Genet.* **43,** 150–156.

25. Bloch, M. and Hayden, M. R. (1990) Opinion: presymptomatic testing for Huntington disease in childhood: challenges and implications. *Am. J. Hum. Genet.* **46,** 1–4.

26. Harper, P. S. and Clarke, A. (1990) Should we test children for "adult" genetic diseases? *Lancet* **335,** 1205–1206.

27. Tyler, A., Quarell, O. W. J., Lazarou, L. P., Meredith, A. L., and Harper P. S. (1990) Exclusion testing in pregnancy for Huntington's disease. *J. Med. Genet.* **27,** 488–499.

28. Tolmie, J. L., Davidson, H. R., May, H. M., McIntosh, K., Paterson, J. S., and Smith, B. (1995) The pre-natal exclusion test for Huntington's disease: experience in the west of Scotland, 1986–1993. *J. Med. Genet.* **32,** 97–101.

29. Martindale, B. and Bottomley, V. (1980) The management of families with Huntington's chorea: a case study to illustrate some recommendations. *J. Child Psychol. Psychiat.* **21,** 343–351.

30. Walker, D. A., Harper, P. S., Wells, C. E., Tyler, A., Davies, K., and Newcombe, R. G. (1981) Huntington's chorea in South Wales. A genetic and epidemiological study. *Clin. Genet.* **19,** 213–221.

31. Harper, P. S., Tyler, A., Smith, S., Jones, P., Newcombe, R. G., and McBroom, V. (1982) A genetic register for Huntington's chorea in South Wales. *J. Med. Genet.* **19,** 241–245.

12

Genotyping and Susceptibility of Sheep to Scrapie

Nora Hunter

1. Introduction

Scrapie, as a disease of sheep and goats, has been recognized for over 300 yr and was first documented in England in 1730 *(1)*. Sheep breeders petitioned the House of Commons of King George II in 1754 requesting the enforcing of regulations governing the sale and distribution of sheep. This is believed to have been prompted by a major scrapie epizootic. However, the petition was referred to "the consideration of a Committee" and vanished without a trace *(2)*—not so different from the present day!

Despite the disease being made notifiable in the European Community in January 1993, the true incidence of scrapie in Great Britain is unknown, partly because of the stigma attached to it. Prior to notification legislation, however, questionnaire surveys were carried out in the hope of establishing the frequency of scrapie cases. One report on two self-administered surveys suggested that one-third of British sheep flocks are infected with scrapie, with mean incidence figures of 0.5 and 1.1 scrapie cases/100 sheep/yr *(3)*. A second report of a postal survey of farmers who had requested Veterinary Investigation Centre (VIC) diagnosis to confirm scrapie in their sheep found median incidence rates ranging from 0.49–1.02 cases/100 sheep/yr but also discussed sources of bias in such studies *(4)*. The cost of VIC diagnosis may have meant that the farmers or the sheep were not typical. The animal may have been a recent expensive purchase and indeed, the proportion of rams among the VIC cases in the second survey was very much higher than that found in the general sheep population. The secrecy surrounding scrapie in Great Britain may lead, in random surveys, to misleading conclusions.

1.1. Clinical Signs

Clinical signs of natural scrapie in an outbreak currently occurring in the flock of Cheviots at the Neuropathogenesis Unit (NPU Cheviots, described in

From: *Methods in Molecular Medicine: Prion Diseases*
Edited by: H. Baker and R. M. Ridley Humana Press Inc., Totowa, NJ

Table 1
Incubation Periods in NPU Cheviot Sheep
Following Subcutaneous Injection with SSBP/1 Scrapie

Sip genotype	sAsA	sApA	pApA
PrP genotype	Val/Val_{136}	Val/Ala_{136}	Ala/Ala_{136}
Mean incubation period, d	167	315	Survive

the following) can last up to 3 or 4 mo, whereas following experimental challenge of the sheep, the course is much shorter—2–3 wk, and sometimes less. With natural scrapie, NPU Cheviots invariably show signs of a progressive pruritus, with hyperesthesia and incoordination of gait. Histopathology shows some variation in intensity of vacuolation of neuroanatomical regions but the brainstem is most frequently affected, e.g., dorsal vagus and thalamic nuclei. If vacuolation of the cerebellum occurs, it is a focal lesion and cortical vacuolation is not common. With experimentally induced disease in the NPU flock, ataxia always is present but pruritus is hardly ever seen. The distribution of vacuolation sometimes can vary depending on route of challenge and source of inoculum. However, following subcutaneous injection of the SSBP/1 source of scrapie, vacuolation of the diencephalon either is very mild or absent altogether *(5)*.

Detection of the abnormal disease-related form of the PrP protein (PrP[Sc]) either confirms the histopathology or is used alone to diagnose scrapie postmortem. Scrapie-associated fibril (SAF) detection and extraction of PrP protein from brain tissues, followed by immunoblotting to reveal the partial proteinase resistance that is a characteristic of PrP[Sc] are techniques covered in other chapters in this volume.

1.2. Sheep PrP Gene Variants and Association with Scrapie

In 1960, a flock of Cheviot sheep (NPU Cheviots) was founded and then selected into two lines based on susceptibility to scrapie challenge. Response to SSBP/1 scrapie (an A group scrapie isolate) in these sheep was controlled by a single gene, *Sip* (Scrapie Incubation Period) with two alleles, sA and pA *(6)*, standing for "short with A group scrapie" and "prolonged with A group scrapie," respectively. Positive line sheep, susceptible to subcutaneous (sc) injection with SSBP/1 are *Sip*[sAsA] or *Sip*[sApA], with the sA allele being dominant. Negative line sheep are *Sip*[pApA] and resist sc injection with SSBP/1. PrP codon 136 variants are tightly linked to *Sip* in that *Sip*[sAsA] animals encode valine (Val or V) and *Sip*[pApA] encode alanine (Ala or A) on both PrP alleles *(8)*. Heterozygous *Sip*[sApA] sheep therefore are Val/Ala_{136} (Table 1). *Sip* genotype groups in the NPU Cheviot flock now are divided into subgroups based on their encoding PrP variant genes and on incubation period groups *(5)*. Variants

Table 2
Sheep PrP Gene Variants

Sheep PrP gene codon polymorphisms			
Codon	Amino acid change	Breed	Ref.
112	Methionine (Met)		
	Threonine (Thr)	Ile de France	*8*
136	Alanine (Ala)		
	Valine (Val)	Many	*7*
141	Leucine (Leu)		
	Phenylalanine (Phe)	NPU Cheviots	*5*
154	Histidine (His)		
	Arginine (Arg)	Many	*7*
171	Glutamine (Gln)		
	Arginine (Arg)	Many	*5, 9, 10*
	Histidine (His)	Texel	
		Awassi	

Sheep protein variants				
Codon	112	136	141	154	171
	Met	Ala	Leu	Arg	Gln
	Met	Val	Leu	Arg	Gln
	Met	Ala	Leu	His	Gln
	Met	Ala	Leu	Arg	Arg
	Met	Ala	Leu	Arg	His
	Met	Ala	Phe	Arg	Gln
	Thr	Ala	Leu	Arg	Gln

at other codons are not linked to *Sip* but form subgroupings of the Ala$_{136}$ encoding PrP allele. Polymorphic codons frequently are discovered and those currently known are shown in Table 2 *(5,7–10)*. The codon variants discussed in this chapter are at codons **136** (Val and Ala), at codon **154**: arginine (Arg or R) and histidine (His or H) and at codon **171** arginine, glutamine (Gln or Q), and histidine.

The ranking of incubation periods with respect to *Sip* genotype seen with SSBP/1 (Table 1) is reversed partially with the C group scrapie source CH1641 that by interacerebral (ic) challenge gives shorter incubation periods in negative line animals than in positive line. There were also some survivors of ic CH1641 challenge in both lines, which was not fully understood until PrP genetics entirely explained the results by the association of incubation period with variation at codon 171 *(11)*. Both positive and negative line animals that

Table 3
Incubation Periods in NPU Cheviot Sheep
Following Intracerebral Injection with BSE

Sip genotype	pApA	sApA	sApA	pApA
PrP genotype	Ala/Ala$_{136}$	Val/Ala$_{136}$	Val/Ala$_{136}$	Ala/Ala$_{136}$
	Gln/Gln$_{171}$	Gln/Gln$_{171}$	Arg/Gln$_{171}$	Arg/Gln$_{171}$
Mean incubation period, d	680	800	Survive >1800	Survive >1800

are Gln/Gln$_{171}$ succumb to injection with CH1641, whereas Arg/Gln$_{171}$ and Arg/Arg$_{171}$ sheep have very much longer survival times, with some animals apparently resistant. This also is true with BSE when transmitted to these sheep (Table 3). The genotype at codon 136, so important with SSBP/1, is of minor relevance with BSE transmission in that Val$_{136}$ is associated with a prolongation of incubation period (Table 3), thus explaining the longer incubation periods in the positive line animals that do succumb to experimental challenge. Codon 136 and 171 variants also have association with incidence of natural scrapie. Most animals with scrapie encode Val$_{136}$ on at least one allele and are also Gln/Gln$_{171}$ (unpublished observations).

Codon 154 is also associated with scrapie incidence. In NPU Cheviots, some positive line animals are affected by the natural disease. All Val/Val$_{136}$ animals die at 700–900 d of age if allowed to live that long. A proportion of heterozygous animals succumb to natural scrapie at over 1100 d of age (those that are Val/Ala$_{136}$:Gln/Gln$_{171}$) but Val/Ala$_{136}$:Arg/Gln$_{171}$ animals survive. Some Val/Ala$_{136}$:Gln/Gln$_{171}$ sheep do survive, but this depends on the genotype at codon 154 in that His/Arg$_{154}$ animals survive, but Arg/Arg$_{154}$ sheep develop scrapie. It seems that Val$_{136}$ is necessary and sufficient for the development of natural scrapie in the homozygous genotype. Val$_{136}$ heterozygotes also succumb if homozygous at codons 154 and 171, but apparently are resistant if heterozygous at either codon 154 or 171. It is uncertain what this means in terms of disease etiology, but it is reminiscent of the human PrP genetics work, which has suggested that heterozygosity gives some protection against the development of Creutzfeldt-Jakob disease (CJD) *(12)*.

2. Methods of Genotyping Sheep

2.1. DNA Extraction

DNA can be extracted from any available tissue sample using methods detailed in ref. *14*. Usually it is most convenient to use blood because it does not have to be homogenized prior to extraction. (Blood samples should be collected in an anticoagulent, such as EDTA, but not heparin, because it inhibits

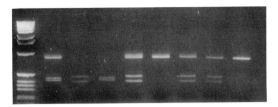

Fig. 1. Genotyping at codon 136 by *Bsp*HI digestion of sheep PrP gene PCR products. Tracks (1) marker; (2) Val/Ala$_{136}$:Arg/Arg$_{154}$; (3) Ala/Ala$_{136}$:His/His$_{154}$; (4) Val/Val$_{136}$:Arg/Arg$_{154}$; (5, 7, 8) Ala/Ala$_{136}$: His/Arg$_{154}$; (6, 9) Ala/Ala$_{136}$:Arg/Arg$_{154}$.

some of the enzymes used in genotyping analysis.) However, any other tissue (brain, liver, spleen, and so on), if removed immediately after death and used immediately or quickly frozen first, also is satisfactory. Large-scale DNA preparations are more reliable than those from commercial kits but are not always necessary.

2.2. Detection of Variants at Codons 136 and 154

Detection of the sequence differences that give rise to the codon variations is generally carried out on the product (amplimer) of polymerase chain reaction (PCR) amplification of the PrP gene. The amplimer can be analyzed in a number of different ways, for example, digestion with restriction enzymes, allele-specific oligonucleotide hybridization, direct sequencing, cloning and sequencing, or analysis on denaturing gradient gels. The polymorphisms at codons 136 and 154 both can be detected by digestion with the restriction enzyme *Bsp*HI (New England Biolabs, Beverly, MA) or its isoschisomers, since at both sites, one allele encodes the enzyme recognition sequence, whereas the other allele does not. One method for this analysis is detailed herein and a set of typical sheep genotype results are shown in Fig. 1.

Using the nucleotide numbering system in which the first base of the initiating methionine is numbered 1, two PCR primers were used as follows: oligo 314 (GGTGAAAAGCCACATAGGCAGTTGG) nucleotides 4–28 and oligo 315 (CTCACAGATGGACGTCGGGACATCA), which is complementary to nucleotides 826–850. Using any standard PCR method *(13)* the amplimer generated by PCR therefore will be 846 bp.

When using *Bsp*HI, it is sometimes necessary to purify the PCR product before digestion, although this is not always the case with every enzyme. This can be done by electrophoresing half of the PCR product (25 µL of a 50 µL reaction) on a 1% Tris-borate agarose gel *(14)* and collecting the amplimer on DEAE membrane (Schleicher and Schuell, Dassel, Germany) inserted into a cut in the gel just "in front" of the DNA fragment, eluting in 1.5M NaCl, 10 mM

Tris-HCl, 1 mM EDTA, pH 8.0 and ethanol precipitating prior to digestion with *Bsp*HI following the instructions of the manufacturer. At codon 136, the Val allele sequence is <u>GTC</u>ATGA, which includes the *Bsp*HI recognition site (TCATGA, cut after the first T leaving a four-base single-stranded overhang on each fragment). The Ala$_{136}$ sequence is <u>GCC</u>ATGA, which destroys the enzyme site. Successful restriction is likely to mean that the Val$_{136}$ allele is present. However, a failed restriction just means the site is no longer encoded, which could be the result of a mutation in any one of the six bases that form the restriction site. Because these alleles have been fully sequenced many times, the assumption is usually made that failed restriction means the genotype is Ala$_{136}$, but it may not be the case. Because of this, the genotypes are sometimes represented with Val/Val$_{136}$ as (+ +), Val/Ala$_{136}$ as (+ −), and Ala/Ala$_{136}$ as (− −).

At codon 154, the His allele has the sequence T<u>CA</u>TGA and is fully digested with *Bsp*HI, whereas the Arg$_{154}$ allele (T<u>CG</u>TGA) is not. The genotypes are described as (+) for His$_{154}$ and (−) for Arg$_{154}$ in a similar manner to codon 136. However, the codon identity is usually assumed.

Restriction with *Bsp*HI at Val$_{136}$ gives rise to a fragments of 403 and 443 bp and at His$_{154}$ generates fragments of 457 and 389 bp. The Val$_{136}$ allele is linked to Arg$_{154}$ and so the doubly digested amplimer is not seen. Following digestion, the DNAs are electrophoresed through a 1.8% Tris-borate agarose gel resulting in sharper ethidium-stained fragments than in a Tris-acetate gel *(14)* that might be used as an alternative. Examples of typical digestion products are given in Fig. 1. The use of different PCR priming oligonucleotides will result in differently sized amplimers and thus differently sized restriction digest products. It is a good idea to digest control samples of known genotype on the same gel both to assist in assessing unknown genotypes and to be sure that the restriction reaction has actually worked. Depending on the PCR primers used, the digestion products from the Val$_{136}$ and from His$_{154}$ alleles could be very similar in size, and it is sometimes necessary to use two controls: one Val/Val$_{136}$ and the other encoding His$_{154}$. The use of a sheep of genotype Val/Ala$_{136}$:Arg/His$_{154}$ will work well for some sizes of fragments, but not all.

2.3. Detection of Variants at Codon 171

For routine analysis, one PCR reaction can be split into two—one-half for codon 136 and 154 analysis and the other for codon 171 analysis. It is possible to divide one PCR reaction product even further and to carry out additional analyses with care.

Codon 171 variants cannot be detected by restriction analysis, but one method that has been used successfully is that of allele-specific differential hybridization. This can be carried out using dot-blots or slot-blots of the PCR product itself *(15)*. Taking the time to run the product on a 1% TBE gel and carrying out standard Southern analysis is worthwhile because every individual DNA does not produce the same amount of PCR product and there are times when it is really necessary to see how well the PCR has worked.

To carry out the procedure, a 1% Tris-borate agarose gel is made with two sets of wells, several centimeters apart (upper and lower wells), and 12.5 µL of each PCR (one quarter of the total) in loading buffer is loaded into each of a pair consisting of one upper well and one lower well. The aim is to get two "copies" of the PCR products on the gel. The DNA is electrophoresed into the gel to a distance of 1–2 cm and a Southern transfer *(14)* is carried out onto a nylon membrane that will allow reprobing if necessary. The membrane then is cut into two pieces, each with its "copy" of the transferred PCR amplimers, and each piece is hybridized to a different allele-specific (AS) radioactive probe.

Oligonucleotide hybridization with radioactively labeled AS oligonucleotides (oligos) *(16)* gives good hybridization (detected using X-ray film) of an oligo to its complementary sequence in homozygotes and heterozygotes. The oligos are 15-mers with the single base mismatch in the middle, i.e., at nucleotide number 7 or 8. There is little or no hybridization to a homozygote-mismatched sequence if the washing temperature and salt concentration (sequence-dependent) are optimized. This means that by making pairs of duplicated filters and using an oligo complementary to the Gln_{171} sequence to probe one "copy" and a second oligo complementary to the Arg_{171} allele to probe the second "copy," an animal can be genotyped in the following way. The Gln_{171} AS oligo will hybridize to Gln/Gln_{171} DNA and to Arg/Gln_{171} DNA but not to Arg/Arg_{171} DNA. Therefore, an animal with DNA that hybridizes with only one AS oligo is a homozygote and an animal with DNA hybridizing to both oligos is a heterozygote. It is not always possible to see the expected reduction in signal in heterozygotes compared with homozygotes. Carrying out this procedure with only one AS oligo is unsatisfactory because the meaning of a negative result will not be clear and using one membrane "copy" of the PCR products and reprobing with the second oligo is not ideal because of the difficulty of first removing the hybridization from the first oligo.

Nonradioactive labeling methods also give good results but are less flexible in terms of rewashing if, for some reason, the wash temperature is not exactly right the first time. The membrane washing temperature depends on the AT content of the oligo—the more As and Ts in the sequence, the lower the wash temperature. Examples of wash temperatures, e.g., with a 15-mer Gln_{171} AS oligo (5'TGGAT<u>CAG</u>TATAGTA, with 66% AT) washes are in 2X SSC or SSPE *(14)* at 37°C, whereas the Arg_{171} AS oligo (5'TGGAT<u>CGG</u>TATAGTA, with 60% AT) washes are at 42–45°C. This method does not work very well for the codon 136 position because of the sequence composition at this point. In this case, in the 15 bases around the single-base difference, the AT content is <50% and wash temperatures are elevated to 55°C or more. This is much more difficult to control and inconsistent results are obtained.

When working with pairs of oligos it is essential to be sure which membrane "copy" is hybridized with which AS oligo. This step can be controlled by making use of known homozygote DNA controls.

2.4. Problems Encountered in Genotyping

Problems encountered when using these techniques for genotyping are largely those with which any molecular biologist will be familiar. There are some real minefields, however, that relate to the intrinsic problems of PCR and the highly polymorphic nature of the sheep PrP gene. The human gene is also very polymorphic *(17)* although the cattle PrP gene is unusually invariant *(18)*. For PCR to work, it is necessary for oligo primers to bind to complementary sequences in the genomic DNA. If there happens to be an unsuspected sequence polymorphism in the region of the genomic DNA covered by the oligo primer, the PCR may not work properly and on heterozygote DNA may amplify only one of the alleles. This will give a misleading result in subsequent digestion or sequencing. Scientists all recognize the necessity of repeating results for confirmation, but with the sheep PrP gene, repeat PCRs should be carried out using priming oligo pairs from different parts of the sequence in order to minimize the risks of hidden mutations resulting in incorrect genotyping. There are many anecdotal stories of such problems.

Hidden sequence differences in the middle of the PCR amplimer can also lead to spurious results. One of these has already been mentioned. Digestion with *Bsp*HI at codons 136 and 154 will identify correctly the codon present. Failure to digest may result from any sequence change that alters the restriction site and may not mean that the expected alternate codon is present. If possible, the genotypes of a number of animals should be checked by sequencing.

Differential oligohybridization also has its pitfalls. For example, there are three codon 171 variants, Gln_{171} and Arg_{171}, which are described in the preceding but there is also an allele that encodes His_{171} (ref. *10*; and Hunter, unpublished). One of the discoveries of this polymorphism came about because of very low levels of hybridization of the Gln_{171} and Arg_{171} oligos to a normal amount of PCR product as seen on ethidium-stained gels. Sequencing revealed the unexpected finding of a third codon variant at this position. Screening by use of a His_{171} AS oligo (5'TGGATCATTATAGTA, 73% AT content, washes at room temperature) works well and an example of this oligo used in conjunction with Gln_{171} and Arg_{171} on three "copies" of a PCR reaction is shown in Fig. 2.

Finally, there is the perennial problem of contamination in the PCR reactions. This possibility must always be remembered and controlled. The best defense against contamination as a source of error is obsessive paranoia and

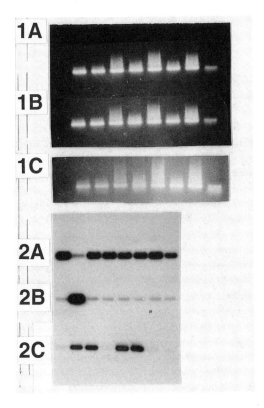

Fig. 2. Genotyping at codon 171 by AS oligo hybridization. (1) Eight sheep PrP gene PCR amplimers, divided into three, on 1% TBE gels. (2) Filters made from **(A)** probed with (A) Gln_{171}; **(B)** Arg_{171}; **(C)** His_{171}. Genotypes are (left to right) Gln/Gln, Arg/His, Gln/His, Gln/Gln, Gln/His, Gln/His, Gln/Gln, Gln/Gln.

suspicion! Control PCR reactions containing all constituents except genomic DNA should reveal any problems.

3. Use of Genotype Information

Once the genotype of an animal is assigned, what use can be made of it? There are breed differences in genotypes of animals affected by natural scrapie. Some breeds, for example Suffolks, do not encode PrP Val_{136} alleles and yet they do contract scrapie *(15,19,20)*. In Suffolks, scrapie occurs in animals that are Gln/Gln_{171}. These animals are also Ala/Ala_{136}, a genotype that in Cheviots, Swaledales, Shetlands, and many other breeds would be expected to be resistant to natural scrapie *(20)*. It is, therefore, important to know the breed of any tested sheep and whether it is Suffolk-like or Cheviot-like before deciding it is scrapie resistant.

Breeding sheep for resistance to scrapie, by elimination of the $Val_{136}:Gln_{171}$ PrP allele, may now be possible. However, because the etiology of scrapie is unknown, it is not clear that this is a good idea in the long term. Scrapie may be the result simply of a faulty gene—a genetic disease or trait—in which case breeding to reduce the frequency of disease-associated alleles of the PrP gene should eradicate the disease. If scrapie is an infectious agent, however, then breeding for resistance to the common natural strains may simply allow selection of rare scrapie mutant strains that could infect the so-called resistant sheep.

Despite this uncertainty, the association of sheep PrP genotype with scrapie susceptibility is now so well understood that the answer to the genetic disease vs genetic susceptibility debate is within reach at last.

References

1. Parry, H. B. (1983) *Scrapie.* Academic, London.
2. Journals of the House of Commons (1754).
3. Morgan, K. L., Nicholas, K., Glover, M. J., and Hall, A. P. (1990) A questionnaire survey of the prevalence of scrapie in sheep in Britain. *Vet. Rec.* **127,** 373–376.
4. Wooldridge, M. J. A., Hoinville, L. J., and Wilesmith, J. W. (1992) A scrapie survey by postal questionnaire: aims, problems and results. *Proc. Soc. Vet. Epid. Prev. Med.,* **1–3 April 1992,** pp. 78–89.
5. Hunter, N., Foster, J. D., Goldmann, W., Stear, M., Hope, J., and Bostock, C. (1995) Natural scrapie in a closed flock of Cheviot sheep occurs only in specific PrP genotypes, in preparation.
6. Dickinson, A. G. and Outram, G. W (1988) Genetic aspects of unconventional virus infections: the basis of the virino hypothesis, in *Novel Infectious Agents and the Central Nervous System* (Bock, J. and Marsh, J., eds.), Ciba Foundation Symposium vol. 135. Wiley-Interscience, London, pp. 63–83.
7. Goldmann, W., Hunter, N., Benson, G., Foster, J. D., and Hope, J. (1991) Different scrapie-associated fibril proteins (PrP) are encoded by lines of sheep selected for different alleles of the *Sip* gene. *J. Gen. Virol.* **72,** 2411–2417.
8. Laplanche, J. L., Chatelain, J., Westaway, D., Thomas, S., Dussaucy, M., Brugere-Picoux, J., et al. (1993) PrP polymorphisms associated with natural scrapie discovered by denaturing gradient gel electrophoresis. *Genomics* **15,** 30–37.
9. Goldmann, W., Hunter, N., Foster, J. D., Salbaum, J. M., Beyreuther, K., and Hope, J. (1990) Two alleles of a neural protein gene linked to scrapie in sheep. *Proc. Natl. Acad. Sci. USA* **87,** 2476–2480.
10. Belt, P. B. G. M., Muileman, I. H., Schreuder, B., Bos-deRuijter, J., Gielkens, A. L. J., and Smits, M. A. (1995) Identification of five allelic variants of the sheep PrP gene and their association with natural scrapie. *J. Gen. Virol.* **76,** 509–517.
11. Goldmann, W., Hunter, N., Smith, G., Foster, J. D., and Hope, J. (1994) PrP genotype and agent effects in scrapie: change in allelic interaction with different isolates of agent in sheep, a natural host of scrapie. *J. Gen. Virol.* **75,** 989–995.

12. Collinge, J., Palmer, M. S., and Dryden, A. J. (1991) Genetic predisposition to iatrogenic Creutzfeldt-Jakob disease. *Lancet* **337,** 1441–1442.

13. McPherson, M. J., Hames, B. D., and Taylor, G. R. (1995) *PCR2: A Practical Approach.* IRL Press at Oxford University Press, Oxford.

14. Sambrook, J., Fritsch, E. F., and Maniatis, T. (1989) *Molecular Cloning: A Laboratory Manual.* Cold Spring Harbor Laboratory, Cold Spring Harbor, NY.

15. Westaway, D., Zuliani, V., Mirenda-Cooper, C., Da Costa, M., Neuman, S., Jenny, A. L., et al. (1994) Homozygosity for prion protein alleles encoding glutamine-171 renders sheep susceptible to natural scrapie. *Gen. Dev.* **8,** 959–969.

16. Deslys, J. P., Marce, D., and Dormont, D. (1994) Similar genetic susceptibility in iatrogenic and sporadic Creutzfeldt-Jakob disease. *J. Gen. Virol.* **75,** 23–27.

17. Gajdusek, D. C. (1994) Spontaneous generation of infectious nucleating amyloids in the transmissible and nontransmissible cerebral amyloidoses. *Mol. Neurobiol.* **8,** 1–13.

18. Hunter, N., Goldmann, W., Smith, G., and Hope, J. (1994) Frequencies of PrP gene variants in BSE affected and healthy cattle in Scotland. *Vet. Rec.* **135,** 400–403.

19. Laplanche, J.-L., Chatelain, J., Beaudry, P., Dussaucy, M., Bounneau, C., and Launay, J.-M. (1993) French autochthonous scrapied sheep without the 136Val PrP polymorphism. *Mam. Gen.* **4,** 463–464.

20. Hunter, N., Goldmann, W., Smith, G., and Hope, J. (1994) The association of a codon 136 PrP gene variant with the occurrence of natural scrapie. *Arch. Virol.* **137,** 171–177.

13

Strain Typing Studies of Scrapie and BSE

Moira E. Bruce

1. Introduction

The basis of strain variation in scrapie and other transmissible spongiform encephalopathies is a crucial issue in the ongoing debate about the nature of the infectious agent. The clear evidence for the existence of multiple strains leads us to conclude that these agents carry some form of strain-specific information that determines the characteristics of the disease. Any valid molecular model must, therefore, include an informational component that can be replicated in the infected host. Further, the behavior of strains when serially passaged in different host species or genotypes limits the range of possible models for the informational component of the agent. In this chapter the methods used for agent strain discrimination are described. The strategies used for the isolation and passaging of different strains are presented and the implications of the results of these simple experiments are discussed.

2. Agent Strain Discrimination
2.1. Disease Characteristics in Mice

Most research into strain variation in the spongiform encephalopathies has been carried out in mouse models of these diseases. From long-term studies of mice infected with many different isolates of scrapie, started by Alan Dickinson in the late 1960s, it has become clear that there are numerous strains of agent that produce distinct patterns of disease in the infected host (1,2). Because there are no serological tests and the molecular basis of strain variation is not yet known, strain discrimination relies on simple measurements of disease characteristics. The major criteria are the incubation periods produced in mice of defined genotypes and the severity and distribution of pathological

From: Methods in Molecular Medicine: Prion Diseases
Edited by: H. Baker and R. M. Ridley Humana Press Inc., Totowa, NJ

changes seen in the brains of these mice. Using these criteria, about 20 pheno-typically distinct strains of scrapie and BSE have been isolated by serial pas-sage in mice.

2.2. Incubation Period Measurement

The incubation period is the interval between initial infection and a standard clinical endpoint *(3)*. For strain-typing studies of mouse-passaged isolates at the Neuropathogenesis Unit in Edinburgh, mice are injected intracerebrally with 20 μL of a 1% homogenate of brain from a clinically affected donor. Injected mice are coded and scored weekly for neurological signs, when they are classified as "unaffected," "possibly affected" or "definitely affected." The endpoint is the date on which the animal receives an unambiguous score of "definitely affected." Slightly different methods of clinical assessment are used in other laboratories.

For people outside the field, it is often difficult to appreciate the precision of the incubation period measurement. Inoculum containing a high dose of a single scrapie strain, injected intracerebrally into genetically uniform mice, will generally give a very tight grouping of incubation periods, with standard errors of <2% of the mean. For a single scrapie strain, the incubation period also is remarkably repeatable using inocula prepared from different brains and remains constant over many serial passages in a single mouse genotype.

2.3. Effect of the Sinc Gene on Incubation Period

The host *Sinc* gene (short for *scrapie incubation*) exerts a major influence on the incubation period of experimental scrapie and related diseases in mice. The action of the *Sinc* gene was first recognized in mice infected with the ME7 strain of scrapie *(3)*. Two alleles of this gene were identified, designated s7 and p7, which gave, respectively, *s*hort or *p*rolonged incubation periods with ME7. Later it was shown that the *Sinc* gene controls the incubation period of all other strains of agent *(1,4)*. It is now clear that the *Sinc* gene encodes PrP; $Sinc^{s7}$ and $Sinc^{p7}$ mice differ consistently in the sequence of the protein, by two amino acids *(5,6)*.

For routine agent strain typing, an isolate is injected intracerebrally into groups of C57BL ($Sinc^{s7}$), VM ($Sinc^{p7}$), and C57BL × VM F_1 ($Sinc^{s7p7}$) mice. Each scrapie or BSE strain has a characteristic and highly reproducible pattern of incubation periods in these three mouse genotypes *(1,2)* (*see* Fig. 1). Agent strains differ:

1. In their incubation periods within a single *Sinc* genotype;
2. In their relative incubation periods in $Sinc^{s7}$ and $Sinc^{p7}$ homozygotes; and
3. In the apparent dominance characteristics shown by the two alleles in the F_1 heterozygote.

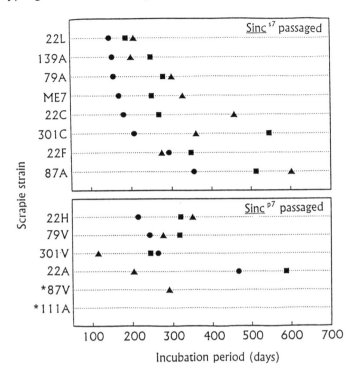

Fig. 1. Incubation periods following ic injection of 1% brain homogenates for 14 scrapie/BSE strains in C57BL (*Sinc*ˢ⁷) (●), VM (*Sinc*ᵖ⁷) (▲), and C57BL × VM (*Sinc*ˢ⁷ᵖ⁷) (■) mice. *The incubation periods for C57BL and C57BL × VM mice with 87V and for all three genotypes with 111A are longer than 700 d and some individuals do not develop clinical disease within their lifespan.

Thus, some agent strains produce shorter incubation periods in *Sinc*ˢ⁷ mice than in *Sinc*ᵖ⁷ mice, but this ranking is reversed for other strains. Depending on the agent strain, the incubation period in the F_1 heterozygote lies either between or beyond those in the two parental genotypes. No agent strain has been identified with an incubation period in the heterozygote that is shorter than in both parents.

2.4. Neuropathological Differences: Lesion Profiles

Scrapie and BSE strains also can be distinguished on the basis of the pathological changes they produce in the brains of infected mice. The most obvious change seen in routine histological sections is a vacuolation of the neuropil. Agent strains show dramatic and reproducible differences in the severity and distribution of this vacuolar degeneration in the brains of genetically uniform mice. This forms the basis of a semiquantitative method of strain discrimina-

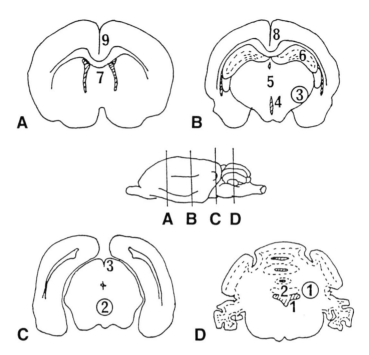

Fig. 2. Standard brain levels and areas in which vacuolation is scored to produce the lesion profile. Circled numbers indicate white matter areas.

tion in which the severity of pathology is scored from coded sections in nine gray matter and three white matter brain areas *(7)* (*see* Fig. 2). Gray matter vacuolation is scored on a scale of 0–5 and white matter vacuolation is scored on a scale of 0–3. The vacuolation score has been shown to bear an approximately logarithmic relationship to the number of vacuoles in a standard field *(7)*. For each group of mice the mean vacuolation score in each area is calculated to construct the "lesion profile."

Each combination of agent strain and mouse genotype has a characteristic lesion profile *(1)* (*see* Fig. 3). Unlike the incubation period, the lesion profile is insensitive to the initial infecting dose and can be used to identify strains in samples containing low levels of infectivity. This method can also be used to determine which strain kills the mouse following injection with a mixture of strains.

The distribution of vacuolar degeneration correlates closely with the distribution of abnormal PrP accumulation in the neuropil, demonstrated by immunostaining with PrP-specific antisera *(8,9)*. These studies have demonstrated clearly that pathological changes are targeted precisely to particular groups

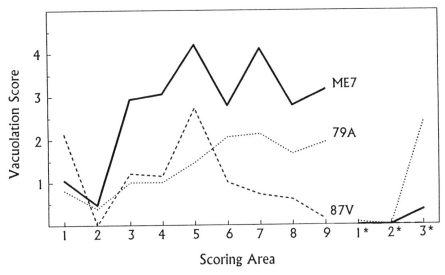

Fig. 3. Examples of lesion profiles in VM mice for three strains of scrapie. One to nine are gray matter areas, 1*–3* are white matter areas.

of neurons and that different strains of agent select different neuronal populations.

3. Isolation of Strains of Spongiform Encephalopathies in Mice

3.1. Primary Transmissions to Mice

The strain-typing work at the Neuropathogenesis Unit has involved transmissions to mice from a wide range of natural and experimental cases of spongiform encephalopathy in sheep, goats, cattle, and a number of other species. Because these diseases are more difficult to transmit between species than within species (the "species barrier" effect) *(10)*, the inoculum is prepared from a 10% rather than a 1% brain homogenate. Also, it has been found that intraperitoneal injection is more effective than intracerebral injection in some primary transmissions *(11)*. In order to maximize the chances of a successful transmission, mice therefore are injected by a combination of routes: 20 µL intracerebrally and 100 µL intraperitoneally. A larger panel of mouse strains is usually included in primary transmission experiments than in subsequent mouse-to-mouse passages because genes other than *Sinc* may have a considerable effect on incubation period at this stage *(11) (see the following)*. In general, the incubation periods in these experiments are very long and there may be survivors. Some sources fail completely to transmit to mice of any genotype.

3.2. Serial Passage in Mice

Following a successful primary transmission, serial mouse-to-mouse passage lines are set up routinely in C57BL ($Sinc^{s7}$) and VM ($Sinc^{p7}$) mice, using 1% brain homogenates from clinically affected individuals. At each passage the incubation period and pathological characteristics are tested in mice of the three $Sinc$ genotypes. On passage in a single inbred mouse strain, the incubation period usually shortens and stabilizes after a few passages to give a strain with characteristic properties. These properties appear to be stable indefinitely on further mouse-to-mouse passage, as long as the conditions of passage, particularly the mouse genotype, remain constant. In Edinburgh alone, 14 unequivocally distinct strains of scrapie and BSE have been isolated in mice (see Fig. 1). A further six isolates have unique disease characteristics, indicative of new strains, but are not yet fully stable; there is no indication that we have reached the limit of this variation. It should be stressed that an agent strain is defined in terms of its characteristics, rather than its passage history. Thus, the same agent strain may be identified in a number of different isolates.

Most primary sources give two different agent strains when passaged in $Sinc^{s7}$ and $Sinc^{p7}$ mice. These differences cannot be imposed by the host: A number of different strains have been isolated in each mouse genotype, and the same strain (e.g., ME7) sometimes has been isolated in both genotypes. However, the resolution of isolates into two distinct stable strains always is consistent with the selection of strains that replicate more efficiently in the mouse genotype used for passage. For example, 301C and 301V were isolated from BSE, by passage in $Sinc^{s7}$ and $Sinc^{p7}$ mice, respectively; 301C has a shorter incubation period than 301V in $Sinc^{s7}$ mice and 301V has a shorter incubation period than 301C in $Sinc^{p7}$ mice (11) (see Fig. 1).

3.3. Biological Cloning of Strains

Even when the disease characteristics have stabilized, it cannot be assumed that the isolate contains only a single agent strain. There is clear evidence that minor strains can be passaged together with the major strain as a stable mixture (2). For many experimental purposes this may not be important. However, for some studies, particularly those concerned with the stability of strains when passaged in different hosts, the presence of minor strains may become a serious complicating factor. To remove minor strains, it is necessary biologically to clone the isolate, by performing three or more sequential passages using the minimum infecting dose. This is achieved by injecting groups of mice with serial tenfold dilutions of brain homogenate and passaging from an affected individual injected with the highest dilution of inoculum to produce clinical cases. For most agent strains, this limiting dilution is approx 10^{-7}. This procedure has been shown to lead to a permanent change in the characteristics or behavior of several isolates.

4. Stability of Mouse-Passaged Strains: Detection of "Mutant" Strains

4.1. Stability on Passage in a Single Mouse Genotype

All but one of the scrapie and BSE strains listed in Fig. 1 appear to be stable indefinitely when serially passaged at high dose in the mouse genotype in which they were originally isolated. Seven of these strains (ME7, 22A, 22C, 22L, 79A, 87V, and 139A) have maintained their characteristics throughout ten or more serial passages in the appropriate mouse genotype. The one exception is 87A. The characteristics of 87A are stable when it is passaged at low dose in $Sinc^{s7}$ mice, but often change suddenly when it is passaged at higher dose in the same mouse genotype, to give a strain with much shorter incubation periods *(12)*. This new strain is stable, even when passaged at high dose, and always is identical to ME7.

87A and ME7 produce very different patterns of pathology in the brain and it is possible to classify each individual mouse on the basis of this pathology. In experiments using 1 or 10% 87A brain homogenates there is a clear bimodal distribution of incubation periods, with the ME7 type of pathology associated with the shorter group *(12)*. The changeover point is at approx 300 d, with little or no overlap between the two groups. The proportion of mice in the shorter group decreases with dilution of the inoculum, suggesting that the starting homogenate contains a mixture of the two strains. Analysis of a large number of experiments using different small brain samples to prepare the inocula has shown that the ratio of ME7 to 87A is, on average, approx $1:10^4$.

The 87A isolate contains ME7 as a minor component, even after it has been cloned. This suggests that ME7 is a shorter incubation period variant strain, derived from 87A by a process analogous to mutation in conventional microorganisms. 87A has been isolated independently from six different natural sheep scrapie sources and in each case has behaved in the same way. This is further evidence that 87A and ME7 are closely related at the molecular level.

4.2. Stability on Changing the Mouse Genotype Used for Passage

From the preceding observations, it is clear that scrapie-like agents carry some form of strain-specific information that is independent of the host. Both the "prion" and "virino" hypotheses for agent structure include the host component, PrP. Therefore, an important question is whether the PrP genotype of the host in which the isolate is passaged directly influences strain characteristics. This question has been addressed by changing the *Sinc* or PrP genotype of mouse used for serial passage, for several well-characterized scrapie strains.

Table 1
Summary of Experiments in Which Well-Characterized Scrapie Strains
Were Passaged Through Different *Sinc* Genotypes of Mouse

Strains originally isolated in *Sinc*^{s7} mice					
Uncloned 22C	→	*Sinc*^{p7} mice	→	Gradual change to 22H	
Cloned 22C	→	*Sinc*^{p7} mice	→	Unchanged 22C	
Cloned ME7	→	*Sinc*^{p7} mice	→	Unchanged ME7	
Uncloned 139A	→	*Sinc*^{p7} mice	→	Unchanged 139A	
Strains originally isolated in *Sinc*^{p7} mice					
Uncloned 22A	→	*Sinc*^{s7} mice	→	Gradual change to 22F	
Cloned 22A	→	*Sinc*^{s7} mice	→	Gradual change to 22F	

At each passage in the new host, the strain characteristics have been tested in the three *Sinc* genotypes. Scrapie strains have been found to differ in the stability of their properties when passaged in this way (*see* Table 1).

The characteristics of some scrapie strains (cloned ME7, cloned 22C, and uncloned 139A) have been completely unchanged by passage in the alternate mouse genotype. Other isolates (uncloned 22C, cloned or uncloned 22A) have changed. This type of experiment must be interpreted carefully to distinguish between the three possible explanations for a change in properties:

1. The selection of a minor agent strain that was present already in the isolate;
2. The selection of a variant strain, derived by "mutation" from the parental strain; and
3. A direct modification of strain determinants by the host.

When such changes occur using uncloned isolates it is not possible to determine their basis, unless other information is available. For example, when uncloned 22C is passaged in *Sinc*^{p7} mice its properties change gradually over several passages to give a new strain, 22H *(2)*. This can be interpreted only with the knowledge that this change is not seen in the equivalent experiment starting with cloned 22C *(2)*. The conclusion is that 22C and 22H coexisted in the early mouse passages of the isolate and that 22H was removed from the mixture by cloning. These results emphasize the need to use cloned strains in studies seeking host modifications of strain characteristics.

On the other hand, when cloned 22A is serially passaged in *Sinc*^{s7} mice, the disease characteristics gradually change over several passages, eventually stabilizing to give the new strain, 22F *(2)*. This suggests that 22F has been generated from 22A, either by a host-induced modification or by a "mutation" in the informational component of the agent. The fact that the change is gradual makes it unlikely that the phenomenon is simply owing to a modification of the agent by the host. The results are more consistent with the gradual selection of a mutant strain, 22F, which has a shorter incubation period in the new passaging genotype than the parental strain, 22A. So far, there is no clear experimental evidence that the *Sinc* or PrP genotype of the mouse can actively modify the characteristics of any strain of agent.

4.3. Stability on Changing the Species Used for Passage

Strain characteristics can also be maintained on passage through another species with a different PrP amino acid sequence. Kimberlin and coworkers have serially passaged a number of mouse scrapie strains in rats or hamsters and then repassaged the isolates in mice *(13,14)*. They found that cloned 22A and cloned ME7 were completely unchanged after serial passage in hamsters and subsequent reisolation in mice. Cloned 139A was unchanged by passage through rats. In contrast, the properties of cloned 139A and cloned 22C were changed permanently by passage in hamsters, giving rise to new strains that were stable on serial passage in mice. The latter results have been interpreted as the selection of strains in the new host species, other than the major strains present in the original host. When this happens, it is likely to contribute to the "species barrier" effect, i.e., the relatively long incubation period seen on primary transmission to a new species, compared with the shorter incubation periods seen on subsequent passage within this new species.

5. Epidemiological Applications of Strain Typing

5.1. Strain Variation in Natural Scrapie

Because of the possibility of the selection of minor variant strains, it is not clear to what extent the mouse-passaged strains isolated from natural scrapie cases are representative of field strains. However, strain typing in mice can give some information about the extent of strain variation in the natural disease. In primary transmissions to mice, the success rate, incubation periods, and pathology have varied enormously between sources. However, on further serial passage in mice most UK isolates have given varying combinations of the same three strains, 87A and ME7 in $Sinc^{s7}$ mice and 87V in $Sinc^{p7}$ mice. Therefore, these sources may not be as diverse as they appear from the primary transmission results. A series of transmissions from Icelandic sheep *(15)* have given at least three strains in mice; these are not yet fully characterized but clearly differ from strains from UK cases. In the United States, the mouse-passaged strains isolated from five natural scrapie sheep have been found to differ between sources *(16)*.

5.2. Strain Typing of BSE Isolates

BSE has been transmitted to mice from a series of unrelated cattle sources, collected at different times during the epidemic, from widely separated loca-

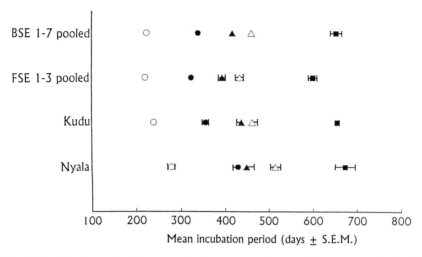

Fig. 4. Incubation periods in a panel of mouse strains on primary transmission from seven cattle with BSE (pooled data), three cats with feline spongiform encephalopathy (FSE) (pooled data), one greater kudu, and one nyala. The mouse strains are RIII (*Sinc*^{s7}) (O), C57BL (*Sinc*^{s7}) (●), VM (*Sinc*^{p7}) (▲), IM (*Sinc*^{p7}) (△), and C57BL × VM (*Sinc*^{s7p7}) (■).

tions within the United Kingdom *(11,17)*. Seven such transmissions have been completed. The results of these experiments were remarkably similar to each other and differed from those in all previous and contemporary transmissions of scrapie. All seven BSE sources have produced a characteristic pattern of incubation periods and pathology in a standard panel of inbred mouse strains and crosses (*see* Figs. 4 and 5). There are large and consistent differences in incubation period between mouse strains of different *Sinc* genotypes and also, surprisingly, between mouse strains of the same *Sinc* genotype (e.g., C57BL and RIII). Further passages in *Sinc*^{s7} and *Sinc*^{p7} mice have produced two strains, 301C and 301V, that differ from all strains derived from sheep or goat scrapie *(11)* (*see* Fig. 1).

The uniformity of the BSE transmissions shows that each cow was infected with the same major strain of agent. The consistency of the pathology reported in cattle with BSE also suggests the involvement of a single or a limited number of strains *(18)*. The major strain of BSE in cattle appears to be different from the strains causing scrapie in sheep. This does not necessarily undermine the widely held assumption that BSE originated from rendered sheep scrapie offal in feed; a possible explanation is that the high temperatures involved in rendering and subsequent passage through cattle have selected variant strains from sheep scrapie.

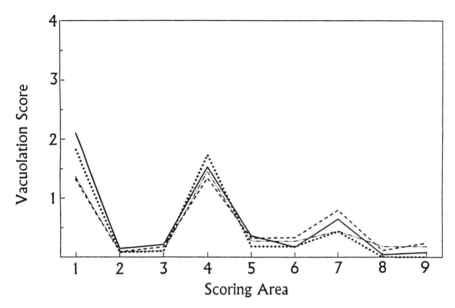

Fig. 5. Lesion profiles in RIII mice in transmissions from cattle, cats, greater kudu, and nyala, using data from the experiments shown in Fig. 4. No white matter vacuolation was seen in any of these experiments.

5.3. Strain Typing
of Novel Spongiform Encephalopathies from Other Species

Recently, transmissions to mice have been achieved from other species with novel spongiform encephalopathies, suspected to be related to the BSE epidemic; the sources were three domestic cats, a greater kudu, and a nyala. The results of all five transmissions were strikingly similar to results from cattle sources, indicating a common source of infection among these species *(11,19)* (Figs. 4 and 5). BSE from cattle also has been transmitted experimentally to sheep, goats, and pigs and then from each of these species to mice; again, the results were similar to direct transmissions of BSE *(11)*. These studies show, first, that the BSE agent is unchanged when passaged through a range of species and, second, that the donor species has little specific influence on the disease characteristics of BSE on transmission to mice.

It is clear from the preceding that transmission and strain typing studies can be used to answer epidemiological questions, for example, to establish links between the natural diseases in different species or in different countries. One obvious application would be to test any future suspicion that BSE has spread to another species, for example, to humans. However, it should be stressed that the methods are cumbersome and slow and that such studies can be undertaken only on a limited scale. Also, this approach can confirm but not refute a suspected link, because there is always the possibility that passage through a new species has selected a variant strain.

6. The Molecular Basis of Strain Variation

The molecular nature of the agent is still a matter for speculation. There are three main hypothetical models:

1. A "prion" *(20)*, consisting solely of modified forms of the host protein, PrP;
2. A "virino" *(21)*, consisting of an infection-specific informational molecule (possibly a nucleic acid) that is closely associated with and protected by abnormal PrP; and
3. An unusual but conventional virus *(22)*.

The existence of multiple strains of agent dictates that any proposed structure should have a replicable informational component. Furthermore, the stability of strains on passage through different host genotypes or species sets a number of conditions for the validity of any particular model. If the agent contains its own nucleic acid, strain variation would be analogous to that seen in conventional microorganisms. It is more difficult to envisage how a protein alone could specify strain diversity.

According to the "prion" hypothesis, the pathogen is PrP, which has been modified in some specific way *(23)*. This abnormal protein is suggested to induce the same modification in new host PrP molecules, by direct interaction. The modification currently is thought to be conformational *(24)*, with specific conformations determining strain properties. There is recent evidence from in vitro studies of two hamster-passaged strains of transmissible mink encephalopathy that such a model might be feasible *(25)*.

If strain specificity resides solely in PrP structure, there must be as many specific modifications as there are distinct strains and each must be able to "replicate" itself accurately over many serial passages, apart from predictably generating other specific modifications. Multiple forms of modified PrP must be capable of retaining their separate identities when passaged as mixtures and such mixtures must be resolvable by biological cloning. Furthermore, it must be possible to reproduce strain-specific modifications faithfully in PrP molecules with different amino acid sequences. It remains to be seen whether PrP alone can fulfill all these criteria, or whether a separate informational molecule is required, such as a nucleic acid.

Acknowledgments

The author would like to thank Irene McConnell, Patricia McBride, and their staff for all their hard work over the years, and Aileen Chree and Laurence Doughty for help in preparing the manuscript.

References

1. Bruce, M. E., McConnell, I., Fraser, H., and Dickinson, A. G. (1991) The disease characteristics of different strains of scrapie in *Sinc* congenic mouse lines: implications for the nature of the agent and host control of pathogenesis. *J. Gen. Virol.* **72,** 595–603.
2. Bruce, M. E., Fraser, H., McBride, P. A., Scott, J. R., and Dickinson, A. G. (1992) The basis of strain variation in scrapie, in *Prion Diseases of Humans and Animals* (Prusiner, S. B., Collinge, J., Powell, J., and Anderton, B., eds.), Ellis Horwood, Chichester, pp. 497–508.
3. Dickinson, A. G., Meikle, V. M. H., and Fraser, H. (1968) Identification of a gene which controls the incubation period of some strains of scrapie in mice. *J. Comp. Pathol.* **78,** 293–299.
4. Dickinson, A. G. and Meikle, V. M. H. (1971) Host-genotype and agent effects in scrapie incubation: change in allelic interaction with different strains of agent. *Mol. Gen. Genet.* **112,** 73–79.
5. Westaway, D., Goodman. P. A., Mirenda, C. A., McKinley, M. P., Carlson, G. A., and Prusiner, S. B. (1987) Distinct prion proteins in short and long scrapie incubation period mice. *Cell* **51,** 651–662.
6. Hunter, N., Dann, J. C., Bennett, A. D., Somerville, R. A., McConnell, I., and Hope, J. (1992) Are *Sinc* and the PrP gene congruent? Evidence from PrP gene analysis in *Sinc* congenic mice. *J. Gen. Virol.* **73,** 2751–2755.
7. Fraser, H. and Dickinson, A. G. (1968) The sequential development of the brain lesions of scrapie in three strains of mice. *J. Comp. Pathol.* **78,** 301–311.
8. Bruce, M. E., McBride, P. A., and Farquhar, C. F. (1989) Precise targeting of the pathology of the sialoglycoprotein, PrP, and vacuolar degeneration in mouse scrapie. *Neurosci. Lett.* **102,** 1–6.
9. Bruce, M. E., McBride, P. A., Jeffrey, M., and Scott, J. R. (1994) PrP in pathology and pathogenesis in scrapie-infected mice. *Mol. Neurobiol.* **8,** 105–112.
10. Dickinson, A. G. (1976) Scrapie in sheep and goats, in *Slow Virus Diseases of Animals and Man* (Kimberlin, R. H., ed.), North Holland, Amsterdam, pp. 209–241.
11. Bruce, M. E., Chree, A., McConnell, I., Foster, J., Pearson, G., and Fraser, H. (1994) Transmission of bovine spongiform encephalopathy and scrapie to mice: strain variation and the species barrier. *Phil. Trans. R. Soc. Lond. B.* **343,** 405–411.
12. Bruce, M. E. and Dickinson, A. G. (1987) Biological evidence that scrapie agent has an independent genome. *J. Gen. Virol.* **68,** 79–89.
13. Kimberlin, R. H., Walker, C. A., and Fraser, H. (1989) The genomic identity of different strains of mouse scrapie is expressed in hamsters and preserved on reisolation in mice. *J. Gen. Virol.* **70,** 2017–2025.
14. Kimberlin, R. H., Cole, S., and Walker, C. A. (1987) Temporary and permanent modifications to a single strain of mouse scrapie on transmission to rats and hamsters. *J. Gen. Virol.* **68,** 1875–1881.
15. Fraser, H. (1983) A survey of primary transmission of Icelandic scrapie (rida) to mice, in *Virus Non Conventionnels et Affections du Système Nerveux Central* (Court, L. A., ed.), Masson, Paris, pp. 34–46.

16. Carp, R. I. and Callahan, S. M. (1991) Variation in the characteristics of 10 mouse-passaged scrapie lines derived from five scrapie-positive sheep. *J. Gen. Virol.* **72,** 293–298.

17. Fraser, H., Bruce, M. E., Chree, A., McConnell, I., and Wells, G. A. H. (1992) Transmission of bovine spongiform encephalopathy and scrapie to mice. *J. Gen. Virol.* **73,** 1891–1897.

18. Wells, G. A. H., Hawkins, S. A. C., Hadlow, W. J., and Spencer, Y. I. (1992) The discovery of bovine spongiform encephalopathy and observations on the vacuolar changes, in *Prion Diseases of Humans and Animals* (Prusiner, S. B., Collinge, J., Powell, J., and Anderton, B., eds.), Ellis Horwood, Chichester, pp. 256–274.

19. Fraser, H., Pearson, G. R., McConnell, I., Bruce, M. E., Wyatt, J. M., and Gruffydd-Jones, T. J. (1994) Transmission of feline spongiform encephalopathy to mice. *Vet. Rec.* **134,** 449.

20. Prusiner, S. B. (1982) Novel proteinaceous infectious particles cause scrapie. *Science* **216,** 136–144.

21. Dickinson, A. G. and Outram, G. W. (1988) Genetic aspects of unconventional virus infections: the basis of the virino hypothesis, in *Ciba Foundation Symposium 135: Novel Infectious Agents and the Central Nervous System* (Bock, G. and Marsh, J., eds.), Wiley, Chichester, pp. 63–83.

22. Rohwer, R. G. (1991) The scrapie agent: "a virus by any other name." *Curr. Top. Microbiol. Immunol.* **172,** 195–232.

23. Prusiner, S. B. (1992) Prion biology, in *Prion Diseases of Humans and Animals* (Prusiner, S. B., Collinge, J., Powell, J., and Anderton, B., eds.), Ellis Horwood, Chichester, pp. 533–567.

24. Baldwin, M. A., Pan, K.-M., Nguyen, J., Huang, Z., Groth, D., Serban, A., et al. (1994) Spectroscopic characterization of conformational differences between PrP^c and PrP^{Sc}. *Phil. Trans. R. Soc. Lond. B.* **343,** 435–441.

25. Bessen, R. A., Kocisko, D. A., Raymond, G. J., Nandan, S., Lansbury, P. T., and Caughey, B. (1995) Non-genetic propagation of strain-specific properties of scrapie prion protein. *Nature* **375,** 698–700.

14

PrP-Deficient Mice in the Study of Transmissible Spongiform Encephalopathies

Jean C. Manson

1. Introduction

The Transmissible Spongiform Encephalopathies (TSEs), such as scrapie, BSE, and Creutzfeldt-Jakob disease, are associated with alterations in the neural membrane protein or prion protein (PrP). This chapter will outline the gene targeting approaches that have been used to mutate the murine PrP gene, resulting in mice with reduction or absence of the PrP protein. It will then describe how these transgenic animals can contribute to our understanding of the role of PrP in agent replication, the pathology of the TSEs, and the normal function of PrP.

1.1. PrP Gene Expression in Infected and Uninfected Animals

The PrP protein is a protease-sensitive cell surface glycoprotein anchored in the membrane by a glycoinositol phospholipid (PrP^C) *(1)*, but during the course of scrapie infection the PrP protein aggregates and accumulates in and around the cells of the brain as protease-resistant deposits (PrP^{Sc}). The distribution of the PrP protein in infected brain is dependent on both the host genotype and the strain of scrapie *(2)*. The amount of PrP^{Sc} detected in the brains of mice or hamsters infected with scrapie is tenfold greater than the amount of PrP^c detected in the uninfected brain *(3)*.

PrP mRNA is detected in neuronal cells throughout the brain with different amounts of mRNA being detected in different populations of neuronal cells *(4)*. In contrast to the alterations in PrP protein detected during disease, there is no difference in the amount or localization of PrP mRNA in brains either uninfected or infected with different strains of scrapie, as detected by Northern analysis in hamsters *(5)* or by *in situ* hybridization in mice *(4)*.

From: *Methods in Molecular Medicine: Prion Diseases*
Edited by: H. Baker and R. M. Ridley Humana Press Inc., Totowa, NJ

1.2. Allelic Variants of PrP

The *Sinc* gene has been shown to be the major gene controlling survival time of mice exposed to scrapie *(6,7)* and animals homozygous for the alleles of *Sinc* s7 and *Sinc* p7, have allelic forms of the PrP gene. The PrP allele with amino acid 108 Leu and 189 Thr is associated with short incubation times when infected with Chandler isolate of scrapie (Prn-p^a), whereas the allele with 108 Phe and 189 Val (Prn-p^b) is associated with long incubation periods with the same isolate of scrapie *(8)*. The *Sinc* gene has been shown to be linked closely to the PrP gene by RFLP analysis *(9)* and *Sinc* congenic mice encode different PrP proteins *(10)*, suggesting PrP may indeed be the product of the *Sinc* locus.

Allelic forms of PrP have also been linked to incubation periods of scrapie in hamsters *(11)* and sheep *(12)*, and are associated with the incidence of Gerstmann-Sträussler syndrome (GSS) and Creutzfeldt-Jakob disease in humans *(13–15)*.

1.3. PrP Gene Dosage and Scrapie Disease

Transgenic models have been produced to investigate the mechanisms of these diseases. Introduction of hamster PrP genes into mice has shown that increasing the copy number of the PrP gene reduces the incubation period of the disease and that the species type of PrP expressed alters susceptibility of the mice to specific isolates of scrapie *(16)*. Transgenic mice with high copy numbers of the Prn-p^b allele of the murine PrP gene also were shown to have shorter incubation periods when injected with the Chandler scrapie isolate than their nontransgenic littermates *(17)*. Transgenic mice with high copy numbers of the murine PrP gene containing a codon 101 proline to leucine mutation spontaneously develop neurodegeneration, spongiform changes in the brain, and astrogliosis *(18)*. However, overexpression of the wild-type PrP gene has also been show to lead to a lethal neurological disease involving spongiform changes in the brain and muscle degeneration *(19)*.

Although these experiments have shown that increasing the copy number of the PrP gene leads to shortening in incubation periods of the disease, random integration of multiple copies presents with clinical artifacts that may not accurately reflect the disease process. More appropriate models to examine the effect of PrP gene dosage and the role of mutations in the PrP gene may be produced by the introduction of specific alterations into the endogenous PrP gene using gene targeting techniques.

2. Methods to Produce PrP-Deficient Mice

Introduction of foreign DNA into cells can lead to recombination of the DNA with homologous sequences of the endogenous DNA. This homologous recombination event, known as gene targeting, has been used to introduce mutations into genes in murine embryonic stem cells. These cells then can be used to produce mice with one or two copies of the mutated gene. This allows the assessment of the specific mutations within a gene in the mouse.

This method of production of transgenic mice has enabled the function of many genes to be assessed by deletion of the gene products. As a tool in the field of the TSEs, gene targeting has already addressed a number of important questions through the introduction of mutations in the PrP gene that have lead to the production of PrP-deficient mice.

2.1. Targeting Vectors

Two different strategies have been used to introduce mutations into the PrP gene, leading to the production of PrP-deficient mice *(20,21)*. Both approaches have used a replacement targeting vector with a positive selection marker for DNA integration, the neomycin phosphotransferase gene *(neo)* (Fig. 1). In the first approach PrP sequences used in the vector were designed to replace 552 bp of the coding sequences of PrP (positions 10–562) with *neo* under control of the thymidine kinase (tk) promoter (Fig. 1A) *(20)*. An alternative strategy was to insert *neo* under control of the metallothioneine promoter into the PrP coding sequence. This led to homologous recombination at the PrP locus without deletion of any PrP sequences (Fig. 1B). In this approach the tk gene was used as a negative selection against random integration of the DNA *(21)*.

2.2. Targeting Mouse Embryonic Stem Cells

The targeting vectors were linearized outside the PrP sequences and electroporated into embryonic stem cells derived from (A) agouti 129/SV(ev) or (B) 129/Ola mice. Selection was carried out and targeted clones were selected in (A) by PCR and in (B) by Southern analysis of genomic DNA. The frequency of homologous recombination in (A) was 1 in 5000 G418-resistant colonies compared with 1 in 800 in (B). These differences may reflect the difference in the PrP sequences used in the targeting vectors.

2.3. Production of Mice Heterozygous and Homozygous for the Mutant Allele

The mutant cells lines produced by Büeler and coworkers *(20)* were injected into 4-d-old blastocysts from C57BL/6J. The chimeric male mice produced were crossed with C57BL females to produce an outbred line of mice both heterozygous (PrP$^{0/+}$) and homozygous (PrP$^{0/0}$) for the mutant allele.

In the experiments carried out by Manson et al. *(21)* the mutant cell lines were injected into 4-d-old blastocysts from C57BL/CBA mice and the male chimeric mice bred with 129/Ola females to produce an inbred line of mice heterozygous (PrP$^{-/+}$) and homozygous (PrP$^{-/-}$) for the mutant allele. I29/Ola

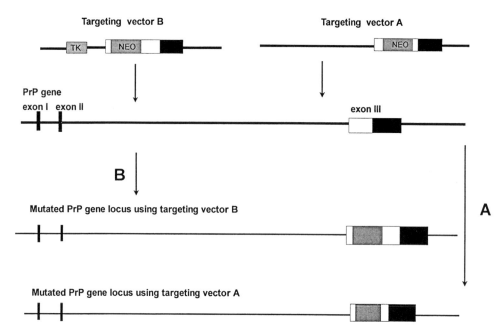

Fig. 1. Targeting strategies for the production of PrP null mice. Vector A. Neomycin phosphotransferase gene *(neo)* under the control of the thymidine kinase promoter, was inserted into exon 3 of the PrP coding region replacing 552 bp of PrP sequence that had been removed (positions 10–562). The vector was linearized outside the PrP sequence and electroporated into 129/SV//ev mouse embryonic stem cells. Homologous recombination of the targeting vector with the endogenous PrP locus resulted in a mutated PrP gene locus lacking 552 bp of PrP and with *neo* inserted into exon 3. Vector B. *Neo* under control of the metallothioneine promoter was inserted into exon 3 of the PrP gene. The thymidine kinase gene *(tk)* was inserted 5' of the PrP sequences to select against random integration. The vector was linearized outside the PrP coding region and electroporated into 129/Ola mouse embryonic stem cells. Homologous recombination of this vector with the endogenous PrP locus results in the loss of the *tk* and a mutated PrP gene with *neo* gene inserted into exon 3. No PrP sequences have been removed in this strategy.

mice are poor breeders and this approach is considerably slower than outbreeding. Nevertheless, it was considered important for the scrapie experiments in these animals to produce an inbred line of mice. In this way differences between wild-type and mutant mice can be attributed solely to the PrP gene.

2.4. PrP mRNA Analysis of Mutant Mice

PrP mRNA analysis of PrP mutant mice by Northern blot has shown both the inbred (PrP$^{+/-}$) and outbred (PrP$^{+/0}$) heterozygous lines have approx 50%

of the wild-type PrP mRNA. *In situ* hybridization studies of the inbred mice have shown the reduction in PrP mRNA is throughout the brain *(22)*. No PrP mRNA was detected in the brains of the inbred mice using probes either 3' or 5' of the *neo* gene. The outbred mice were shown to contain substantial quantities of a chimeric mRNA containing the *neo* and residual PrP sequences. Reduced levels of the PrP-neo RNA molecules were also detected in the outbred heterozygous mice *(20)*.

2.5. PrP Protein Analysis of Mutant Mice

Protein extracted from membrane fractions of mouse brains was analyzed by Western blotting. No PrP protein was detected in either of the homozygous mutant lines. The PrP protein detected in the heterozygous lines was qualitatively identical to that detected in the wild-type mice but was significantly reduced in amount. Quantitation of the PrP protein in the outbred heterozygous mice estimated it to be 50% of that detected in the wild-type mice *(20)* whereas in the inbred heterozygous line the estimated amount of PrP protein was approx 70% of the wild-type mice *(22)*. These differences may simply reflect the problems of accurate quantitation by this technique or alternatively may reflect actual differences in protein amount owing to the different targeting strategies used.

3. Use of PrP-Deficient Mice in Scrapie Research

3.1. The Role of PrP in Disease Pathology

3.1.1. Onset of Disease

Mice with no functional copies of the PrP gene have been shown to survive scrapie infection when injected with Chandler isolate of scrapie *(23)* or the ME7 strain *(22)*. No disease symptoms were observed in these mice over 500 d after inoculation, whereas terminal stages of the disease are detected in wild-type mice around 150 d. These results have clearly demonstrated that PrP gene expression is required for the development of scrapie disease.

In mice with one functional copy of the PrP gene, the time of onset of disease is delayed *(22–24)*, but the amount by which the onset is delayed in the inbred heterozygotes appears to differ, depending on the strain of scrapie (Table 1). It has not yet been established if differences between the outbred and inbred PrP-deficient mouse lines also will lead to differences in timing of the onset of disease, when injected with the same strain of scrapie.

Whereas these experiments have shown that decreasing the copy number of the PrP genes lengthens the incubation period, they have also shown that there is no simple formula relating the PrP gene copy number and incubation period of disease. The strains of mice used in the gene targeting experiments contain the Prn-p^a allele of the PrP gene. The incubation period for the 129Ola PrP$^{+/-}$

Table 1
Incubation Period of PrP[+/+], PrP[+/−], and PrP[−/−] Mice Infected with Scrapie[a]

Mouse genotype	PrP[+/+]	PrP[+/−]	PrP[−/−]	*Sinc* s7/p7 (Prn-p[a]/Prn-p[b])
Scrapie isolate				
ME7	154 ± 0 ($n = 4$)	280 ± 4 ($n = 8$)	>530	251 ± 2
	160 ± 4 ($n = 6$)	284 ± 0 ($n = 4$)		
301C	154 ± 1^c ($n = 14$)	230 ± 2^c ($n = 21$)	$>440^c$ ($n = 21$)	505 ± 5^b

Mouse genotype	PrP[+/+]	PrP[0/+]	PrP[0/0]	*Sinc* s7/p7 (Prn-p[a]/Prn-p[b])
Scrapie isolate				
RML	171 ± 11	415 ± 30 ($n = 15$)	>600	268 ± 4

[a]The time-point at the terminal stages of scrapie for a group of animals of each genotype was used to give an estimate of incubation period of ME7, 301C strains in inbred lines of mice (129/Ola PrP[+/+], PrP[+/−], and PrP[−/−]) and chandler isolate in the outbred lines (PrP[+/+], PrP[0/+], and PrP[0/0]). These incubation periods are compared with the incubation period for each scrapie strain/isolate in *Sinc* s7/p7 (Prn-p[a]/Prn-p[b]) mice to determine the role of the different alleles of PrP in determining the incubation period of the disease.

[b]Isolate passaged three times in C57Bl mice.

[c]Isolate passaged four times in C57Bl mice (incubation periods for *Sinc* s7/p7 and *Sinc* p7/p7 identical to passage 3).

infected with the ME7 strain of scrapie is approximately twice that of the 129/Ola PrP[+/+] and is similar to that obtained with a heterozygous mouse containing a Prn-p[a] and a Prn-p[b] allele of the PrP gene. These results suggest that for the ME7 strain the Prn-p[a] allele of the PrP gene is the predominant allele involved in either generating the ME7 agent or interacting with it. This is not, however, the case with other strains of scrapie. PrP[0/+] mice injected with the RML scrapie isolate had incubation periods (400–465), more prolonged than the heterozygous Prn-p[a]/Prn-p[b] mice (268 d) *(23,24)* and PrP[−/+] infected with 301C isolate have incubation periods considerably shorter than the Prn-p[a]/Prn-p[b] mice (Table 1).

Gene dosage is clearly not the only factor involved in the incubation period of disease. The interaction of the different alleles of the PrP gene with specific strains of scrapie and also genes other than PrP can be investigated using the PrP-deficient mice, into which multiple copies of the Prn-p[a] or Prn-p[b] allele have been inserted. This type of approach has been used to show that the Prn-p[a] allele of PrP is the major determinant of incubation period in mice infected with the RML isolate of scrapie since increasing copy numbers of the Prn-p[a] allele leads to a reduction in incubation periods. However, expression of the Prn-p[b] allele appears to confer resistance to disease in mice infected with 87V strain of scrapie and increases the incubation time with 22A strain infection *(25)*.

3.1.2. PrP Accumulation in Deficient Mice

In PrP$^{+/-}$ mice infected with the ME7 strain of scrapie, PrP deposition starts in the same brain area as the wild-type mice and can be detected as early as 50 d. The pattern of PrP deposition in the brain of heterozygotes follows an identical course, but builds up more slowly than in the wild-type mice. By the terminal stages of disease the amount detected in the brain by immunohistochemical techniques is identical in the PrP$^{+/-}$ and PrP$^{+/+}$ mice *(22)*. These results contrast with the PrP$^{0/+}$ and PrP$^{+/+}$ mice injected with the RML isolate in which the heterozygote animals were shown to have levels of PrP^{Sc} detected by immunoblotting, comparable to their terminally ill counterparts for many months prior to clinical disease and death *(26)*. These experiments indicate that PrP^{Sc} accumulation is not limited by the levels of PrP^{C} and raise questions as to whether there is a causal relationship between PrP^{Sc} accumulation in the brain and the development of the clinical disease.

3.1.3. PrP Gene Dosage and Severity of Disease

The timing of disease onset and progression of disease symptoms is delayed in the PrP-deficient mice *(22,26)*. However, there appears to be no difference in the severity of the clinical signs and the pathology in the brain at the terminal stages of diseases *(22)*. In PrP$^{+/-}$ mice injected with ME7 the pattern and degree of vacuolation was similar to that detected in the wild-type mice and disease symptoms, although delayed, were no less severe than those seen in the wild-type mice *(22)* (Fig. 2).

Suggestions have been put forward for the use of therapies that regulate the production of PrP as a means of controlling these disease *(23)*. These findings suggest that the use of such measures may delay the onset and progression of the disease but have no effect on the final severity of disease.

3.2. The Role of PrP in Agent Replication

PrP$^{0/0}$ mice inoculated with the RML isolate of scrapie were resistant to disease *(23,24)*. Brain homogenates from these animals at 2, 8, 12, and 25 wk after infectivity showed no infectivity. A sample from 20 wk showed low titer infectivity in a group of mice and in a repeat experiment in only one of a group of mice *(23,27)*. Although these results indicate that PrP is necessary for agent replication, whether contamination, residual infectivity, or low level propagation account for the occasional case of scrapie remains to be established.

Initial experiments with the PrP$^{+/+}$ and PrP$^{0/+}$ mice infected with the RML isolate of scrapie indicate that infectivity in the wild-type and PrP-deficient mice reaches similar maximal levels by 20 wk postinfection *(26)*. These results suggest that the reduced levels of PrP^{C} in the heterozygous mice do not limit agent replication, but whether there is any alteration in the rate of replication requires more detailed analysis.

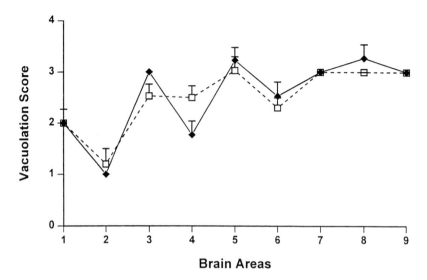

Fig. 2. Lesion profile for 129/Ola PrP$^{+/+}$ and PrP$^{+/-}$ mice infected with ME7 strain of scrapie. The amount of vacuolation in nine areas of the brain at the terminal stages of disease were used to calculate the lesion profile. A lesion profile was produced for the terminal stage of disease in the 129/Ola PrP$^{+/-}$ mice (280 d) and compared to the terminal 129/Ola PrP$^{+/+}$ profile (150 d). The brain areas assessed were (1) medulla, (2) cerebellum, (3) superior colliculus, (4) hypothalamus, (5) thalamus, (6) hippocampus, (7) septum, (8) thalamic cortex, (9) forebrain cortex. Dashed line; endpoint PrP$^{+/+}$. Solid line; end point PrP$^{+/-}$.

4. Normal Function of PrP

One of the major reasons for the numerous gene targeting experiments undertaken in the last few years has been to investigate gene function. The normal function of PrP has not yet been established. The PrP gene is expressed at high levels in neuronal cells of the adult brain *(4,28)* and recently has been detected in astrocytes and oligodendrocytes *(29)*. Lower levels of PrP mRNA can be detected in other tissues, such as heart, lung, and spleen *(5,30)*. PrP mRNA has also been detected during mouse embryogenesis. By 6.5 d gene expression can be detected in the extraembryonic tissue and by 11.5 d in the developing central and peripheral nervous system *(31)*. These studies have suggested a role for PrP in promoting neuronal cell differentiation and in maintaining neuronal function in the differentiated neurons. However, the expression of the PrP gene is not limited to neuronal cell populations since it has been detected in astrocytes and oligodendrocytes, during kidney development, in the developing tooth bud, and in placenta amnion and yolk sac, suggesting the role of PrP may be more widespread, perhaps as part of a cell signaling system required for differentiation of specific cells or as a cell adhesion molecule.

It was hoped that the generation of PrP null mice would address the question of the normal function of PrP. The initial analysis of the PrP null mice *(20,21)*, however, showed that without PrP, mice appeared to develop and reproduce normally. It has now been shown there is little or no phenotypic effect in many null mutant mice produced by gene targeting. This lack of phenotype is thought to be owing to the organism compensating for the loss of a gene by alteration in expression of other genes or use of alternative developmental pathways.

More recently, however, $PrP^{0/0}$ mice were shown to display weakened long-term potentiation (LTP) in the hippocampal CA1 subfield *(32)*. Abnormalities in the LTP were also detected in the inbred PrP null mice that consistently showed an absence of LTP in the CA1 region of the hippocampus and in its place a short-term potentiation, which decays to control levels within 1 h *(33)*. Since the two lines of mice have been produced by different targeting strategies, these results indicate that the impairment of synaptic plasticity is a result of the loss of PrP and not an artifact of gene targeting and subsequent mouse production. Reintroduction of the high copy numbers of the human PrP gene into PrP null mice has been shown to restore the LTP response to that seen in the wild-type controls *(34)*. The effect on LTP therefore can be attributed to an absence of PrP, but it remains difficult to ascribe the differences between knockout and wild-type mice to a specific function of PrP or to compensatory effects of the organism in the absence of PrP.

5. Introduction of Point Mutations into the Endogenous Murine PrP Gene by Gene Targeting

Variant forms of the PrP protein have been shown to be associated with human TSE *(13–15)* and amino acid differences have been shown to be associated with different incubation periods of scrapie in mice *(8)*, hamsters *(10)*, and sheep *(12)*.

Recent advances in the gene targeting technology now have made it possible for the role of the specific alterations in the PrP gene to be analyzed by introducing point mutations into the endogenous PrP gene *(35)*. In this way the relationship between *Sinc* and PrP can be clearly defined by introducing the 108 and 189 sequences associated with the *Sinc* p7 allele into the PrP gene from a *Sinc* s7 mouse.

This approach also can be used to produce transgenic models for the human TSEs. Standard transgenic approaches have been used to produce mice expressing high levels of a chimeric hamster murine PrP gene with the codon 101 proline to leucine mutation *(18)*. The equivalent mutation (codon 102) in humans has been shown to be associated with Gerstmann-Sträussler Syndrome *(13,14)*. Transgenic mice overexpressing the mutant PrP protein have been shown to spontaneously develop neurodegeneration, spongiform changes in the brain, and astrogliosis *(18)*. Whether this disease can be attributed specifically to the PrP mutation or to overexpression of the PrP gene is difficult to establish. More definitive transgenic models have now been produced by introducing specific alterations into the endogenous PrP gene using gene targeting techniques. Using a double replacement strategy, mice have been produced in which amino acid 101 of the endogenous PrP gene has been mutated from Pro to Leu. Both heterozygous and homozygous lines of mice carrying this mutation have been produced *(36)*. No spontaneous neurodegenerative disease has been observed in these animals in up to 250 d of age.

6. Future Prospects

Gene targeting has provided us with an extremely powerful approach to analyze these diseases. The introduction of specific mutations will allow the role of these mutations in PrP to be defined. Gene targeting technology is now being developed through bacteriophage P1 cre/lox site-specific recombinase systems *(37–39)* to allow inducible gene expression, both tissue specifically and temporally.

Inducible expression will allow the development of more specific models that ablate PrP gene expression at different time-points in development. These models will be able to differentiate between normal function of the PrP protein in the adult mouse and compensatory effects resulting from the complete ablation of PrP during development.

Tissue-specific expression will allow the role of PrP gene expression in specific cell types and specific tissues to be analyzed. These models will be able to address questions relating to agent replication, transfer of infectivity from the periphery to central nervous system, and the role of PrP expression in specific neuronal cell populations in the development of the disease pathology.

Transgenic models can also be developed using this approach, which will allow expression of mutant and wild-type PrP genes at different time-points during the disease process. These models will allow in vivo analysis of interactions between different PrP molecules and the precise mechanisms leading to the development of the pathology of the disease to be defined.

References

1. Stahl, N., Borchelt, D. R., Hsiao, K., and Prusiner, S. B. (1987) Scrapie prion protein contains a phosphoinositol glycolipid. *Cell* **51**, 229–240.
2. Bruce, M., McBride, P. A., and Farquhar, C. F. (1989) Precise targeting of the pathology of the sialoglycoprotein, PrP, and vacuolar degeneration in mouse scrapie. *Neurosci. Lett.* **102**, 1–6.
3. DeArmond, S. J., Mobley, W. C., DeMott, D. L., Barry, R. A., Beckstead, J. H., and Prusiner, S. B. (1987) Changes in the localisation of the brain prion proteins during scrapie infection. *Neurology* **37**, 1271–1281.

4. Manson, J., McBride, P., and Hope, J. (1992) Expression of the PrP gene in the brain of Sinc congenic mice and its relationship to the development of scrapie. *Neurodegeneration* **1,** 45–52.

5. Oesch, B., Westaway, D., Wälchi, M., McKinley, M. P., Kent, S. B. H., Aebersold, R., et al. (1985) A cellular gene encodes scrapie PrP 27–30 protein. *Cell* **40,** 735–746.

6. Dickinson, A. G., Meikle, M. V., and Fraser, H. (1968) Identification of a gene which controls incubation period of some strains of scrapie in mice. *J. Comp. Pathol.* **78,** 293–299.

7. Carlson, G. A., Kingsbury, D. T., Goodman, P. A., Coleman, S., Marshall, S. T., DeArmond, S., et al. (1986) Linkage of prion protein and scrapie incubation time genes. *Cell* **46,** 503–511.

8. Westaway, D., Goodman, P. A., Mirenda, C., McKinley, M. P., Carlson, G. A., and Prusiner, S. B. (1987) Distinct prion proteins in short and long incubation period mice. *Cell* **51,** 651–662.

9. Hunter, N., Hope, J., McConnell, I., and Dickson, A. G. (1987) Linkage of the scrapie associated fibril protein(PrP) gene and *Sinc* using congenic mice and restriction length fragment polymorphisms analysis. *J. Gen. Virol.* **68,** 2711,2712.

10. Hunter, N., Dann, J. C., Bennett, A. D., Somerville, R. A., McConnell, I., and Hope, J. (1992) Are *Sinc* and the PrP gene congruent? Evidence from PrP gene analysis in *Sinc* congenic mice. *J. Gen. Virol.* **73,** 2751–2755.

11. Lowenstein, D. H., Butler, D. A., Westaway, D., McKinley, M. P., DeArmond, S. J., and Prusiner, S. B. (1990) Three hamster species with different incubation times and neuropathological features encode different prion proteins. *Mol. Cell Biol.* **10,** 1153–1163.

12. Goldmann, W., Hunter, N., Benson, G., Foster, J., and Hope, J. (1991) Scrapie-associated fibril proteins are encoded by lines of sheep selected for different alleles of the *Sip* gene. *J. Gen. Virol.* **72,** 2411–2417.

13. Doh-ura, K., Tateishi, J., and Sasaki, H. (1989) Pro-Leu change at position 102 of prion protein is the most common but not the sole mutation related to Gerstmann-Sträussler syndrome. *Biochem. Biophys. Res. Commun.* **163,** 974–979.

14. Collinge, J., Harding, A. E., Owen, F., Poulter, M., Lofthouse, R., Boughey, A. M., et al. (1989) Diagnosis of Gerstmann-Sträussler syndrome in familial dementia with prion protein gene analysis. *Lancet* **2,** 15–17.

15. Owen, F., Poulter, M., Shah, T., Collinge, J., Lofthouse, R., Baker, H., et al. (1990) An in-frame insertion in the prion protein gene in familial Creutzfeldt-Jakob disease. *Mol. Brain Res.* **7,** 273–276.

16. Scott, M., Foster, D., Mirenda, C., Serban, D., Coufal, F., Wälchli, M., et al. (1989) Transgenic mice expressing hamster prion protein produce species-specific scrapie infectivity and amyloid plaques. *Cell* **59,** 847–857.

17. Westaway, D., Mirenda, C. A., Foster, D., Zebarjadian, Y., Scott, M., Torchia, M., et al. (1991) Paradoxical shortening of scrapie incubation times by expression of prion protein transgenes derived from long incubation period mice. *Neuron* **7,** 59–68.

18. Hsiao, K. K., Scott, M., Foster, D., Groth, D. F., DeArmond, S. J., and Prusiner, S. B. (1990) Spontaneous neurodegeneration in transgenic mice with mutant prion protein. *Science* **250,** 1587–1590.

19. Westaway, D., DeArmond, S. J., Cayetano-Canlas, J., Groth, D., Foster, D., Yang, S.-L., et al. (1994) Degeneration of skeletal muscle, peripheral nerves and central nervous system in transgenic mice overexpressing wild type prion proteins. *Cell* **76,** 117–129.

20. Büeler, H., Fischer, M., Lang, Y., Bluthmann, H., Lipp, H.-L., DeArmond, S. J., et al. (1992) The neuronal cell surface protein PrP is not essential for normal development and behaviour in mice. *Nature* **356,** 577–582.

21. Manson, J. C., Clarke, A. R., Hooper, M., Aitchison, L., McConnell, I., and Hope, J., et al. (1994) 129/ola mice carrying a mutation in PrP that abolishes mRNA production are developmentally normal. *Mol. Neurobiol.* **8,** 1–5.

22. Manson, J. C., Clarke, A. R., McBride, P. A., McConnell, I., and Hope, J. (1994) PrP gene dosage determines the timing but not the final intensity or distribution of lesions in scrapie pathology. *Neurodegeneration* **3,** 311–340.

23. Büeler, H., Aguzzi, A., Sailer, A., Greiner, R.-A., Autenreid, P., Aguet, M., et al. (1993) Mice devoid of PrP are resistant to scrapie. *Cell* **73,** 1339–1347.

24. Prusiner, S. B., Groth, D., Serban, A., Koehler, R., Foster, D., Torchia, M., et al. (1993) Ablation of the prion protein (PrP) gene in mice prevents scrapie and facilitates the production of anti-PrP antibodies. *Proc. Natl. Acad. Sci. USA* **90,** 10,608–10,612.

25. Carlson, G. A., Ebeling, C., Yang, S.-L., Telling, G., Torchia, M., Groth, D., et al. (1994) Prion protein specified allotypic interactions between cellular and scrapie prion proteins in congenic and transgenic mice. *Proc. Natl. Acad. Sci. USA* **91,** 5690–5694.

26. Büeler, H., Raeber, A., Sailer, A., Fischer, M., Aguzzi, A., and Weissmann, C. (1994) High prion and PrP^{Sc} levels but delayed onset of disease in scrapie-inoculated mice heterozygous for a disrupted PrP gene. *Mol. Med.* **1,** 19–30.

27. Sailer, A., Bueler, H., Fischer, M., Aguzzi, A., and Weissmann, C. (1994) No propagation of prions in mice devoid of PrP. *Cell* **77,** 967,968.

28. Kretzchmar, H. A., Prusiner, S. B., Stowring, L. E., and DeArmond, S. J. (1986) Scrapie prion proteins are synthesised in neurones. *Am. J. Pathol.* **122,** 1–5.

29. Moser, M., Colello, R. J., Pott, U., and Oesch, B. (1995) Developmental expression of the prion protein gene in glial cells. *Neuron* **14,** 509–517.

30. Caughey, B., Race, R., and Chesebro, B. (1988) Detection of prion protein mRNA in normal and scrapie-infected tissues and cell lines. *J. Gen. Virol* **69,** 711–716.

31. Manson, J., West, J. D., Thomson, V., McBride, P., Kaufman, M. H., and Hope, J. (1992) The prion protein gene: a role in mouse embryogenesis? *Development* **115,** 117–122.

32. Collinge, J., Whittington, M., Sidle, K., Smith, C., Palmer, M., Clarke, A., et al. (1994) Prion protein is necessary for normal synaptic function. *Nature* **370,** 295–297.

33. Manson, J., Hope, J., Clarke, A., Johnston, A., Black, C., and MacLeod, N. (1995) PrP gene dosage and long term potentiation. *Neurodegeneration* **4,** 113–115.

34. Whittington, M., Sidle, K., Gowland, I., Meads, J., Hill, A., Palmer, M., et al. (1995) Rescue of neurophysiological phenotype seen in PrP null mice by transgene encoding human prion protein. *Nature Genet.* **9**, 197–201.

35. Stacey, A., Schnieke, A., McWhir, J., Cooper, J., Colman, A., and Melton, D. W. (1994) Use of double replacement gene targeting to replace the murine alpha lactalbumin gene with its human counterpart in embryonic stem cells in mice. *Mol. Cell Biol.* **14**, 1009–1016.

36. Moore, R., Redhead, N., Selfridge, J., Hope, J., Manson, J., and Melton, D. (1995) Double replacement gene targeting for the production of a series of mouse strains with different prion protein gene alterations. *Biotechnology* **13**, 999–1004.

37. Orban, P. C., Chui, D., and Marth, J. D. (1992) Tissue- and site-specific DNA recombination in transgenic mice. *Proc. Natl. Acad. Sci. USA* **89**, 6861–6865.

38. Lasko, M., Sauer, B., Mosinger, B., Lee, E., Manning, R., Yu, S.-H., et al. (1992) Targeted oncogene activation by site-specific recombination in transgenic mice. *Proc. Natl. Acad. Sci. USA* **89**, 6232–6236.

39. Gu, H., Marht, J., Orban, P., Mossman, H., and Rajewsky, K. (1994) Deletion of a DNA polymerase B gene segment in T cells using cell type specific gene targeting. *Science* **265**, 103–106.

15

Transgenic Approaches to Prion "Species-Barrier" Effects

David Westaway

1. Introduction

Like conventional viruses, prion isolates exhibit distinctive, and often restricted host-ranges. However, the molecular events that shape the host-ranges of these two classes of pathogen are dissimilar, reflecting their fundamentally different life cycles. As discussed by Ridley and Baker (Chapter 1), many lines of experimentation indicate that PrP^{Sc}, an aberrant form of a host-encoded neuronal sialoglycoprotein PrP^{C}, is the major constituent of the scrapie prion. Resistance to protease digestion in vitro is a convenient hallmark of PrP^{Sc}. PrP^{C} or a closely related protease-sensitive macromolecule is converted to PrP^{Sc} in a posttranslational event: This event may correspond to a conformational change templated by PrP^{Sc} molecules *(1,2)*. Whereas the host-range and cellular tropism of mammalian viruses frequently reflects their binding to cell-surface proteins to gain entry into cells via endocytosis, PrP^{Sc}-receptors have yet to be identified: Although PrP^{C} plays a critical role in determining prion host-range *(see the following)* and is displayed on the cell surface via a glycolipid anchor *(3)*, three observations suggest that it cannot be considered a "receptor" in the conventional sense, e.g., in the sense that CD4 is the receptor for HIV. First, the cell surface molecules co-opted by mammalian viruses cycle within the endocytic pathway and are destined for degradation: During prion replication PrP^{C} is converted to PrP^{Sc} and changes a variety of biophysical properties *(4)*. It is noteworthy that PrP^{Sc} has a half-life of >48 h in pulse-chase experiments and may thus be protease-resistant in vivo, as well as in vitro *(5)*. Second, PrP^{C} that is not displayed on the cell-surface to a significant degree (owing to ablation of Asn-linked glycosylation sites) can also become pro-

From: *Methods in Molecular Medicine: Prion Diseases*
Edited by: H. Baker and R. M. Ridley Humana Press Inc., Totowa, NJ

tease-resistant *(6)*. Third, aggregates of purified PrP^{Sc} are taken into cultured cells, presumably by phagocytosis *(7)*.

It is also unlikely that the immune system plays any substantial role in determining prion host-range. It has long been noted that scrapie infection fails to induce inflammatory *(8)* or humoral immune responses *(9)*. Although one gene affecting Creutzfeldt-Jakob disease (CJD) incubation times was mapped to the major histocompatibility complex of mouse *(10)*, other workers using different congenic strains of inbred mice inoculated with scrapie prions failed to reproduce this finding *(11)*.

Again, although statistically significant changes in scrapie incubation time can be observed when transferring prions between nonisogenic mouse strains, these effects have an explanation that lies outside of the realm of transplantation biology *(12)*. The factors that do control prion host-range will be considered in the following, after a description of the origins and definition of "species barrier" effects.

2. Species Barriers to Prion Transmission

Although first described in rodents, "species barrier" effects have been confirmed in many experimental settings *(13)*. The general observation is that the first transmission of prions from one species to another results in incubation times that are more protracted than subsequent repassages in the recipient species: Adaptation to the new species frequently is complete by the second passage *(14–16)*. There are variations on this theme: Some species barriers are very hard to traverse, with incubation times approaching the natural lifespan of the recipient, often with only a fraction of the inoculated animals exhibiting disease, e.g., transmission of CJD prions to mice *(17,18)* or hamsters *(19)*. Some barriers appear insurmountable, for example, passage of the hamster-adapted scrapie prion isolates 263K and Sc237, or transmissible mink encephalopathy prions into mice *(20–22)*.

In addition to laboratory studies, species barriers operating in natural prion diseases can be tentatively inferred from epidemiological data. For example, even though natural scrapie is endemic in several European countries and the United States, handling of or consumption of meat from affected sheep does not appear to be a significant risk factor for sporadic CJD *(23,24)*. CJD cases have yet to be reported in laboratory workers handling rodent-adapted scrapie isolates but have been described in histopathology technicians handling human biopsy material *(25)*.

3. PrP Sequences and Species Barriers

The hypothesis that primary structure of PrP plays a pivotal role in species barrier effects was derived from several pieces of information that had become available by the late 1980s. These included the observations that PrP is a major component of purified preparations of scrapie prions *(26)* and that different scrapie-susceptible mammals encoded nonidentical PrPs *(27–29)*. It was also known that scrapie incubation time gene alleles in mice were correlated with two alleles of the prion protein gene, *Prnp*: These *a* and *b* alleles diverge at codons 108 (Leu → Phe) and 189 (Thr → Val) *(30)*. Furthermore, isolates of prions derived from *Prnp^a* mice differed in duration and variance of scrapie incubation period (when repassaged in *Prnp^a* mice) from those derived from *Prnp^b* mice *(12)*. Taken together, these data suggested that optimal replication of prions might be dependent on "matching" incoming PrP^Sc with endogenous PrP^C such that both types of molecules shared the same primary structure. Although transgenic studies that directly address this hypothesis will be scrutinized presently, recent compilations of PrP sequences are compatible with this interpretation.

Transmission of CJD prions into nonhuman primates is correlated with the presence of Met at codon 112 *(31)*. Data from a smaller study is in accord with this finding, even though these authors reached the opposite conclusion that "primary structural similarity may not be an important factor in disease transmission": This divergent interpretation may reflect considerations detailed in Section 7., absence of Capuchin monkey and marmoset PrP sequences from the compilation, or presence of PrP-coding polymorphisms in the particular Rhesus monkeys used for inoculations, which exhibited a transmission rate of only 73%, yet (in other individuals of this species) 96.4% amino-acid homology to human PrP *(32)*. Other studies have elaborated on the concept that PrP gene alleles within a species can comprise an "internal" (intra-) species barrier to prion infection. Collinge and coworkers have demonstrated that Val/Met heterozygosity at codon 129 in the human PrP gene may confer a degree of resistance to the development of CJD *(33)*. Similarly, a common Gln/Arg polymorphism at codon 171 in the sheep PrP gene may comprise a major determinant of susceptibility to natural scrapie in Suffolk sheep *(34)*.

4. Studies with Syrian Hamster PrP Transgenes

To address the relationship between PrP primary structure and species barrier phenomena, a cosmid clone encompassing a complete Syrian hamster (SHa) PrP gene was used to construct transgenic mice *(35)*. Four transgenic lines were shown to express SHa PrP mRNA and PrP^C: All exhibited heightened susceptibility to infection with hamster-adapted prions. Furthermore, incubation times were inversely related to the level of SHaPrPC expression: In the cases of the Tg(SHaPrP)81 and Tg(SHaPrP)7 lines these times equalled (75 d) and surpassed (55 d), respectively, those observed in hamsters. Titers of mouse prions in Tg mice inoculated with SHa-adapted prions were low or undetectable. Conversely, SHa prions were undetectable in the same Tg mice when inoculated with mouse-adapted prions. Patterns of spongiform change and amyloid deposition in TgSHaPrP mice inoculated with SHa-adapted prions closely resembled those of hamsters *(21)*. These results indicated that a highly specific interaction between a transgene encoded molecule (presumably PrP^C) and SHaPrP^{PSc} in the inoculum features in prion replication. This hypothesis was appraised in a further set of experiments.

By engineering restriction sites into mouse (Mo) and SHa PrP cDNA clones, a "mix-and-match" approach was used to create chimeric PrP coding cassettes. These were then inserted into a PrP expression vector, cosSHa Tet, where the coding sequences of the SHa PrP cosmid have been replaced by a tetracycline resistance gene flanked by *Sal*I restriction sites *(36)*. One such construct harboring the MH2M cassette, which differs from mouse PrP at residues 108, 111, 138, 154, and 169 (i.e., in the central region of the PrP^C molecule), engendered the Tg(MH2M)92 transgenic line. Interestingly, this line was found to be susceptible to both SHa- and mouse-adapted prions. Furthermore, prion isolates that transited Tg(MH2M)92 mice gained novel host-range properties. For example, subsequent to passage in Tg(MH2M)92 mice, the Sc237 SHa-adapted isolate was able to infect mice with an incubation period of approx 185 d. Conversely, the mouse-adapted Rocky Mountain Laboratory prion isolate (RML) gained the ability to infect Syrian hamsters with an incubation period of approx 215 d *(37)*.

Transgenic lines expressing a chimeric PrP^C with a smaller hamster-derived coding region (differences only at codons 108 and 111) were not susceptible to Sc237 prions. These experiments demonstrate the crucial importance of certain amino acid side chains in the central region of the PrP in determining host-range and further implicate homophilic PrP^{Sc}/PrP^C interactions in prion replication. Additionally, since the only differences between the constructs in Tg lines, such as Tg(MH2M)92 and Tg(MHM2)285, lie in the PrP coding region, these experiments dispense with an earlier caveat that other genes or sequences within the SHaPrP cosmid control dissolution of the barrier to infection with Sc237 prions *(35)*.

5. Studies with Human and Sheep PrP Transgenic Mice

Extrapolating from the precedent established by Tg(SHaPrP) mice, attempts have been made to transmit human and sheep prions to cognate TgPrP mice. Transgenic lines Tg(HuPrP)110 and Tg(HuPrP)152 were constructed by inserting wild-type human PrP coding sequences encoding Val at codon 129 into cosSHa Tet. For the Tg(HuPrP)152 line levels of human PrP^C were approximately four- to eightfold higher than that found in human brain. These Tg mice were challenged with inocula from cases of Gerstmann-Sträussler syndrome (GSS), as well as sporadic and iatrogenic CJD. With the exception of transmission from iatrogenic CJD case #170 into Tg(HuPrP)152 (positive at approx 240 d, J. Collinge, personal communication), incubation times were protracted (≥589 d ± SE), with only 14/169 inoculated mice exhibiting scrapie-like clinical symptoms. Notably, a similar (10.3%) disease incidence was encountered in non-Tg controls *(18)*.

Difficulties have also been encountered transmitting sheep prions to Tg(PrP) mice. Thus, the Tg(ShePrP)217 line expressing wild-type sheep PrP^C (corresponding to the Gln 171 allele) did not exhibit heightened susceptibility to natural scrapie prions from Suffolk sheep, since both transgenic and nontransgenic animals became sick at ≥360 d after inoculation *(34,38,39)*, (D. Westaway, D. Foster, and S. B. Prusiner, in preparation). Thus, experiences with Tg(HuPrP) and Tg(ShePrP) mice suggest that expression of a donor-derived PrP transgene is not always sufficient to erase species barriers to prion infection.

6. Enhancing Transmission of Exogenous Prions into TgPrP Mice

One explanation for the "failures" with human and sheep transgenes detailed above focuses on the contribution of endogenous PrP genes. Incubation times to illness in TgSHaPrP lines inoculated with RML mouse prions range from 18–38 (Tg69) to 46–66 (Tg81) d longer than in non-Tg littermates. This effect crudely parallels levels of transgene expression, suggesting that the presence of SHaPrP^C can interfere with the conversion of Mo PrP^C to Mo PrP^{Sc} *(21)*. This explanation is further supported by:

1. Approximately 40 d prolongations of incubation times with RML prions in Tg(MHu2M)PrP mice compared to non-Tg mice *(see the following [18])*;
2. Studies in scrapie-infected mouse neuroblastoma cells, where expression of heterologous PrP sequences can attenuate production of protease-resistant PrP *(36)*; and
3. Deletion of endogenous PrP coding sequences.

Thus, introduction of the Tg81 SHaPrP transgene array onto a Prn-p$^{0/0}$ ablated background led to successive reductions in the time from inoculation with Sc237 prions to onset of clinical disease from 75 ± 1 d (Prnp$^{+/+}$) to 67 ± 4 d (Prnp$^{+/0}$) to 56 ± 3 d (Prnp$^{0/0}$) *(40,41)*. Analogous experiments to assess the behavior of Tg(HuPrP)152 and Tg(ShePrP)217 transgenes within a Prnp$^{0/0}$ background are in progress.

Since transgene expression levels have a potent effect on incubation times *(21,42)*, another maneuver is to construct Tg lines with yet higher levels of expression. Unfortunately, the highest levels of PrP^C expression obtained to date are associated with a spontaneous neuromuscular disease. In the case of Tg(SHaPrP)7$^{+/-}$ heterozygous mice this syndrome has an onset after 700 d, and does not affect the execution or interpretation of experiments where animals are inoculated with Sc237 SHa-adapted prions. However, there is a more tangible problem in the case of Tg(ShePrP) mice. Although several Tg(ShePrP) lines equal the levels of ShePrP mRNA seen in Tg(ShePrP)217 mice, founder mice with higher expression levels and copy numbers are rare ($n = 2$) and engender unstable lines that "revert" to a Tg(ShePrP)217-like copy number *(39)*. Thus, the caveats of spontaneous disease and genetic instability associated with high copy-number transgene arrays—with PrP^C-associated cytotoxicity perhaps acting as the selection pressure for the emergence of lower copy-number revertants—limit the practicality of this approach.

A third enhancement to transgenic strategies involves use of chimeric coding sequences. Insertion of the central region of human PrP into a mouse coding cassette to create a "MHu2M" transgenic line resulted in incubation times to illness of ≤240 d subsequent to inoculation with CJD prions *(18)*. All the inoculated mice developed disease ($n = 24$) and incubation times in individual experiments were clustered (standard error of the mean ± 4.6 d or less), indicative of a nonstochastic process. Demonstration of spongiform degeneration typical of prion diseases and accumulation of transgene-encoded PrP^{Sc} detected by a monoclonal antiserum argues that the Tg(MHu2M)PrP mice are more susceptible to CJD prions than their nontransgenic relatives.

7. Beyond Primary Structure: Other Components of Species Barriers

Although the usefulness of Tg mice susceptible to heterologous prions needs no elaboration, it should be clearly understood that parameters other than PrP primary structure can affect transit across a species barrier.

7.1. "Strain"-Type

Different prion isolates or "strains" propagated within the same inbred host may have different abilities to cross a species barrier, e.g., hamster-adapted scrapie strains 263K and 431K *(20)*. Originally attributed to the presence of a nucleic acid genome, the molecular basis of "strainness" is unresolved but in the case of scrapie and TME "strains" is associated with PrP^{Sc} molecules with distinct physiochemical properties *(43,44)*.

7.2. Chaperones/Foldases

Failure of Tg(HuPrP) mice to succumb to CJD prions has been interpreted to reflect divergence in a gene encoding a chaperone or foldase involved in the conversion of PrP^C to PrP^{Sc} *(18)*. Although aspects of induction of the heat-shock proteins (hsps) hsp72, hsp28, and hsp73 are now known to differ between prion-infected and uninfected neuroblastoma cells *(45)*, these hsps seem unlikely candidates for PrP-specific chaperones: They are not known to have affinities for particular mature (completely folded) proteins, instead having a catholic affinity for denatured proteins *(46)*. Nonetheless, because there is at least one example of a "dedicated" foldase in a eukaryote, a putative rhodopsin-specific proline isomerase from *D. melanogaster (47)*, this line of reasoning may warrant further attention.

7.3. Carbohydrate Modifications

PrP^{Sc} is known to be decorated with two complex Asn-linked carbohydrate trees *(48)* and contains a glycan core within the phosphatidylinostol glycolipid anchor *(3)*. Since mammalian species differ in their repertoire of glycosyl transferases *(49)* it is possible that species-specific patterns of glycosylation of PrP^{Sc} also affect the ability to infect a new host. Indeed, the related notion that prion strains could correspond to different PrP^{Sc} glycoforms is quite widely held *(43,50,51)*.

8. Reprise: General Approaches to Creating Tg(PrP) Mice

Experiences set out herein can be distilled into the following recommendations: Most or all of these points should be considered before embarking on transgenic studies.

1. Probes. Nucleic acid and antibody probes directed against the heterologous PrP gene should be available. Ideally, these should exhibit no crossreactivity with mouse PrP mRNA or PrP^C. In the case of cosTet.SHaPrP the diverged 3' untranslated region has proven a useful probe for DNA and RNA analyses *(36)*.
2. The transgene construct. Because expression of "minigene" transgene constructs is weak or unpredictable *(35,52)*, molecular clones encompassing introns and many kilobases of flanking sequence are recommended. A variety of PrP cosmids has been used to date: These have important advantages in that most stable Tg lines express PrP^C, i.e., expression of the cosmids appears independent of chromosomal context *(21,39,42)*, and since they use PrP promoters, they direct expression of PrP^C to the correct cellular lineages. Procedures for the production of transgenic mice are detailed by Hogan et al. *(53)*.
3. Allele-matching. If the species under study exhibits PrP coding sequence polymorphisms, care should be taken to derive the transgene construct from the appropriate allele of the donor species.
4. Genetic background. Use of Prnp$^{0/0}$ (Prnp "ablated") mice is recommended *(40)*. This gene knockout was originally created on the 129/Sv background but because this strain has low fecundity and oocytes that are difficult to microinject, repeated backcrossing to transfer the ablated Prnp locus to PrP Tg mice within the FVB/N or C57Bl6 backgrounds should be considered.
5. Expression levels. In general, stable Tg lines with high levels of expression ($\geq 4 \times$ endogenous PrP^C) are recommended for bioassays, because they exhibit the most rapid onset of clinical disease. Uninoculated Tg mice should be set aside to moni-

tor for the appearance of spontaneous (as well as intercurrent) disease *(39)*. Breeding homozygotes for the transgene array offers the possibility of obtaining yet higher expression levels but there are attendant complications:

a. Since transgenes can be insertional mutagens there is a possibility of creating homozygotes for irrelevant recessive lethal mutations *(54)*;

b. DNA or segregation analyses will have to be performed to distinguish heterozygotes from homozygotes;

c. There is the possibility of creating a spontaneous PrP transgene-mediated neurodegenerative disease *(see earlier)*; and

d. Unequal crossing over may destabilize the transgene arrays (which usually correspond to tandem repeat units *[54]*).

6. Chimeric or nonchimeric PrP coding sequences? Tg(MHu2M)PrP mice have established the utility of chimeric coding sequences. Such mice might be considered for use in a sensitive bioassay for CJD prions but use in other contexts might be inappropriate. For example, the biophysical properties of "artificial" prions containing MHu2M.PrP[CJD] may differ from those of authentic CJD prions.

7. Disease diagnosis. Procedures for inoculations and the clinical diagnosis of scrapie-like disease in rodents have been described elsewhere *(12,16,18,55)*. "Positive" diagnoses typically should be verified by histopathology and the presence of PrP[Sc]. Mouse colonies should be assessed for and protected against extraneous pathogens *(53)*.

8. Biohazards. Last, but not least, prions cause fatal neurologic disease in humans and animals and should be handled with due respect. Safety issues are considered by Taylor (Chapter 6). Laboratory studies should be undertaken under the auspices of local biohazard and recombinant DNA advisory committees and readers are also referred to refs. *56* and *57*.

9. Concluding Remarks

Because parameters other than primary structure above may modulate species barrier effects (*see earlier*; and ref. *31*), recommendations listed here enhance, but do not guarantee, success in constructing mice sensitive to heterologous prions. Also it is worth noting that TgPrP mice may have an intrinsic limit in their susceptibility to prion disease, corresponding to an incubation time of ca. 50 d *(21,40,58)*: This behavior is exemplified by inoculation of Sc237 prions into Tg(SHaPrP)7[+/+] mice within the Prnp[0/0] background, and by inoculation of mouse-adapted RML prions into Tg(Mo-PrP-A)B4053 mice *(18)*. Although these transgenic mice represent a considerable advance over transmissions into nontransgenic hosts, it should be borne in mind that animal bioassays ultimately will be supplanted by faster (and most probably cheaper) in vitro systems for the conversion of PrP[C] to PrP[Sc] *(59,60)*.

Note Added in Proof

Telling et al. (*Cell* **83,** 79–90, 1995) have documented an inhibitory effect of mouse PrP[C] on transmission of human prions into Tg(Hu) PrP mice. The authors hypothesize that interactions between PrP^C and a macromolecule designated "protein X" provide an additional level of specificity in species-barrier effects.

Acknowledgments

I am indebted to George Carlson, Richard Rubenstein, and John Collinge for their comments, and the Alzheimer's Association of Ontario for support.

References

1. Pan, K.-M., Baldwin, M., Nguyen, J., Gasset, M., Serban, A., Groth, D., et al. (1993) Conversion of α-helices into β-sheets features in the formation of the scrapie prion proteins. *Proc. Natl. Acad. Sci. USA* **90,** 10,962–10,966.
2. Caughey, B. (1994) Scrapie-associated PrP accumulation and agent replication: effects of sulphated glycosaminoglycan analogues. *Phil. Trans. R. Soc. Lond. B* **343,** 399–404.
3. Stahl, N., Borchelt, D. R., Hsiao, K., and Prusiner, S. B. (1987) Scrapie prion protein contains a phosphatidylinositol glycolipid. *Cell* **51,** 229–240.
4. Prusiner, S. B. (1992) Chemistry and biology of prions. *Biochemistry* **31,** 12,278–12,288.
5. Borchelt, D. R., Scott, M., Taraboulos, A., Stahl, N., and Prusiner, S. B. (1990) Scrapie and cellular prion proteins differ in their kinetics of synthesis and topology in cultured cells. *J. Cell Biol.* **110,** 743–752.
6. Taraboulos, A., Rogers, M., Borchelt, D. R., McKinley, M. P., Scott, M., Serban, D., et al. (1990) Acquisition of protease resistance by prion proteins in scrapie-infected cells does not require asparagine-linked glycosylation. *Proc. Natl. Acad. Sci. USA* **87,** 8262–8266.
7. Taraboulos, A., Serban, D., and Prusiner, S. B. (1990) Scrapie prion proteins accumulate in the cytoplasm of persistently infected cultured cells. *J. Cell Biol.* **110,** 2117–2132.
8. Zlotnik, I. (1962) The pathology of scrapie: a comparative study of lesions in the brain of sheep and goats. *Acta Neuropathol. (Berl.)* **1(Suppl.),** 61–70.
9. Porter, D. D., Porter, H. G., and Cox, N. A. (1973) Failure to demonstrate a humoral immune response to scrapie infection in mice. *J. Immunol.* **111,** 1407–1410.
10. Kingsbury, D. T., Kasper, K. C., Stites, D. P., Watson, J. D., Hogan, R. N., and Prusiner, S. B. (1983) Genetic control of scrapie and Creutzfeldt-Jakob disease in mice. *J. Immunol.* **131,** 491–496.
11. Mohri, S. and Tateishi, J. (1989) Host genetic control of incubation periods of Creutzfeldt-Jakob disease in mice. *J. Gen. Virol.* **70,** 1391–1400.
12. Carlson, G. A., Westaway, D., DeArmond, S. J., Peterson-Torchia, M., and Prusiner, S. B. (1989) Primary structure of prion protein may modify scrapie isolate properties. *Proc. Natl. Acad. Sci. USA* **86,** 7475–7479.
13. Pattison, I. H. (1965) Experiments with scrapie with special reference to the nature of the agent and the pathology of the disease, in *Slow, Latent and Temperate Virus Infections, NINDB Monograph 2* (Gajdusek, D. C., Gibbs, C. J., Jr., and Alpers, M. P., eds.), US Government Printing Office, Washington, DC, pp. 249–257.

14. Kimberlin, R. H., Cole, S., and Walker, C. A. (1987) Temporary and permanent modifications to a single strain of mouse scrapie on transmission to rats and hamsters. *J. Gen. Virol.* **68,** 1875–1881.

15. Lowenstein, D. H., Butler, D. A., Westaway, D., McKinley, M. P., DeArmond, S. J., and Prusiner, S. B. (1990) Three hamster species with different scrapie incubation times and neuropathological features encode distinct prion proteins. *Mol. Cell. Biol.* **10,** 1153–1163.

16. Carp, R. I. and Callahan, S. M. (1991) Variation in the characteristics of 10 mouse-passaged scrapie lines derived from five scrapie-positive sheep. *J. Gen. Virol.* **72,** 293–298.

17. Gibbs, C. J., Jr., Gajdusek, D. C., and Amyx, H. (1979) Strain variation in the viruses of Creutzfeldt-Jakob disease and kuru, in *Slow Transmissible Diseases of the Nervous System,* vol. 2 (Prusiner, S. B. and Hadlow, W. J., eds.), Academic, New York, pp. 87–110.

18. Telling, G. C., Scott, M., Foster, D., Yang, S.-L., Torchia, M., Sidle, K. C. L., et al. (1994) Transmission of Creutzfeldt-Jakob disease from humans to transgenic mice expressing chimeric human-mouse prion protein. *Proc. Natl. Acad. Sci. USA* **91,** 9936–9940.

19. Brown, P., Gibbs, C. J., Jr., Rodgers-Johnson, P., Asher, D. M., Sulima, M. P., Bacote, A., et al. (1994) Human spongiform encephalopathy: the National Institutes of Health series of 300 cases of experimentally transmitted disease. *Ann. Neurol.* **35,** 513–529.

20. Kimberlin, R. H. and Walker, C. A. (1978) Evidence that the transmission of one source of scrapie agent to hamsters involves separation of agent strains from a mixture. *J. Gen. Virol.* **39,** 487–496.

21. Prusiner, S. B., Scott, M., Foster, D., Pan, K.-M., Groth, D., Mirenda, C., et al. (1990) Transgenetic studies implicate interactions between homologous PrP isoforms in scrapie prion replication. *Cell* **63,** 673–686.

22. Taylor, D. M., Dickinson, A. G., Fraser, H., and Marsh, R. F. (1986) Evidence that transmissible mink encephalopathy agent is biologically inactive in mice. *Neuropathol. Appl. Neurobiol.* **12,** 207–215.

23. Masters, C. L. (1987) Epidemiology of Creutzfeldt-Jakob disease: studies on the natural mechanisms of transmission, in *Prions—Novel Infectious Pathogens Causing Scrapie and Creutzfeldt-Jakob Disease* (Prusiner, S. B. and McKinley, M. P., eds.), Academic, Orlando, FL, pp. 511–522.

24. Harries-Jones, R., Knight, R., Will, R. G., Cousens, S., Smith, P. G., and Matthews, W. B. (1988) Creutzfeldt-Jakob disease in England and Wales, 1980–1984: a case-control study of potential risk factors. *J. Neurol. Neurosurg. Psychiat.* **51,** 1113–1119.

25. Miller, D. C. (1988) Creutzfeldt-Jakob disease in histopathology technicians. *N. Engl. J. Med.* **318,** 853–854.

26. McKinley, M. P., Bolton, D. C., and Prusiner, S. B. (1983) A protease-resistant protein is a structural component of the scrapie prion. *Cell* **35,** 57–62.

27. Basler, K., Oesch, B., Scott, M., Westaway, D., Wälchli, M., Groth, D. F., et al. (1986) Scrapie and cellular PrP isoforms are encoded by the same chromosomal gene. *Cell* **46,** 417–428.

28. Locht, C., Chesebro, B., Race, R., and Keith, J. M. (1986) Molecular cloning and complete sequence of prion protein cDNA from mouse brain infected with the scrapie agent. *Proc. Natl. Acad. Sci. USA* **83**, 6372–6376.

29. Kretzschmar, H. A., Stowring, L. E., Westaway, D., Stubblebine, W. H., Prusiner, S. B., and DeArmond, S. J. (1986) Molecular cloning of a human prion protein cDNA. *DNA* **5**, 315–324.

30. Westaway, D., Goodman, P. A., Mirenda, C. A., McKinley, M. P., Carlson, G. A., and Prusiner, S. B. (1987) Distinct prion proteins in short and long scrapie incubation period mice. *Cell* **51**, 651–662.

31. Schätzl, H. M., Da Costa, M., Taylor, L., Cohen, F. E., and Prusiner, S. B. (1994) Prion protein gene variation among primates. *J. Mol. Biol.* **245**, 362–374.

32. Cervenakova, L., Brown, P., Goldfarb, L. G., Nagel, J., Pettrone, K., Rubenstein, R., et al. (1994) Infectious amyloid precursor gene sequences in primates used for experimental transmission of human spongiform encephalopathy. *Proc. Natl. Acad. Sci. USA* **91**, 12,159–12,162.

33. Palmer, M. S., Dryden, A. J., Hughes, J. T., and Collinge, J. (1991) Homozygous prion protein genotype predisposes to sporadic Creutzfeldt-Jakob disease. *Nature* **352**, 340–342.

34. Westaway, D., Zuliani, V., Cooper, C. M., Da Costa, M., Neuman, S., Jenny, A. L., et al. (1994) Homozygosity for prion protein alleles encoding glutamine-171 renders sheep susceptible to natural scrapie. *Gene Dev.* **8**, 959–969.

35. Scott, M., Foster, D., Mirenda, C., Serban, D., Coufal, F., Wälchli, M., et al. (1989) Transgenic mice expressing hamster prion protein produce species-specific scrapie infectivity and amyloid plaques. *Cell* **59**, 847–857.

36. Scott, M. R., Köhler, R., Foster, D., and Prusiner, S. B. (1992) Chimeric prion protein expression in cultured cells and transgenic mice. *Prot. Sci.* **1**, 986–997.

37. Scott, M., Groth, D., Foster, D., Torchia, M., Yang, S.-L., DeArmond, S. J., et al. (1993) Propagation of prions with artificial properties in transgenic mice expressing chimeric PrP genes. *Cell* **73**, 979–988.

38. Westaway, D., Neuman, S., Zuliani, V., Mirenda, C., Foster, D., Detwiler, L., et al. (1992) Transgenic approaches to experimental and natural prion diseases, in *Prion Diseases of Humans and Animals* (Prusiner, S. B., Powell, J., Collinge, J., and Anderton, B., eds.), Ellis Horwood, London, pp. 474–482.

39. Westaway, D., DeArmond, S. J., Cayetano-Canlas, J., Groth, D., Foster, D., Yang, S.-L., et al. (1994) Degeneration of skeletal muscle, peripheral nerves, and the central nervous system in transgenic mice overexpressing wild-type prion proteins. *Cell* **76**, 117–129.

40. Büeler, H., Aguzzi, A., Sailer, A., Greiner, R.-A., Autenried, P., Aguet, M., et al. (1993) Mice devoid of PrP are resistant to scrapie. *Cell* **73**, 1339–1347.

41. Prusiner, S. B., Groth, D., Serban, A., Koehler, R., Foster, D., Torchia, M., et al. (1993) Ablation of the prion protein (PrP) gene in mice prevents scrapie and facilitates production of anti-PrP antibodies. *Proc. Natl. Acad. Sci. USA* **90**, 10,608–10,612.

42. Westaway, D., Mirenda, C. A., Foster, D., Zebarjadian, Y., Scott, M., Torchia, M., et al. (1991) Paradoxical shortening of scrapie incubation times by expression

of prion protein transgenes derived from long incubation period mice. *Neuron* **7,** 59–68.

43. Kascsak, R. J., Rubenstein, R., Merz, P. A., Carp, R. I., Robakis, N. K., Wisniewski, H. M., et al. (1986) Immunological comparison of scrapie-associated fibrils isolated from animals infected with four different scrapie strains. *J. Virol.* **59,** 676–683.

44. Bessen, R. A. and Marsh, R. F. (1992) Identification of two biologically distinct strains of transmissible mink encephalopathy in hamsters. *J. Gen. Virol.* **73,** 329–334.

45. Tatzelt, J., Zuo, J., Voellmy, R., Scott, M., Hartl, U., Prusiner S. B., et al. (1995) Scrapie prions selectively modify the stress response in neuroblastoma cells. *Proc. Natl. Acad. Sci. USA* **92,** 2944–2948.

46. Morimoto, R. I., Tissieres, A., and Georgeopoulos, C. (1994) *The Biology of Heat Shock Proteins and Molecular Chaperones* Cold Spring Harbor Laboratory, Cold Spring Harbor, NY.

47. Stamnes, M. A., Shieh B.-H., Chuman, L., Harris, G. L., and Zuker, C. S. (1991) The cyclophilin homolog ninaA is a tissue-specific integral membrane protein required for the proper synthesis of a subset of Drosophila rhodopsins. *Cell* **65,** 219–227.

48. Haraguchi, T., Fisher, S., Olofsson, S., Endo, T., Groth, D., Tarantino, A., et al. (1989) Asparagine-linked glycosylation of the scrapie and cellular prion proteins. *Arch. Biochem. Biophys.* **274,** 1–13.

49. Rademacher, T. W., Parekh, R. B., and Dwek, R. A. (1988) Glycobiology. *Annu. Rev. Biochem.* **57,** 785–838.

50. Somerville, R. A. and Ritchie, L. A. (1990) Differential glycosylation of the protein (PrP) forming scrapie-associated fibrils. *J. Gen. Virol.* **71,** 833–839.

51. Hecker, R., Taraboulos, A., Scott, M., Pan, K.-M., Torchia, M., Jendroska, K., et al. (1992) Replication of distinct prion isolates is region specific in brains of transgenic mice and hamsters. *Genes Dev.* **6,** 1213–1228.

52. Brinster, R. L., Allen, J. M., Behringer, R. R., Gelinas, R. E., and Palmiter, R. D. (1988) Introns increase transcriptional efficiency in transgenic mice. *Proc. Natl. Acad. Sci. USA* **85,** 836–840.

53. Hogan, B., Beddington, R., Costantini, F., and Lacy, E. (1994) *Manipulating the Mouse Embryo, a Laboratory Manual*, 2nd ed., Cold Spring Harbor Laboratory, Cold Spring Harbor, NY.

54. Palmiter, R. D. and Brinster, R. L. (1986) Germ-line transformation of mice. *Annu. Rev. Genet.* **20,** 465–499.

55. Bruce, M. E. and Fraser, H. (1991) Scrapie strain variation and its implications. *Curr. Top. Microbiol. Immunol.* **172,** 125–138.

56. Chatigny, M. A., and Prusiner, S. B. (1979) Biohazards and risk assessment of laboratory studies on the agents causing the spongiform encephalopathies, in *Slow Transmissible Diseases of the Nervous System*, vol. 2 (Prusiner, S. B. and Hadlow, W. J., eds.), Academic, New York, pp. 491–514.

57. Prusiner, S. B. (1987) The biology of prion transmission and replication, in *Prions—Novel Infectious Pathogens Causing Scrapie and Creutzfeldt-Jakob Disease* (Prusiner, S. B. and McKinley, M. P., eds.), Academic, Orlando, FL, pp. 83–112.

58. Carlson, G. A., Ebeling, C., Yang, S.-L., Telling, G., Torchia, M., Groth, D., et al. (1994) Prion isolate specified allotypic interactions between the cellular and scrapie prion proteins in congenic and transgenic mice. *Proc. Natl. Acad. Sci. USA* **91,** 5690–5694.

59. Butler, D. A., Scott, M. R. D., Bockman, J. M., Borchelt, D. R., Taraboulos, A., Hsiao, K. K., et al. (1988) Scrapie-infected murine neuroblastoma cells produce protease-resistant prion proteins. *J. Virol.* **62,** 1558–1564.

60. Kocisko, D. A., Come, J. H., Priola, S. A., Chesebro, B., Raymond, G. J., Lansbury, P. T., Jr., et al. (1994) Cell-free formation of protease-resistant prion protein. *Nature* **370,** 471–474.

16

Methods for Studying Prion Protein Amyloid

Fabrizio Tagliavini, Frances Prelli, Giorgio Giaccone,
Gianluigi Forloni, Mario Salmona, Pedro Piccardo,
Bernardino Ghetti, Blas Frangione, and Orso Bugiani

1. Introduction

Prion encephalopathies are neurodegenerative diseases characterized by the accumulation of abnormal isoforms of the prion protein (PrP) and the deposition of PrP amyloid in the central nervous system (CNS) *(1,2)*. The disease-specific PrP molecules are distinguishable from their normal homologs by their relative resistance to proteinase K digestion *(1,2)*; they are thought to be derived from protease-sensitive precursors by a posttranslational process that involves a conformational change with a shift from α-helix to β-sheet structure *(3–5)*.

Amyloid formation occurs to the highest degree in Gerstmann-Sträussler-Scheinker (GSS) disease (Fig. 1), an autosomal dominant disorder that exhibits a wide spectrum of clinical presentations, including ataxia, spastic paraparesis, parkinsonism, and dementia, and is associated with variant *PRNP* genotypes resulting from the combination of a pathogenic mutation with a common polymorphism (ATG → GTG, M/V) at codon 129 (Fig. 2) *(6)*. To date, point mutations at codons 102 (CCG → CTG, P → L) and 145 (TAT → TAG, Y → stop) have been found on a M129 allele, whereas mutations at codons 105 (CCA → CTA, P → L), 117 (GCG → GTG, A → V), 198 (TTC → TCC, F → S), and 217 (CGG → CAG, Q → R) have been detected on a V129 allele *(6)*.

The clinical variability of GSS disease is related to the distribution and extent of amyloid deposition as well as the occurrence of associated lesions *(6)*. Unicentric and multicentric amyloid deposits may be found throughout the CNS. In most cases they are particularly abundant in the cerebellar cortex,

From: *Methods in Molecular Medicine: Prion Diseases*
Edited by: H. Baker and R. M. Ridley Humana Press Inc., Totowa, NJ

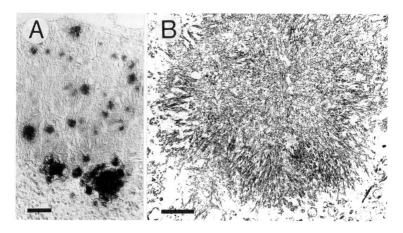

Fig. 1. Amyloid deposits in the cerebellar cortex of a patient with GSS disease. **(A)** Amyloid cores are labeled by an antiserum to a synthetic peptide homologous to residues 90–102 of PrP. **(B)** At the electron microscopic level, amyloid deposits are composed of radially oriented, 8–10-nm diameter fibrils. Bars: (A), 50 μm; (B), 2 μm.

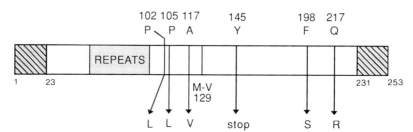

Fig. 2. Schematic representation of human PrP showing the mutations associated with GSS disease and PrP-CAA. The protein consists of 253 residues. A signal peptide of 22 amino acids is cleaved at the N terminus during biosynthesis and a hydrophobic signal sequence of 23 amino acids is removed from the C terminus on addition of a glycosyl phosphatidylinositol anchor (shaded areas). A polymorphism at codon 129 (M/V) is common in the Caucasian population. Pathogenic mutations at codon 102 (P → L) and codon 145 (Y → stop) are in phase with M129, whereas mutations at codons 105 (P → L), 117 (A → V), 198 (F → S), and 217 (Q → R) are in phase with V129.

although in a few instances (e.g., some patients with the codon 117 mutation) the cerebellum is not involved. Amyloid deposition usually occurs in the neuropil. However, in the case of the 145 mutation it takes place primarily in the wall of parenchymal and leptomeningeal vessels. This neuropathologic phenotype is distinct from that of GSS disease and has been designated as PrP cerebral amyloid angiopathy (PrP-CAA) *(7)*. In addition to amyloid deposits, many patients with P → L substitution at codon 102 show severe spongiform changes in the cerebral cortex, whereas patients with codon 145, 198, or 217 mutation display neurofibrillary tangles composed of paired helical filaments in neo- and archicortex as well as several subcortical nuclei *(6–10)*.

Amyloid deposition is consistently accompanied by hypertrophy and proliferation of astroglial and microglial cells, neuritic abnormalities, and neuronal loss leading to a variable degree of atrophy of affected regions *(6)*. The close topographical relationship between amyloid deposits and tissue changes suggests that PrP amyloid plays a role in the pathogenesis of nerve cell degeneration and glial cell reaction. On this ground, we carried out studies aiming to define the composition of amyloid fibrils in GSS families with different *PRNP* mutations, to identify the PrP sequences that are critical for amyloid formation, and to evaluate the biological effects of PrP peptides on nerve and glial cells in vitro.

2. The Amyloid Protein in GSS Disease

The biochemical composition of PrP amyloid was first determined in brain tissue samples obtained from patients of the Indiana kindred of GSS *(11,12)*, carrying a F → S substitution at PrP residue 198 in coupling phase with V129 *(13,14)*. Amyloid cores were isolated by a procedure combining buffer extraction, sieving, collagenase digestion, and sucrose density gradient (Fig. 3). Proteins were extracted from amyloid fibrils with formic acid, purified by gel filtration chromatography and reverse-phase high-performance liquid chromatography (HPLC), analyzed by sodium dodecyl sulfate-polyacrylamide gel electrophoresis (SDS-PAGE) and immunoblot, and sequenced as described in detail below. The amyloid preparations contained two major peptides of ~11 and ~7 kDa (Fig. 4) spanning residues 58–150 and 81–150 of PrP, respectively *(11,12)*. Sequence analysis showed that these peptides had ragged N- and C-termini.

The finding that the amyloid protein was an N- and C-terminal truncated fragment of PrP was verified by immunostaining brain sections with antisera raised against synthetic peptides homologous to residues 23–40, 90–102, 127–147, and 220–231 of human PrP, and residues 15–40, 90–102, and 220–232 of hamster PrP. The amyloid cores were strongly immunoreactive with the antisera to the mid-region of the molecule, whereas only the periphery of the cores was immunostained by antibodies to N- or C-terminal domains *(6,15)*.

In GSS 198, the amyloid protein does not include the region containing the amino acid substitution. To establish whether amyloid peptides originate from mutant or both mutant and wild-type PrP, we analyzed patients heterozygous at codon 129 (M/V) and used V129 as a marker of the mutant allele, since in this family V129 is in phase with mutant S198. Amino acid sequencing and electrospray mass spectrometry of peptides generated by digestion of the amyloid protein with

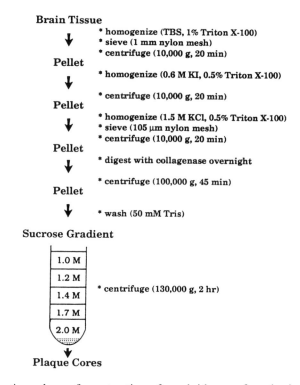

Fig. 3. Preparative scheme for extraction of amyloid cores from brain tissue.

endoproteinase Lys-C showed that the samples contained only peptides with V129, suggesting that only mutant PrP was involved in amyloid formation (Fig. 5A) *(12)*.

Subsequently, we characterized the amyloid protein in GSS kindreds with other *PRNP* mutations (i.e., A → V at codon 117 and Q → R at codon 217) and found that the smallest amyloid subunit was a ~7 kDa N- and C-terminal trun- cated fragment of PrP of similar size and sequence and was derived from the mutant allele (Fig. 5B) *(12)*. In patients with GSS 117, the amyloid protein contained the mutant V117 *(16)*.

3. Assembly and Conformation of PrP Peptides In Vitro

To determine which residues are important for polymerization of PrP peptides into amyloid fibrils and which conditions promote peptide assembly, we investi- gated fibrillogenesis in vitro using synthetic peptides homologous to consecutive segments of the amyloid protein purified from GSS brains (i.e., the octapeptide repeat region, residues 89–106, 106–126, and 127–147). We found that peptides within the PrP sequence 106–147 readily assembled into fibrils, whereas pep- tides corresponding to the N-terminal segment of the amyloid protein did not

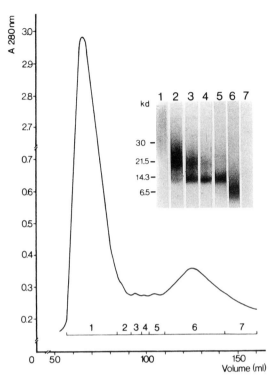

Fig. 4. Partial purification of GSS amyloid proteins by gel filtration chromatography. Elution profile of proteins extracted by formic acid from amyloid cores and fractionated on a Sephadex G-100 column. Protein peaks were collected as indicated (fractions 1–7) and subjected to immunoblot analysis with the monoclonal antibody 3F4 to residues 109–112 of human PrP (inset). Gel filtration yielded two major peaks: the void volume (fraction 1) and a low-mol-wt peak (fraction 6) that was present as a broad band centered at 7 kDa on SDS-polyacrylamide minigels. In addition, minor intermediate peaks (fractions 2–5) were present in the chromatograms. Fractions 4 and 5 contained an 11-kDa PrP fragment.

(17). In particular, the peptide homologous to residues 106–126 was extremely fibrillogenic and formed dense meshworks of straight filaments ultrastructurally similar to those observed in GSS patients (Fig. 6A). The fibrillary assemblies were partially resistant to proteinase K digestion, exhibited tinctorial and optical properties of *in situ* amyloid (i.e., birefringence under polarized light after Congo red staining and yellow fluorescence after thioflavine S treatment), and showed an X-ray diffraction pattern consistent with that of native amyloid fibrils, with reflections corresponding to H-bonding between antiparallel β-sheets *(17,18)*.

Circular dichroism (CD) spectroscopy revealed that peptide 106–126 is able to adopt different conformations in relation to the microenvironment (Fig.

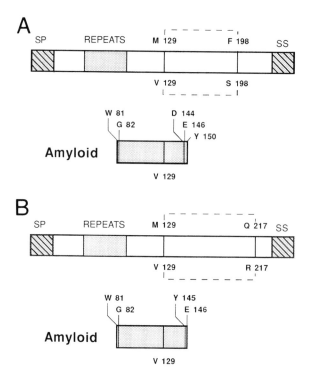

Fig. 5. Schematic representation of the PrP molecule and the 7-kDa amyloid protein in GSS patients with mutations at codon 198 **(A)** or codon 217 **(B)** and heterozygous M/V at codon 129. The amyloid peptides have ragged N and C termini, and contain only V129, suggesting that only mutant PrP is involved in amyloid formation.

6B,C) *(19)*. It showed primarily a β-sheet secondary structure in phosphate buffer, pH 5.0, a combination of β-sheet and random coil in phosphate buffer, pH 7.0, a random coil conformation in deionized water, and an α-helical structure in the presence of micelles formed by a 5% SDS solution. The addition of α-helix stabilizing solvents (e.g., trifluoroethanol or hexafluoropropanol) to a solution of peptides in deionized water induced a conformational shift from random coil to α-helix, but did not modify the β-sheet structure of the peptide previously suspended in phosphate buffer, pH 5.0 *(19)*. These data suggest that the PrP region including residues 106–126 may feature in the conformational transition from normal to abnormal PrP.

4. Biological Effects of PrP Peptides In Vitro

To test the hypothesis that the accumulation of PrP amyloid protein may be the direct cause of nerve cell degeneration and glial cell reaction in GSS disease and PrP-CAA, we investigated the effects of the exposure of primary

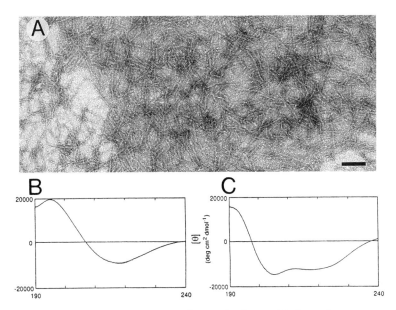

Fig. 6. Assembly and conformation of a synthetic peptide homologous to residues 106–126 of human PrP. **(A)** Electron micrograph of fibrils generated in vitro by the peptide. The fibrils are straight, unbranched, and have a diameter of ~8 nm. Bar: 100 nm. **(B,C)** Circular dichroism spectra of PrP 106–126. The peptide adopts predominantly a β-sheet conformation when suspended in 200 mM phosphate buffer, pH 5.0 (B), and an α-helical structure when dissolved in 50% trifluoroethanol (C).

cultures of neurons and astrocytes to the synthetic peptides used for fibrillogenesis studies.

The prolonged exposure of hippocampal neurons to micromolar concentrations of PrP 106–126 resulted in marked neuronal loss (Fig. 7A,B) *(20)*. Conversely, the other PrP peptides and a scrambled sequence of PrP 106–126 were not effective. The neurotoxic effect of PrP 106–126 on cultured neurons was dose-dependent; the toxic response was first detected at a concentration of 20 μM, was statistically significant at 40 μM, and resulted in virtually complete neuronal loss at 60 or 80 μM. Fluorescence microscopy following treatment with DNA-binding fluorochromes (e.g., Hoechst 33258; Sigma, St. Louis, MO) as well as electron microscopy showed that neurons chronically exposed to PrP 106–126 presented a typical apoptotic morphology, with condensation of the chromatin and fragmentation of the nucleus. Agarose gel electrophoresis of DNA extracted from cultured cells after 7-d treatment with the peptide showed an apoptotic pattern of DNA fragmentation, resulting from cleavage of nuclear DNA in internucleosomal regions *(20)*.

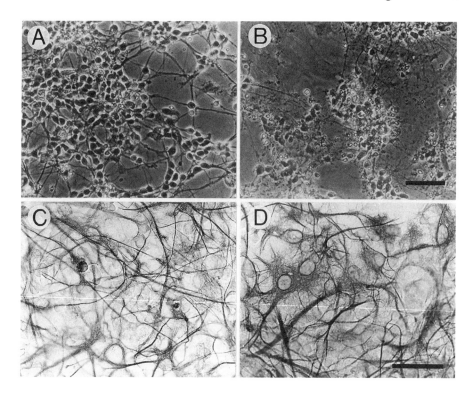

Fig. 7. **(A,B)** Neurotoxicity of a synthetic peptide homologous to residues 106–126 of human PrP. Photomicrographs of primary hippocampal neurons exposed for 10 d to a scrambled peptide (A) or to PrP 106–126 (B) at concentrations of 80 μ*M*. **(C,D)** Growth-promoting activity of peptide PrP 106–126 on glial cells. Photomicrographs of primary astroglial cultures exposed for 14 d to a scrambled peptide (C) or to PrP 106–126 (D) at concentrations of 50 μ*M*. Following treatment, cultures were immunostained with an antiserum to GFAP. Bars: (B), 200 μm; (D), 100 μm.

The prolonged exposure of primary astroglial cultures to the peptide PrP 106–126 resulted in a remarkable increase in size and density of astroglial processes (Fig. 7C,D) *(21)*. Conversely, no effects were observed following treatment of cultures with the other PrP peptides. The hypertrophy of astrocytes was associated with a striking increase in glial fibrillary acidic protein (GFAP) transcripts as revealed by Northern blot analysis. Densitometric quantitation of GFAP mRNA normalized for the level of β-actin message showed that this increment was dependent on peptide concentration, being significant at 10 μ*M* and resulting in three- and fivefold increase above control values at 25 and 50 μ*M*, respectively. The rise of GFAP transcripts was accompanied by a substantial increase in GFAP, as determined by Western blot analysis *(21)*. The hypertrophy of astroglial cells was associated with 1.5-fold increase in cell number at a peptide concentration of 50 μ*M*. However, the proliferation rate of astrocytes was much higher when cultures were kept in serum-free medium for the duration of the experiment. Under these conditions, the uptake of thymidine after 9-d treatment was 50 times higher than the basal values *(22)*. The proliferative effect of PrP 106–126 was abolished by cotreatment of astroglial cultures with nicardipine, a blocker of L-type voltage-sensitive calcium channels. Microfluorimetric analysis of intracellular calcium levels in single astrocytes showed that PrP 106–126 induced a rapid increase in cytosolic calcium concentrations, followed by a slow return to basal levels. This effect was not observed with the scrambled peptide. It was absent when calcium was removed from the medium and was prevented by preincubation of cultures with nicardipine. These data suggest that PrP 106–126 stimulates astroglial proliferation via an increase in intracellular calcium concentration, through the activation of L-type voltage-sensitive calcium channels *(22)*.

In summary, our studies showed that the amyloid protein in GSS disease is an N- and C-terminal truncated fragment of PrP that originates from mutant molecules. This fragment contains a sequence (i.e., residues 106–126) that can adopt different conformations in distinct environments, although it has a high propensity to form stable β-sheet structures. When synthesized as a peptide, this sequence is fibrillogenic and partially resistant to protease digestion; it is toxic to neurons but has a growth-promoting activity on glial cells. This sequence is an integral part of PrP peptides that accumulate in the CNS of patients with prion diseases, and might be a major contributor to the molecular characteristics and the pathogenic properties of disease-specific PrP isoforms and PrP amyloid.

5. Methods
5.1. Buffers Used in the Methods Described
1. Lysis buffer: 100 m*M* NaCl, 10 m*M* EDTA, 0.5% Nonidet P-40, 0.5% sodium deoxycholate in 10 m*M* Tris-HCl, pH 7.4.
2. Buffer A: 10 m*M* Tris-HCl, 150 m*M* NaCl, 2 m*M* EDTA, 2 m*M* EGTA, 0.4 m*M* PMSF, 1% Triton X-100, pH 7.5.
3. Buffer B: 10 m*M* Tris-HCl, 0.6*M* KI, 2 m*M* EDTA, 2 m*M* EGTA, 0.4 m*M* PMSF, 0.5% Triton X-100, pH 7.5.
4. Buffer C: 10 m*M* Tris-HCl, 1.5*M* KCl, 2 m*M* EDTA, 2 m*M* EGTA, 0.4 m*M* PMSF, 0.5% Triton X-100, pH 7.5.
5. Buffer D: 50 m*M* Tris-HCl, 150 m*M* NaCl, pH 7.5.
6. Buffer E: 50 m*M* Tris-HCl, 10 m*M* CaCl$_2$, 3 m*M* NaN$_3$, pH 7.5.
7. Buffer F: 50 m*M* Tris-HCl, pH 7.5.

5.2. Antibodies to PrP
Polyclonal antibodies were raised in New Zealand white rabbits to synthetic peptides homologous to residues 23–40, 58–71, 90–102, 95–108, 127–147,

151–165, and 220–231 of the amino acid sequence deduced from human PrP cDNA. Peptide synthesis and purification were carried out as described in Section 5.5.1. The peptides PrP 23–40, 90–102, 127–147, and 220–231 were coupled to keyhole limpet hemocyanin through a cysteine residue added at the N-terminus (PrP 127–147 and 220–231) or at the C-terminus (PrP 23–40 and 90–102). The peptides PrP 58–71, 95–108, and 151–165 were directly synthesized on a small nonimmunogenic core of branching lysine residues, which provide a scaffolding to support multiple copies of the peptide antigen *(23)*. Polyclonal antibodies were generated by multiple intradermal and/or subcutaneous injections of 100–200 μg peptide at 2–3-wk intervals. The rabbits were bled according to the evolution of the antibody titer against the unconjugated peptide evaluated by enzyme-linked immunosorbent assay (ELISA). Rabbit antisera to synthetic peptides homologous to residues 15–40, 90–102, and 220–232 of hamster PrP (provided by S. B. Prusiner) *(24)* as well as the monoclonal antibody 3F4 (provided by R. J. Kascsak) *(25)* were also used for the studies. The latter was obtained against Syrian hamster PrP 27–30 and its epitope corresponds to the sequence M-K-H-M, i.e., residues 108–111 of hamster PrP and 109–112 of human PrP *(26)*.

5.3. Analysis of Crude Brain Extracts

Brain tissues that had been frozen at –80°C at the time of autopsy were used in the following protocol.

1. Homogenize tissue in 9 vol of cold lysis buffer *(27)* and centrifuge at 10,000*g* for 10 min.
2. Take aliquots of supernatant each equivalent to 100 μg protein and digest with proteinase K (20–100 μg/mL) for 1 h at 37°C.
3. Terminate digestion by adding PMSF to a final concentration of 3 m*M*.

Selected proteinase K-treated samples were subjected to enzymatic deglycosylation with PNGase F (New England Biolabs, Beverly, MA) using reaction conditions recommended by the manufacturer. Proteins were then fractionated on 12.5% tricine-SDS-polyacrylamide minigels (tricine-SDS-PAGE) *(28)* under reducing conditions, electrophoretically transferred to polyvinylidene difluoride membranes (Immobilon, Millipore, Bedford, MA) and probed with anti-PrP antibodies. The immunoreactivity was visualized by enhanced chemiluminescence (Amersham, Arlington Heights, IL). Specificity of the reactions was checked by using normal rabbit or mouse sera as primary antibodies and was confirmed by absorption of the PrP synthetic peptide antisera with the relevant peptides. Each antiserum was incubated with 10 m*M* of the relevant peptide at 37°C for 1 h and then at 4°C overnight. After centrifugation at 10,000*g*, the supernatants were used as primary antibodies.

5.4. Characterization of GSS Amyloid Protein

5.4.1. Protocol for Isolation of Amyloid Fibrils from Brain Tissue

Amyloid plaque cores were isolated from 20–40 g of cerebellum, cerebral cortex, or basal ganglia that had been frozen at –80°C at the time of autopsy. Patients were selected on the basis of the *PRNP* genotype, whereas the brain regions were chosen based on a semiquantitative evaluation of the density of amyloid deposits in 7-μm thick paraffin sections immunostained with anti-PrP antibodies *(see* Section 5.4.7.).

1. Remove leptomeninges and large vessels and homogenize tissue with a Brinkmann homogenizer in buffer A at a sample-to-buffer ratio of 1:5 (w/v).
2. Sieve homogenate through a 1-mm nylon mesh and centrifuge the filtrate at 10,000*g* for 20 min.
3. Rehomogenize pellet in buffer B and centrifuge at 10,000*g* for 20 min, then rehomogenize pellet in buffer C and centrifuge at 10,000*g* for 20 min.
4. Wash pellet three times in buffer D and centrifuge at 70,000*g* for 30 min.
5. Resuspend pellet in buffer E at a sample-to-buffer ratio of 1:25 (w/v) and digest with collagenase (Collagenase 3.4.24.3. Type I, Sigma, St. Louis, MO) for 18 h at 37°C using a 1:100 (w/v) ratio of enzyme to pellet.
6. Centrifuge at 70,000*g* for 60 min, resuspend pellet, and wash three times in buffer F.
7. Load suspension on to a discontinuous sucrose gradient (1, 1.2, 1.4, 1.7, and 2*M* sucrose in 10 m*M* Tris, pH 7.5) and centrifuge at 130,000*g* for 120 min.

Each interface was collected, washed, and pelleted three times in buffer F. Aliquots of each pellet were assessed for the presence of amyloid fibrils by polarized light microscopy after Congo red staining, fluorescence microscopy after thioflavine S treatment, and electron microscopy after negative staining with 2% aqueous phosphotungstic acid or 5% uranyl acetate. Amyloid plaque cores were recovered in the 2*M* sucrose fraction.

5.4.2. Protocol for Gel Filtration Chromatography

1. Suspend amyloid-enriched pellet in 99% formic acid and sonicate four times for 20 s.
2. Add 2 vol of distilled water and centrifuge at 10,000*g* for 10 min.
3. Apply supernatant to a calibrated Sephadex G-100 column (1.2 × 120 cm), equilibrated with 3*M* formic acid.
4. Pool protein peaks and concentrate with a speed vacuum concentrator.

The purity and molecular weight of the fractions were determined by tricine-SDS-PAGE using 12.5% polyacrylamide minigels under reducing conditions and by immunoblot analysis with the antisera to peptides PrP 23–40, 58–71, 90–102, 127–147, 151–165, and 220–231 (1:1,000) and the monoclonal antibody 3F4 (1:50,000).

5.4.3. High-Performance Liquid Chromatography

The fraction obtained by gel filtration that contained the major amyloid subunit (as revealed by SDS-PAGE and immunoblot analysis) was further purified by reverse-phase chromatography on a C4 column (214TP104, Vydac, Alltech, Deerfield, IL) with a 0–80% linear gradient of acetonitrile containing 0.1% (v/v) trifluoroacetic acid, pH 2.5. The column eluents were monitored at 214 nm and protein peaks were lyophilized. Following tricine-SDS-PAGE and immunoblot analysis, aliquots of the HPLC-purified, PrP-immunoreactive peptides were digested with endoproteinase Lys-C *(see* Section 5.4.4.*)*. The peptides generated from enzymatic digests were separated by reverse-phase chromatography on a Delta-Pak C18 column (0.39 × 30 cm, Waters, Milford, MA) with a 0–70% linear gradient of acetonitrile containing 0.1% (v/v) trifluoroacetic acid, pH 2.5. The column eluents were monitored at 214 nm and protein peaks were lyophilized.

5.4.4. Protocol for Enzymatic Digestion and Reduction of Peptides

1. Dissolve HPLC-purified amyloid peptides in 25 mM Tris, 1 mM EDTA, pH 8.5.
2. Incubate for 24 h at 37°C with endoproteinases Lys-C (Boehringer, Mannheim, Germany) at an enzyme-to-substrate ratio of 1:30 (w/w).
3. Terminate proteolysis by rapid freezing.

Following fractionation of enzymatic digests by HPLC *(see* Section 5.4.3.*)*, aliquots of lyophilized peptides were dissolved in 5% acetic acid and reduced with 0.725M dithiothreitol at 37°C for 30 h *(29)*. The reduced peptides were purified by HPLC on a reverse-phase C4 column.

5.4.5. Amino Acid Sequencing

Sequence analyses of the intact amyloid protein as well as of peptides generated by enzymatic digestion were carried out on a 477A microsequencer and the resulting phenylthiohydantoin amino acid derivatives were identified using the on-line 120A PTH analyzer and the standard program (Applied Biosystems, Foster City, CA).

5.4.6. Mass Spectrometry

C-terminal fragments of the amyloid protein generated by endoproteinase Lys-C digestion were reduced, repurified by HPLC on a reverse-phase C4 column, and subjected to electrospray mass spectrometry *(30)*. The analysis was carried out using a VG BioTech Bio-Q mass spectrometer with quadrupole analyzer. The mass scale was calibrated with myoglobin.

5.4.7. Immunocytochemistry

Coronal sections of cerebral hemispheres as well as sections of cerebellum and brain stem were fixed in 4% formaldehyde or Carnoy solution (i.e., ethanol:chloroform:acetic acid 6:3:1) and embedded in paraplast.

For light microscopy, 7-µm thick serial sections were stained with Congo red and thioflavine S, or incubated with the antisera to synthetic peptides corresponding to residues 23–40, 58–71, 90–102, 95–108, 127–147, 151–165, and 220–231 of human PrP, or residues 15–40, 90–102, and 220–232 of hamster PrP. The antisera were used at a dilution of 1:100/1:200 and were detected by the peroxidase-antiperoxidase (PAP) method with swine-antirabbit immunoglobulins, PAP complex (Dako, Santa Barbara, CA) and 3-3'-diaminobenzidine as chromogen. Before immunostaining, sections from formalin-fixed blocks were treated with 98% formic acid at room temperature for 30 min and/or subjected to hydrolytic autoclaving as follows. The sections were immersed in 1 mM hydrochloric acid in deionized water, autoclaved at 121°C for 10 min, and cooled to room temperature in tap water. In addition, selected sections from Carnoy-fixed blocks were digested with proteinase K (25 µg/mL) at 37°C for 10 min.

For electron microscopy, 50-µm thick paraplast-embedded sections were deparaffinized in xylene and rehydrated through graded ethanol solutions. Immunogold labeling with anti-PrP antisera (1:20) was performed by means of goat-antirabbit immunoglobulins conjugated with 10 nm colloidal gold particles (Biocell [Cardiff, UK], 1:20), following a pre-embedding procedure. Specificity of immunoreactions was checked by using normal rabbit serum as primary antibody or by peptide absorption.

5.5. Assembly and Conformation of Synthetic PrP Peptides

5.5.1. Synthesis and Purification of Peptides

The peptides for investigation were selected on the basis of the predicted secondary structure and the hydropathic profile of human PrP. By the analysis of these structural parameters, the 11 kDa amyloid protein purified from GSS 198 was divided into four peptides, corresponding to residues 57–64, 89–106, 106–126, and 127–147 of the amino acid sequence deduced from human PrP cDNA. PrP 57–64 (i.e., the octapeptide repeat) and PrP 89–106 (i.e., the N-terminal region of PrP 27–30) are hydrophilic, whereas PrP 106–126 and PrP 127–147 contain a sequence of hydrophobic amino acids flanked by hydrophilic residues. Scrambled sequences of these peptides were also used for the study.

Stepwise solid-phase peptide synthesis was carried out on a 430A synthesizer (Applied Biosystems) using 9-fluorenyl-methoxycarbonyl as the protective group for aminic residues, and N-hydroxybenzotriazole, *O*-benzotriazol-1-yl-

N,N,N',N'-tetramethyluronium hexafluorophosphate, and *N,N'*-dicyclohexyl-carbodiimide as activators of carboxylic residues. The peptides were purified by reverse-phase HPLC. Identity and purity of all peptides were determined by amino acid analysis using a Beckman (Fullerton, CA) 6300 analyzer and amino acid sequencing by automated Edman degradation.

5.5.2. Morphology and Staining Properties of Peptide Assemblies

To overcome solubility problems, all peptides were dissolved in 30% formic acid or in the α-helix stabilizing solvent 1,1,1,3,3,3-hexafluoro-2-propanol at a concentration of 0.5, 1, 2.5, 5, and 10 mg/mL. The solutions were dialyzed at room temperature for 24 h against deionized water or phosphate-buffered saline, pH 7.4. For light microscopy, 50 μL of each suspension were air dried on gelatin-coated slides, stained with 0.2% Congo red in 80% ethanol saturated with NaCl, and examined under polarized light, or treated with 1% aqueous thioflavine S and observed with fluorescent light (excitation filter 410–490 nm, barrier filter 520 nm). For electron microscopy, 5 μL of each suspension were applied to formvar-coated nickel grids, negatively stained with 5% uranyl acetate, and observed in a Philips EM 410 at 80 kV.

5.5.3. X-Ray Diffraction Analysis

X-ray diffraction was performed on peptides that proved to be fibrillogenic at electron microscopy examination. Peptides were analyzed as a lyophilized powder, allowed to sediment by slow dehydration after suspension in distilled water, or solubilized in either 30% formic acid or 1,1,1,3,3,3-hexafluoro-2-propanol prior to placement into 0.8-mm diameter siliconized Lindeman glass capillary tubes. Diffraction patterns were obtained after exposure to nickel-filtered CuKα radiation from a sealed X-ray tube generator source operated at 40 kV, 16 mA, with exposure times of 1–2 h. Spacings were measured directly on CEA Reflex 25 X-ray films, using a specimen-to-film distance of 7.5 cm.

5.5.4. CD Spectroscopy

The peptides were suspended in 200 m*M* phosphate buffer, pH 5.0 and 7.0, or in deionized water at a concentration of 0.5, 1.5, and 3.0 mg/mL. Following incubation at room temperature for 5 min or 1 h, CD spectra were recorded in quartz cells with an optical path of 0.01 or 0.1 cm, using a Jasco (Toyko, Japan) J-500 dichrograph at a scan speed of 1 nm/min. Mean residue ellipticities were calculated using the following equation:

$$(\theta)_M = A \times 3300 \times M/C \times l, \tag{1}$$

where A = observed dichroism absorbance, l = path length in cm, C = concentration of peptide in g/L, and M = mean residue weight. The percentages of the secondary structure of the peptides were calculated according to Yang et al. *(31)*. To investigate the influence of α-helix stabilizing solvents on the secondary structure of the peptides, 1,1,1-trifluoroethanol was added up to a final concentration of 50% to suspensions of peptides in phosphate buffer or deionized water. To analyze the effects of SDS micelles on peptide conformation, CD spectra were collected from suspensions of peptides in 5% SDS in deionized water or in 5 m*M* phosphate buffer, pH 7.4.

5.6. Biological Effects of PrP Peptides In Vitro

5.6.1. Neuronal Cultures

Hippocampal neuron cultures were prepared using the following protocol.

1. Remove brains from rat fetuses on embryonic d 17.
2. Dissect out hippocampus and dissociate cells in serum-free medium containing 0.1% trypsin (Difco, Surrey, UK) and 25 μg/mL deoxyribonuclease for 5 min at room temperature.
3. Plate cells at 5×10^5 cells/well in Primaria (Falcon, Becton Dickinson, Plymouth, UK) 15-mm wells, precoated with poly-D-lysine (50 μg/mL, Sigma).
4. Culture cells in basal medium Eagle (BME-Hank's salt, Gibco-BRL, Gaithersburg, MD) supplemented with 10% fetal calf serum (FCS, Gibco) and 2 m*M* glutamine. Keep cultures at 37°C in a humidified 5% CO_2 atmosphere.
5. After 5 d in vitro, halt nonneuronal cell division by exposure to $10^{-5}M$ cytosine arabinoside to prevent overgrowth of glial cells.

Additional cultures were prepared in tissue culture chamber slides (Nunc, Naperville, IL) to analyze culture composition immunocytochemically, using antibodies that selectively label astrocytes, microglial cells, or neurons.

Cultures were exposed either acutely or chronically to PrP peptides at a concentration of 20, 40, 60, and 80 μ*M*. For short-term treatment, peptides were applied once at the time of plating or after 9 d of culture; for long-term treatment, the peptides were added to the medium on d 0, 2, 4, 6, and 8 of culture. Control cultures were exposed to scrambled peptides or to vehicle only.

On d 10 cell viability was evaluated by light and electron microscopy or assessed quantitatively by a colorimetric method as follows. Neurons cultured in microtiter dishes were stained with crystal violet (0.5% in water/methanol 4:1). After washing, cells were dried, solubilized in sodium citrate/ethanol 1:1, and spectrophotometrically analyzed at 540 nm with an automated microplate reader (Perkin-Elmer [Norwalk, CT] lambda reader). The absorbance of the solution is proportional to the number of viable cells.

5.6.2. Glial Cell Cultures

Glial cell cultures were prepared using the following protocol.

1. Remove brains from neonatal rats and remove leptomeninges.

2. Dissociate nervous tissue by trituration with a Pasteur pipet.
3. Grow cells in poly-L-lysine coated Primaria (Falcon) dishes containing Dulbecco's modified minimal essential medium (DMEM; Gibco) supplemented with 10% FCS and 2 mM glutamine, at 37°C in a humidified 5% CO_2 atmosphere.
4. Change medium every 3 d.

Additional cultures were prepared in tissue culture chamber slides (Nunc) to analyze culture composition immunocytochemically, using antibodies that selectively label astrocytes, microglial cells, or neurons.

The cultures were treated with the peptides (10, 25, and 50 µM concentration) for 1 or 14 d. For short-term treatment, the PrP peptides were delivered 2 wk after plating, and the cultures were examined 24 h later. For long-term treatment, the PrP peptides were added to the medium every 3 d for 2 wk starting from d 0 or 7, and the cultures were analyzed on d 14 or 21, respectively. Control cultures were exposed to scrambled peptides or to vehicle only.

After treatments, cell viability and growth were estimated by light and electron microscopy, or assessed quantitatively by a colorimetric method as follows. Cultures were supplied with 3-(4,5-dimethylthiazol-2yl)-2,5-diphenyl tetrazolium bromide (MTT; 1.5 mg/mL), which is known to be converted into an insoluble blue formazan product by living cells but not by dying cells or their lytic debris *(32,33)*. The blue product was then solubilized by adding 80 mM HCl in isopropanol (500 µL/well) and the color intensity was measured at 570 nm using an automated microplate reader (Perkin-Elmer lambda reader). The optical density of the solution in the microwell is proportional to the number of viable cells. Cell culture growth was also quantitatively assessed by spectrophotometric determination of DNA *(34)* or [^{3}H]-thymidine uptake *(35)*.

Acknowledgments

This work was supported by the Italian Ministry of Health, Department of Social Services, and by the US National Institutes of Health (Grant NS29822).

References

1. Prusiner, S. B. (1991) Molecular biology of prion diseases. *Science* **252,** 1515–1522.
2. Prusiner, S. B. (1993) Genetic and infectious prion diseases. *Arch. Neurol.* **50,** 1129–1153.
3. Caughey, B. W., Dong, A., Bhat, K. S., Ernst, D., Hayes, S. F., and Caughey, W. S. (1991) Secondary structure analysis of the scrapie-associated protein PrP 27–30 in water by infrared spectroscopy. *Biochemistry* **30,** 7672–7680.
4. Pan, K.-M., Baldwin, M., Nguyen, J., Gasset, M., Serban, A., Groth, D., Mehlhorn, I., Huang, Z., Fletterick, R. J., Cohen, F. E., and Prusiner, S. B. (1993) Conversion of α-helices into β-sheets features in the formation of the scrapie prion protein. *Proc. Natl. Acad. Sci. USA* **90,** 10,962–10,966.

5. Safar, J., Roller, P. P., Gajdusek, D. C., and Gibbs, C. J. (1993) Conformational transitions, dissociation, and unfolding of scrapie amyloid (prion) protein. *J. Biol. Chem.* **268**, 20,276–20,284.

6. Ghetti, B., Dlouhy, S. R., Giaccone, G., Bugiani, O., Frangione, B., Farlow, M. R., and Tagliavini, F. (1995) Gerstmann-Sträussler-Scheinker disease and the Indiana kindred. *Brain Pathol.* **5**, 61–75.

7. Ghetti, B., Piccardo, P., Spillantini, M. G., Ichimiya, Y., Porro, M., Perini, F., Kitamoto, T., Tateishi, J., Seiler, C., Frangione, B., Bugiani, O., Giaccone, G., Prelli, F., Goedert, M., Dlouhy, S. R., and Tagliavini, F. (1996) Vascular variant of prion protein cerebral amyloidosis with τ-positive neurofibrillary tangles: the phenotype of the stop codon 145 mutation in *PRNP*. *Proc. Natl. Acad. Sci. USA*, **93**, 744–748.

8. Ghetti, B., Tagliavini, F., Giaccone, G., Bugiani, O., Frangione, B., Farlow, M. R., and Dlouhy, S. R. (1994) Familial Gerstmann-Sträussler-Scheinker disease with neurofibrillary tangles. *Mol. Neurobiol.* **8**, 41–48.

9. Giaccone, G., Tagliavini, F., Verga, L., Frangione, B., Farlow, M. R., Bugiani, O., and Ghetti, B. (1990) Neurofibrillary tangles of the Indiana kindred of Gerstmann-Sträussler-Scheinker disease share antigenic determinants with those of Alzheimer disease. *Brain Res.* **530**, 325–329.

10. Tagliavini, F., Giaccone, G., Prelli, F., Verga, L., Porro, M., Trojanowski, J., Farlow, M. R., Frangione, B., Ghetti, B., and Bugiani, O. (1993) A68 is a component of paired helical filaments of Gerstmann-Sträussler-Scheinker disease, Indiana kindred. *Brain Res.* **616**, 325–328.

11. Tagliavini, F., Prelli, F., Ghiso, J., Bugiani, O., Serban, D., Prusiner, S. B., Farlow, M. R., Ghetti, B., and Frangione, B. (1991) Amyloid protein of Gerstmann-Sträussler-Scheinker disease (Indiana kindred) is an 11 kd fragment of prion protein with an N-terminal glycine at codon 58. *EMBO J.* **10**, 513–519.

12. Tagliavini, F., Prelli, F., Porro, M., Rossi, G., Giaccone, G., Farlow, M. R., Dlouhy, S. R., Ghetti, B., Bugiani, O., and Frangione, B. (1991) Amyloid fibrils in Gerstmann-Sträussler-Scheinker disease (Indiana and Swedish kindreds) express only PrP peptides encoded by the mutant allele. *Cell* **79**, 695–703.

13. Dlouhy, S. R., Hsiao, K., Farlow, M. R., Foroud, T., Conneally, P. M., Johnson, P., Prusiner, S. B., Hodes, M. E., and Ghetti, B. (1992) Linkage of the Indiana kindred of Gerstmann-Sträussler-Scheinker disease to the prion protein gene. *Nature Genet.* **1**, 64–67.

14. Hsiao, K., Dlouhy, S. R., Farlow, M. R., Cass, C., DaCosta, M., Conneally, P. M., Hodes, M. E., Ghetti, B., and Prusiner, S. B. (1992) Mutant prion proteins in Gerstmann-Sträussler-Scheinker disease with neurofibrillary tangles. *Nature Genet.* **1**, 68–71.

15. Giaccone, G., Verga, L., Bugiani, O., Frangione, B., Serban, D., Prusiner, S. B., Farlow, M. R., Ghetti, B., and Tagliavini, F. (1992) Prion protein preamyloid and amyloid deposits in Gerstmann-Sträussler-Scheinker disease, Indiana kindred. *Proc. Natl. Acad. Sci. USA* **89**, 9349–9353.

16. Tagliavini, F., Prelli, F., Porro, M., Rossi, G., Giaccone, G., Bird, T. D., Dlouhy, S. R., Young, K., Piccardo, P., Ghetti, B., Bugiani, O., and Frangione, B. (1991)

Only mutant PrP participates in amyloid formation in Gerstmann-Sträussler-Scheinker disease with Ala → Val substitution at codon 117. *J. Neuropathol. Exp. Neurol.* **54**, 416.

17. Tagliavini, F., Prelli, F., Verga, L., Giaccone, G., Sarma, R., Gorevic, P., Ghetti, B., Passerini, F., Ghibaudi, E., Forloni, G., Salmona, M., Bugiani, O., and Frangione, B. (1993) Synthetic peptides homologous to prion protein residues 106–147 form amyloid-like fibrils *in vitro*. *Proc. Natl. Acad. Sci. USA* **90**, 9678–9682.

18. Selvaggini, C., De Gioia, L., Cantù, L., Ghibaudi, E., Diomede, L., Passerini, F., Forloni, G., Bugiani, O., Tagliavini, F., and Salmona, M. (1993) Molecular characteristics of a protease-resistant, amyloidogenic and neurotoxic peptide homologous to residues 106–126 of the prion protein. *Biochem. Biophys. Res. Commun.* **194**, 1380–1386.

19. De Gioia, L., Selvaggini, C., Ghibaudi, E., Diomede, L., Bugiani, O., Forloni, G., Tagliavini, F., and Salmona, M. (1994) Conformational polymorphism of the amyloidogenic and neurotoxic peptide homologous to residues 106–126 of the prion protein. *J. Biol. Chem.* **269**, 7859–7862.

20. Forloni, G., Angeretti, N., Chiesa, R., Monzani, E., Salmona, M., Bugiani, O., and Tagliavini, F. (1993) Neurotoxicity of a prion protein fragment. *Nature* **362**, 543–545.

21. Forloni, G., Del Bo, R., Angeretti, N., Chiesa, R., Smiroldo, S., Doni, R., Ghibaudi, E., Salmona, M., Porro, M., Verga, L., Giaccone, G., Bugiani, O., and Tagliavini, F. (1994) A neurotoxic prion protein fragment induces rat astroglial proliferation and hypertrophy. *Eur. J. Neurosci.* **6**, 1415–1422.

22. Florio, T., Grimaldi, M., Scorziello, A., Salmona, M., Bugiani, O., Tagliavini, F., Forloni, G., and Schettini, G. (1995) The prion protein fragment 106–126 increases intracellular calcium levels through a dihydropyridine-sensitive mechanism and induces cortical type I astrocyte proliferation *in vitro*. *Soc. Neurosci. Abstr.* **21(1)**, 494.

23. Posnett, D. N., McGrath, H., and Tam, J. P. (1988) A novel method for producing anti-peptide antibodies. *J. Biol. Chem.* **256**, 495–497.

24. Barry, R. A., Vincent, M. T., Kent, S. B., Hood, L. E., and Prusiner, S. B. (1988) Characterization of prion proteins with monospecific antisera to synthetic peptides. *J. Immunol.* **140**, 1188–1193.

25. Kascsak, R. J., Rubenstein, R., Merz, P. A., Tonna-DeMasi, M., Fersko, R., Carp, R. I., Wisniewski, H. M., and Diringer, H. (1987) Mouse polyclonal and monoclonal antibody to scrapie-associated fibril proteins. *J. Virol.* **61**, 3688–3693.

26. Rogers, M., Serban, D., Gyuris, T., Scott, M., Torchia, T., and Prusiner, S. B. (1991) Epitope mapping of the Syrian hamster prion protein utilizing chimeric and mutant genes in a vaccinia virus expression system. *J. Immunol.* **147**, 3568–3574.

27. Serban, D., Taraboulos, A., DeArmond, S. J., and Prusiner, S. B. (1990) Rapid detection of Creutzfeldt-Jakob disease and scrapie prion proteins. *Neurology* **40**, 110–117.

28. Schagger, H. and von Jagow, G. (1987) Tricine-sodium dodecyl sulfate-polyacrylamide gel electrophoresis for the separation of proteins in the range from 1 to 100 kDa. *Anal. Biochem.* **166**, 368–379.

29. Houghten, R. A. and Li, C. H. (1983) Reduction of sulfoxides in peptides and proteins, in *Methods in Enzymology, Vol. 91: Enzyme Structure Part I* (Hirs, C. H. W. and Timasheff, S. N., eds.), Academic, New York, pp. 549–559.

30. Biemann, K. (1992) Mass spectrometry of peptides and proteins. *Ann. Rev. Biochem.* **61**, 977–1010.

31. Yang, J. T., Wu, C.-S. C., and Martinez, H. M. (1986) Calculation of protein conformation from circular dichroism, in *Methods in Enzymology,* vol. 130, Academic, New York, pp. 208–269.

32. Mosmann, T. (1983) Rapid colorimetric assay for cellular growth and survival: application to proliferation and cytotoxic assay. *J. Immunol. Methods* **65**, 55–63.

33. Manthorpe, M., Fagnani, R., Skaper, S., and Varon, S. (1986) An automated colorimetric microassay for neurotrophic factors. *Dev. Brain. Res.* **25**, 191–198.

34. Maniatis, T., Fritsch, E. F., and Sambrook, J. (1982) *Molecular Cloning: A Laboratory Manual.* Cold Spring Harbor Laboratory. Cold Spring Harbor, NY.

35. Florio, T., Pan, M. G., Newman, B., Hershberger, R. E., Civelli, O., and Stork, P. J. S. (1992) Dopaminergic inhibition of DNA synthesis in pituitary tumor cells is associated with phosphotyrosine phosphatase activity. *J. Biol. Chem.* **267**, 24,169–24,172.

17

Methods for Studying Prion Protein (PrP) Metabolism and the Formation of Protease-Resistant PrP in Cell Culture and Cell-Free Systems

Byron Caughey, David A. Kocisko, Suzette A. Priola,
Gregory J. Raymond, Richard E. Race, Richard A. Bessen,
Peter T. Lansbury, Jr., and Bruce Chesebro

1. Introduction

The pathogenesis of scrapie and other transmissible spongiform encephalopathies (TSEs) appears to be based on the posttranslational conversion of the host's protease-sensitive prion protein (PrP-sen or PrP^c) to abnormal protease-resistant forms (PrP-res or PrP^{Sc}). In vitro studies using both scrapie-infected tissue culture cells and cell-free reactions have been used effectively to help define the cellular and molecular details of this process and how it might be inhibited. Here we discuss experimental approaches that we have used in these studies.

1.1. PrP-Sen and PrP-Res Metabolism

Striking differences in the cellular metabolism of PrP-sen and PrP-res have been observed that may account for the pathogenic accumulation of PrP-res in scrapie-infected animals (reviewed in ref. *1*). The abnormal accumulation of PrP-res in humans with familial TSEs may also be potentiated by abnormal metabolism of mutant PrP-sen molecules. Thus, comparative analyses of the metabolism of PrP forms remain important in understanding the etiology and pathogenesis of TSEs. PrP metabolism has been studied in C127 cells *(2)*, murine neuroblastoma (MNB) cells (reviewed in ref. *3*), hamster HaB cells (reviewed in ref. *4*), fibroblasts *(5)*, PC12 cells (R. Rubenstein and B. Caughey, unpublished data), T-cells *(6)*, and mixed leukocytes *(5)*. The best characterized experimental models of PrP metabolism are the MNB cells and their

From: *Methods in Molecular Medicine: Prion Diseases*
Edited by: H. Baker and R. M. Ridley Humana Press Inc., Totowa, NJ

scrapie-infected counterparts (Sc⁺MNB) *(7–9)*. In this section we will focus on methodologies used in metabolic studies with these cells.

1.1.1. Generation and Maintenance of Sc⁺MNB Cells

The Neuro 2A or C1300 clones of MNB cells can be infected using brain homogenates from Chandler scrapie-infected mice as an inoculum *(7–9)*. Typically, a 10% w/v suspension of scrapie-infected brain in physiological buffer is prepared by Dounce homogenization. The suspension is cleared of large debris by low speed centrifugation, diluted 1:4 in tissue culture medium (MEM with 10% fetal bovine serum [FBS]) and incubated over a 50% confluent cell monolayer with rocking for 4 h. The medium/inoculum is diluted further to 2 mL with medium and the cells are incubated for an additional 4 h. Six milliliters of additional fresh medium is added and the cells are grown another 1–2 d to confluence. To obtain more homogeneously infected cultures, single-cell clones are derived by limiting dilution and analyzed for scrapie infection using PrP-res as a marker. Often only 1–2% of clones derived from the initial bulk culture replicate scrapie infectivity and produce PrP-res *(8)*. The stability of the scrapie infection varies in cultures derived from individual clones. Some have produced PrP-res for over 100 serial passages, whereas others have unpredictably lost scrapie infectivity (or PrP-res production) after only a few passages. Two factors that can be important in maintaining the infection are the culture medium and FBS. We have identified individual lots of each that reproducibly cure the Sc⁺MNB cultures of the scrapie infection within one to two passages. Thus, individual lots of these components should be screened for use with Sc⁺MNB. In this respect, Gibco (Gaithersburg, MD) OptiMEM (supplemented with 10% FBS) has been the least problematic medium that we have used. Cell culture crises, such as overgrowth and loss of CO_2 in the incubator, also can be detrimental to the scrapie infection without permanently affecting the growth of the cells. Once infected clones are identified, aliquots can be frozen in 10% dimethylsulfoxide, 20% OptiMEM, and 70% FBS and stored in liquid nitrogen.

The potential instability of the infection requires that Sc⁺MNB cultures routinely be monitored for the production of PrP-res. This can be done by lysing a monolayer of cells in a lysing buffer (LB) containing 0.5% Triton X-100 and 0.5% sodium deoxycholate (1 mL/25 cm² flask), clearing the lysate of debris with a 5-min centrifugation at 1000g, digesting the supernatant with 50 μg/mL proteinase K for 30 min at 37°C, pelleting the PrP-res at 350,000g for 45 min, and analyzing the pellet for PrP-res by immunoblot *(10)*.

1.1.2. Metabolic Labeling and Immunoprecipitation of PrP

Many types of analyses of PrP metabolism require labeling various forms of PrP with [³⁵S]methionine and/or [³⁵S]cysteine in live cells followed by immunoprecipitation of the protein from cell lysates. Our usual procedure for MNB cells involves preincubating the cells in methionine and cysteine-deficient MEM (supplemented with 1% dialyzed FBS) for 1 h and incubating with Expre^{35}S^{35}S (DuPont NEN, Boston, MA) label (0.1–1 mCi/25 cm^2 flask of 60–80% confluent cells) for various time periods (typically, 2 h) in this same medium. The cells can then be washed and lysed in ice-cold LB immediately or after a chase incubation in full-methionine/cysteine medium.

The lysates are cleared by centrifugation for 5 min at 1000g at 4°C, and the proteins precipitated with 4 vol of methanol at –20°C for ≥1 h. The precipitated proteins are resuspended by cuphorn sonication into DLPC buffer *(11)* (1 mL/25 cm^2 flask equivalent of cell proteins). The appropriate PrP-specific antibody is incubated with the suspension and antibody-antigen complexes plus free antibodies are collected by binding to protein A sepharose beads. The beads are washed prior to elution and SDS-PAGE/fluorography. The stringency of the washes can be adjusted for different antibodies to minimize nonspecific background without destabilizing the protein A–antibody-PrP complex. For the selective immunoprecipitation of PrP-res, the cleared lysate can be treated with PK as described earlier to eliminate PrP-sen prior to methanol precipitation. Alternatively, the PrP-res can be isolated from the PrP-sen by ultracentrifugation of the cleared lysate and suspension of the pellet in DLPC buffer *(12)*.

1.1.3. PrP Metabolism

1.1.3.1. Kinetics of Biosynthesis and Turnover

The rates of biosynthesis and turnover of various PrP molecules can be compared using standard pulse-chase metabolic labeling experiments *(11,13,14)*. The vast difference in the pulse labeling rates of PrP-sen and PrP-res provided the first metabolic indication that PrP-res was derived from PrP-sen *(11,14)*. Similar experiments can also be used to determine if these rates are affected by other non-PrP factors. For instance, pulse-chase labeling experiments showed that PrP-res accumulation can be inhibited by Congo red and certain sulfated glycans without affecting the formation and half-life of its precursor, PrP-sen *(vide infra) (10,15)*.

1.1.3.2. Addition of Glycophosphatidylinositol (GPI)

The addition of a GPI moiety is one of the earliest events in PrP biosynthesis *(13,16)*. Several observations can be used as evidence for the presence of the GPI moiety on PrP *(13,16–19)*:

1. Direct chemical analysis.
2. Metabolic labeling with radioactive precursors of the GPI moiety.

3. Shifts in SDS-PAGE migration (approx 1 kDa) of PrP molecules on treatment with phosphatidylinositol-specific phospholipase C (PI-PLC).
4. Release of PrP from cells or membranes by PI-PLC.

In the last case, it should be emphasized that the lack of release cannot be taken as evidence that the GPI moiety is missing since certain GPI-linked PrP forms (e.g., PrP-res) are known to be resistant to release by PI-PLC *(17–19)*.

1.1.3.3. N-Linked Glycosylation

PrP is variably glycosylated at two Asn-linked glycosylation sites *(13,20–24)* and there is evidence that the extent of glycosylation can affect the efficiency of PrP-sen conversion to PrP-res *(14)*. The N-linked glycosylation of PrP forms can be monitored by specific metabolic labeling with radioactive glycan precursors, such as N-acetylglucosamine or fucose, or by changes in SDS-PAGE migration on treatment of the PrP molecules with endoglycosidases *(13,25)*. Lectins and endoglycosidases with varying specificities can be used to discriminate between various types of glycans *(20,21,26,27)*. For instance, endoglycosidase H will remove only the high mannose glycans present on early PrP precursors produced, whereas endoglycosidase F and peptidyl-N-glycanase can remove the more mature complex and hybrid glycans as well. The glycosylation of PrP can be manipulated metabolically by using specific inhibitors *(28)*. Tunicamycin blocks the formation of the precursor of N-linked glycans, and, when applied to cells making PrP, leads to the formation of only the 25–26 kDa unglycosylated PrP-sen and 19 kDa PrP-res *(13,25)*.

1.1.3.4. Transport Of PrP to and from the Cell Surface

Studies with a variety of cells have indicated that PrP-sen can be labeled on the cell surface by membrane immunofluorescence *(2,11,16,17,29)*, biotinylation *(11)*, radioiodination *(14,29)*, and immunogold *(30–33)*. Cell surface PrP-sen is also susceptible to treatments of intact cells by trypsin, proteinase K, dispase, and PI-PLC *(5,13,16,17,29,34)*. Although trypsin and proteinase K have the added effect of releasing the tissue culture cells from their vessels, PI-PLC and dispase can remove PrP-sen from the plasma membrane without dislodging the cells. The sensitivity of PrP-sen in intact cells to PI-PLC varies even between laboratories working with MNB cells *(11,14,17,34)*, but the reason for this variability is not clear.

The rate at which PrP-sen is translocated to the plasma membrane can be tested by monitoring its exposure to extracellular proteases or PI-PLC in pulse-chase metabolic labeling experiments *(13,34)*. The transport of newly synthesized PrP-sen through the Golgi apparatus and to the plasma membrane can be disrupted by brefeldin A *(35)*.

Once PrP-sen is at the cell surface, its reinternalization and cycling through endocytic processes can be monitored after surface iodination, immunofluorescence, and immunogold labeling *(29,31)*. PrP-res is formed on the plasma membrane or along an endocytic pathway to the lysosomes in Sc$^+$MNB cells *(12,14,34,35)*. The N-terminal truncation of PrP-res by cellular acid proteases that are sensitive to lysosomotropic amines can be used as a marker for the transport of PrP-res to an acidic endosomal or lysosomal compartment *(12,35)*. Immunochemical staining has indicated that PrP-res accumulates in secondary lysosomes in Sc$^+$MNB cells *(36)*.

1.1.3.5. PRECURSOR-PRODUCT RELATIONSHIPS

As noted, pulse-chase labeling experiments can be used to help establish precursor–product relationships between various forms of PrP. More definitive evidence that PrP-res is derived from PrP-sen in Sc$^+$MNB cells has been obtained by combining the pulse-labeling approach with the ability to selectively remove PrP-sen from the cell surface enzymatically *(14,34)*. When PrP-sen was labeled with a short pulse-chase and then selectively removed from the cell surface by PI-PLC or proteases, no subsequent labeling of PrP-res occurred. This established that PrP-res is derived from a protease- and phospholipase-sensitive precursor that is at least transiently exposed on the cell surface. Similar approaches might be useful in determining whether this sequence of events varies in other TSE-infected cell types or in cells expressing mutant PrP molecules.

1.1.4. Inhibitors of PrP-res Accumulation and Scrapie Agent Replication

One useful approach to dissecting the mechanism of PrP-res formation that may also lead to effective drug treatments for the TSEs is the identification of inhibitors of PrP-res formation. Sc$^+$MNB cell cultures can be used to screen for such inhibitors *(10,15)*. Typically, potential inhibitors are added to the medium of cultures seeded at low density. The cultures are grown up to confluence and assayed for effects on the accumulation of PrP-res by immunoblot. Compounds such as Congo red and certain sulfated glycans have been shown to block PrP-res accumulation and scrapie agent replication at concentrations in the nanomolar or subnanomolar range and may act by blocking PrP-glycosaminoglycan interactions *(10,15,37,38)*. These inhibitors also are known to prolong the lives of animals inoculated with scrapie *(39–43)*.

The selectivity of an inhibitor for PrP-res formation, as opposed to PrP biosynthesis generally, can be tested using the described pulse-chase metabolic labeling experiments to look for effects on PrP-sen labeling, transport to the cell surface, and turnover. For instance, Congo red and the sulfated glycan inhibitors block PrP-res accumulation without apparent effects on PrP-sen metabolism *(10,15)*.

1.1.5. Expression of Recombinant PrP Molecules in MNB Cells

Recombinant PrP molecules have been expressed in a variety of eukaryotic vector systems *(2,44–47)*. We have used a retroviral expression vector, pSFF, which is derived from the murine spleen focus-forming retrovirus *(45,46)*. Recombinant PrP is cloned into the pSFF polylinker using standard cloning protocols. The PrP-pSFF clone is transfected into the retroviral packaging cells Ψ2 and PA317 *(45)*. We maintain Ψ2 cells in 10% calf bovine serum in DMEM and PA317 cells in 10% FBS in RPMI. The PrP-pSFF vector is packaged into an infectious retroviral particle and a "ping-pong" effect leads to the spread of the infectious retrovirus throughout the culture *(48)*. Viral spread of the recombinant PrP is monitored at each cell passage by membrane immunofluorescence using an antibody specific to the recombinant PrP or by cytoplasmic immunofluorescence using an antibody to a retroviral gag protein expressed from the pSFF vector *(17,45)*. Depending on the efficiency of the initial transfection, the PrP-containing retrovirus will spread to 100% of the cells within 1–2 wk. Occasionally, when <30–40% of the cells are positive for the recombinant PrP, retroviral spread will stop. This is a result of interference of viral spread by infectious helper virus that has been "rescued" from the Ψ2/PA317 culture *(48)*. If this occurs, the cultures should be discarded, replaced with early passage Ψ2 and PA317 cells, and the transfection repeated.

When 100% of the cells express the recombinant PrP protein, viral supernatants are harvested. The cells are split and when the culture is approx 70% confluent, the media is replaced with fresh media and the cells are incubated a further 24 h. The media is collected the next day, cell debris is centrifuged out at low speed, and the cleared supernatant is stored in aliquots at −70°C. Depending on cell viability, one to three viral supernatants can be harvested over consecutive days. The supernatants can be used directly to infect the desired cell type. Alternatively, the Ψ2/PA317 cells expressing the recombinant PrP can be cloned by standard limiting dilution cell cloning to derive a clonal cell line expressing the PrP gene of interest.

The recombinant PrP-pSFF retroviral supernatants can be used to infect any cell type susceptible to infection by an amphotropic or ecotropic murine retrovirus. We have used both mouse neuroblastoma cells *(46,49,50)* and monkey Vero cells. Briefly, cells are split into a six-well tissue culture plate at $3–5 \times 10^5$ cells/well and incubated overnight at 37°C. The cells should be 70–80% confluent by the next day. Original medium is replaced with fresh medium containing 1 mg/mL polybrene, and a 1-mL aliquot of the retroviral supernatant is added directly to the cells. The culture is incubated overnight at 37°C, split the next day at 1:5 into a new six-well tissue culture plate, and assayed 1–4 d later for expression of the recombinant PrP by immunofluorescence *(17,45)*. Typically, from 50–100% of the cells will be positive for the recombinant PrP protein. The cells are cloned by limiting dilution to derive a clonal cell line expressing high levels of the recombinant PrP protein. One or more copies of the recombinant PrP will be integrated into the cellular DNA, but deletions in the recombinant PrP gene occur with variable frequency in this system and it is important to insure that full-length recombinant PrP protein is expressed in the clonal cell line. The size of the clonal insert(s) can be confirmed by genomic Southern blot while the recombinant PrP protein is assayed by radioimmunoprecipitation as described previously (*see* Section 1.2.). The mouse neuroblastoma clonal cell lines can stably express recombinant PrP protein and be maintained indefinitely in culture or frozen in liquid nitrogen by standard procedures.

2. Materials

2.1. Reagents

1. TEND: (10 mM Tris-HCl, pH 8.3, 1 mM EDTA, 130 mM NaCl, 1 mM dithiothreitol, supplemented with 0.5 μg/mL leupeptin, 1.0 μg/mL aprotinin, 0.7 μg/mL pepstatin, 0.1 mM Petabloc SC, Boehringer, Mannheim (Mannheim, Germany).
2. 10% Sarkosyl (N-lauryl-sarcosine) in TEND.
3. 10% NaCl, 1% sulfobetaine (SB) in TEND.
4. TMCS: (10 mM Tris-HCl, 5 mM MgCl$_2$, 5 mM CaCl$_2$, 100 mM NaCl, pH 7.4).
5. SNSB: (1M sucrose, 100 mM NaCl, 0.5% sulfobetaine).
6. Stock 3M guanidine hydrochloride (GdnHCl)—to be diluted as appropriate.
7. TN (130 mM NaCl, 50 mM Tris-HCl, pH 7.4 at 20°C).
8. Cetylpyridinium chloride (CPC).

3. Methods

3.1. Cell-Free Conversion of PrP-Sen to Protease-Resistant Forms

Until recently, the simplest system capable of generating PK-resistant PrP was an intact scrapie-infected tissue culture cell. Efforts to observe the conversion of PrP-sen to PrP-res in subfractions of such cells have failed *(51)*. To simplify the analysis of PrP-res formation, we recently developed a cell-free system that converts PrP-sen to PK-resistant PrP *(52–54)*. This system facilitates more defined studies of the conversion reaction. The main strategy that we used in developing this PrP-sen-to-PrP-res conversion was to keep the reactants as concentrated as possible to help drive what might otherwise be a slow or inefficient reaction. The major components of the cell-free conversion reaction are PrP-res purified from scrapie-infected brain tissue and metabolically labeled PrP-sen immunoprecipitated from uninfected tissue culture cells.

3.1.1. Isolation of the Cell-Free Conversion Reaction Components

3.1.1.1. PRP-RES

The purification method we have used is a modification of the procedure of Bolton and coworkers *(55,56)*. We will present this protocol in some detail because this particular preparation of PrP-res may be important in getting the conversion reaction to work as we have described *(52–54)*.

1. Thaw brains from scrapie-infected hamsters or mice and rinse in phosphate-buffered saline.
2. Using a Dounce homogenizer, make a 10% (w/v) homogenate in 10% sarkosyl (N-lauryl-sarcosine) in TEND.
3. Centrifuge the homogenate in a Beckman (Fullerton, CA) Ti 50.2 rotor at 28,000*g* at 4°C for 30 min and keep the pellet.
4. Centrifuge the supernatant at 180,000*g* at 4°C for 2.5 h in the same rotor. Discard the supernatant and keep the pellet.
5. Combine the pellets, rinse them, and resuspend in 10% NaCl and 1% sulfobetaine (SB) in TEND using a Dounce homogenizer. Centrifuge the suspension at 250,000*g* for 90 min at 20°C (TL 50.2 rotor). Discard the supernatant.
6. Rinse pellet and resuspend by Dounce homogenization in TMCS. Add DNase I and RNase A to concentrations of 20 and 100 μg/mL, respectively, and agitate the suspension overnight at 4°C.
7. To the suspension add EDTA to 20 m*M*, NaCl to 10% w/v, and SB to 1% and layer the suspension above a cushion of SNSB. Centrifuge at 250,000*g* for 90 min at 20°C.
8. Discard the supernatant, rinse the pellet, and sonicate into 0.5% sulfobetaine in PBS.

The purity of the preparation is checked using SDS-PAGE with silver staining. The yield can be determined by a BCA protein assay (Pierce, Rockford, IL) if it is sufficiently pure and, if not, by quantitative immunoblotting. The PrP-res preparations can be stored frozen indefinitely, but repeated freeze–thaw cycles may affect their behavior.

3.1.1.2. PRP-SEN

The PrP-sen that is used as a substrate in the cell-free reaction is labeled for 90–120 min with ^{35}S in uninfected tissue culture cells (e.g., MNB, Ψ2, or PA317 cells) and immunoprecipitated as described earlier (*see* Section 1.2.), except that after a final water wash of the PrP–antibody–protein A Sepharose complex, the residual liquid is drawn off the bead pellet with a Hamilton syringe and the PrP is eluted in a minimal volume of 3–7.5*M* GdnHCl (e.g., 25 μL/5 mg dry wt equivalent of beads). The beads are agitated in the eluant for 15 min at 37°C. The eluate is drawn off the beads, and the elution step is repeated with fresh eluant. Both eluates usually have enough cpm of ^{35}S-PrP-sen

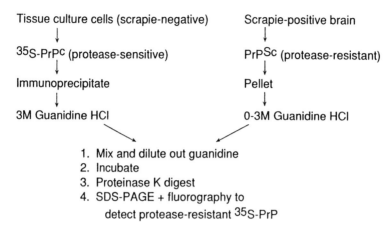

Fig. 1. Flow diagram of the cell-free reaction for converting PrP-sen to PK-resistant forms.

to be useful in conversion reactions. The eluates are stored on ice and, other than radioactive decay, are often stable for weeks.

3.1.2. The Cell-Free Conversion Reaction

The conversion of ^{35}S-PrP-sen to proteinase K-resistant forms can be achieved by incubating ^{35}S-PrP-sen with brain-derived PrP-res that has been pretreated with 0–3M guanidine hydrochloride (GdnHCl) at 37°C for 0.25–24 h (Fig. 1) *(52–54)*. The optimal GdnHCl concentration varies slightly with the PrP-res preparation. For instance, the converting activity of many preparations of hamster PrP-res (263K strain) is enhanced by a 3M GdnHCl pretreatment *(52)*, whereas the activity of some other preparations of 263K PrP-res have been optimal with pretreatments of 2–2.5M GdnHCl. The reason for this variation in GdnHCl sensitivity is unclear but it may be influenced in part by the purity of the preparation or repeated freeze–thaw cycles. A 1–3 μg/μL suspension of PrP-res in the desired GdnHCl concentration is mixed with an equal volume of ^{35}S-PrP-sen eluate prediluted to the same GdnHCl concentration. The mixture is then diluted further to 0.75M GdnHCl using TN supplemented with cetylpyridinium chloride (CPC) to give a final concentration of 1.5 mM CPC and incubated at 37°C for ≥16 h.

The PK-resistance of the ^{35}S-PrP is then tested after a further dilution of the GdnHCl to 0.075 – 0.4M by digesting with 20–500 μg/mL of PK for 1 h at 37°C *(54)*. The PK is inhibited using Pefabloc SC (Boehringer Mannheim) as recommended by the manufacturer. Twenty micrograms of a carrier protein (thyroglobulin) is added, and the proteins are precipitated with methanol. The resulting pellet is boiled in SDS-PAGE sample buffer and electrophoresed. Radiolabeled PrP-res was visualized by fluorography with Entensify (DuPont, Wilmington, DE) or Phosphorimager analysis (Molecular Dynamics, Sunnyvale, CA).

In summary, a typical reaction would be:

1. 2 μL of 1 mg/mL hamster PrP-res in $3M$ GdnHCl.
2. 2 μL hamster ^{35}S-PrP-sen with 30,000 cpm/μL in $3M$ GdnHCl.
3. 12 μL TN + CPC.
4. Mix these constituents, sonicate briefly, and incubate for 1 d at 37°C.
5. Dilute with 64 μL TN and then add 4 μL of 1 mg/mL proteinase K.
6. After the 37°C incubation, add 20 μL of 5 mM Pefabloc SC, 4 μL 5 mg/mL thyroglobulin, and 4 vol methanol.
7. After 1 h at –20°C, the tube is centrifuged for 15 min at 11,000g.
8. Sonicate and boil the pellet into 20 μL of sample buffer. Analyze by SDS-PAGE-fluorography.

Using these conditions, newly converted PK-resistant ^{35}S-PrP bands often can be detected with an overnight fluorographic exposure of the gel. Since PK treatment of scrapie brain-derived PrP-res generally results in a 6–7 kDa reduction in apparent molecular mass, we look for PK-resistant ^{35}S-PrP bands that are 6–7 kDa smaller than the corresponding untreated ^{35}S-PrP-sen precursor *(52–54)*. The presence of ^{35}S-PrP bands that are the same size as the ^{35}S-PrP-sen precursor would suggest that some nonspecific trapping or protection of ^{35}S-PrP-sen precursor from PK may have occurred. We have also observed PK-resistant ^{35}S-PrP bands that are >6–7 kDa smaller than the ^{35}S-PrP-sen precursors, which has suggested that a partial conversion occurred *(52,53)*. The relationship between these incomplete cell-free conversion products and brain-derived PrP-res molecules is not clear. To control for the role of PrP-res in the conversions, comparisons can be made to samples of ^{35}S-PrP-sen incubated without PrP-res or samples to which PrP-res is added immediately before the PK digestion *(52,53)*. The efficiency of conversion of hamster ^{35}S-PrP-sen to PK-resistant species using the above protocol with hamster PrP-res has been variable in our hands. Usually 5–30% of the ^{35}S-PrP-sen precursor in the reaction becomes PK-resistant but, on occasion, as much as 100% of the label is converted. The reason for this variability is not yet understood.

3.1.3. Species and Strain Specificities in the Cell-Free Conversion Reaction

The cell-free conversion reaction can be used to investigate the species specificity of PrP-res-PrP-sen interactions. For instance, mouse and hamster PrP-res have been combined in conversion reactions with mouse, hamster, and chimeric PrP-sen molecules *(53)*. The species specificities in the conversion reactions correlated with the relative transmissibilities of these scrapie strains in vivo. The hamster PrP-res (263K strain) would not convert mouse PrP-sen to PK-resistant forms just as the hamster scrapie is not infectious to mice. Chandler mouse PrP-res converted both mouse and hamster PrP-sen molecules to PK-resistant forms that correlated with the ability of the Chandler mouse scrapie strain to infect both mice and hamsters.

Conversion experiments performed with chimeric mouse/hamster PrP-sen precursors identified certain species-specific residues that affected the conversion efficiency and the size of the resultant protease-resistant PrP species. These studies indicated that incompatibilities in direct PrP-res-PrP-sen interactions may be a molecular basis for the barriers to interspecies transmissions of TSEs in vivo.

The observation of this type of specificity in the cell-free conversion reaction and its correlation with in vivo species barrier effects suggests that this experimental system may be useful for quick in vitro tests of the susceptibility of various species to TSE agents of other species. Of particular interest at present is the issue of whether humans are likely to be susceptible to bovine spongiform encephalopathy (BSE) agent. Tests of the ability of PrP-res from BSE cattle brain to convert human PrP-sen to PK-resistant forms might shed light on this subject. However, given the newness of the cell-free conversion reaction and the likelihood that other factors profoundly influence TSE species barrier effects in vivo, caution should be used in extrapolating between the in vitro and in vivo situations. The conditions for optimal cell-free conversion may differ from those described herein when PrP-res and PrP-sen derived from various species and strains of TSEs are used.

The scrapie-strain specificity of the cell-free conversion reaction also can be investigated under circumstances in which there is no difference in the amino acid sequence between PrP-res and PrP-sen *(54)*. This type of experiment can be used to model the strain-specific differences in PrP-res formation that occur in a single host species. For example, two hamster adapted strains of transmissible mink encephalopathy (TME) lead to the in vivo formation of PrP-res forms that are cleaved differently by PK. When these two forms of PrP-res were reacted with the same labeled PrP-sen precursor in the cell-free conversion reaction, the resulting conversion products exhibited a similar difference in PK cleavage. These data provided evidence that the strain-specific properties (or phenotypes) of PrP-res can be transmitted to PrP-sen as a result of direct PrP-res-PrP-sen interactions.

References

1. Caughey, B. (1991) Cellular metabolism of normal and scrapie-associated forms of PrP. *Sem. Virol.* **2,** 189–196.

2. Caughey, B., Race, R. E., Vogel, M., Buchmeier, M. J., and Chesebro, B. (1988) In vitro expression in eukaryotic cells of the prion protein gene cloned from scrapie-infected mouse brain. *Proc. Natl. Acad. Sci. USA* **85**, 4657–4661.

3. Caughey, B. (1993) Scrapie associated PrP accumulation and its prevention: insights from cell culture. *Br. Med. Bull.* **49**, 860–872.

4. Taraboulos, A., Borchelt, D. R., McKinley, M. P., Raeber, A., Serban, D., DeArmond, S. J., et al. (1992) Dissecting the pathway of scrapie prion synthesis in cultured cells, in *Prion Diseases of Humans and Animals* (Prusiner, S. B., Collinge, J., Powell, J., and Anderton, B., eds.), Ellis Horwood, Chichester, pp. 435–444.

5. Meiner, Z., Halimi, M., Polakiewicz, R. D., Prusiner, S. B., and Gabizon, R. (1992) Presence of prion protein in peripheral tissues of Libyan Jews with Creutzfeldt-Jakob disease. *Neurology* **42**, 1355–1360.

6. Cashman, N. R., Loertscher, R., Nalbantoglu, J., Shaw, I., Kascsak, R. J., Bolton, D. C., et al. (1990) Cellular isoform of the scrapie agent protein participates in lymphocyte activation. *Cell* **61**, 185–192.

7. Race, R. E., Fadness, L. H., and Chesebro, B. (1987) Characterization of scrapie infection in mouse neuroblastoma cells. *J. Gen. Virol.* **68**, 1391–1399.

8. Race, R. E., Caughey, B., Graham, K., Ernst, D., and Chesebro, B. (1988) Analyses of frequency of infection, specific infectivity, and prion protein biosynthesis in scrapie-infected neuroblastoma cell clones. *J. Virol.* **62**, 2845–2849.

9. Butler, D. A., Scott, M. R. D., Bockman, J. M., Borchelt, D. R., Taraboulos, A., Hsiao, K. K., et al. (1988) Scrapie-infected murine neuroblastoma cells produce protease-resistant prion proteins. *J. Virol.* **62**, 1558–1564.

10. Caughey, B. and Raymond, G. J. (1993) Sulfated polyanion inhibition of scrapie-associated PrP accumulation in cultured cells. *J. Virol.* **67**, 643–650.

11. Borchelt, D. R., Scott, M., Taraboulos, A., Stahl, N., and Prusiner, S. B. (1990) Scrapie and cellular prion proteins differ in the kinetics of synthesis and topology in cultured cells. *J. Cell Biol.* **110**, 743–752.

12. Caughey, B., Raymond, G. J., Ernst, D., and Race, R. E. (1991) N-terminal truncation of the scrapie-associated form of PrP by lysosomal protease(s): implications regarding the site of conversion of PrP to the protease-resistant state. *J. Virol.* **65**, 6597–6603.

13. Caughey, B., Race, R. E., Ernst, D., Buchmeier, M. J., and Chesebro, B. (1989) Prion protein (PrP) biosynthesis in scrapie-infected and uninfected neuroblastoma cells. *J. Virol.* **63**, 175–181.

14. Caughey, B. and Raymond, G. J. (1991) The scrapie-associated form of PrP is made from a cell surface precursor that is both protease- and phospholipase-sensitive. *J. Biol. Chem.* **266**, 18,217–18,223.

15. Caughey, B. and Race, R. E. (1992) Potent inhibition of scrapie-associated PrP accumulation by Congo red. *J. Neurochem.* **59**, 768–771.

16. Stahl, N., Borchelt, D. R., Hsiao, K., and Prusiner, S. B. (1987) Scrapie prion protein contains a phosphatidylinositol glycolipid. *Cell* **51**, 229–240.

17. Caughey, B., Neary, K., Buller, R., Ernst, D., Perry, L., Chesebro, B., et al. (1990) Normal and scrapie-associated forms of prion protein differ in their sensitivities to phospholipase and proteases in intact neuroblastoma cells. *J. Virol.* **64**, 1093–1101.

18. Stahl, N., Borchelt, D. R., and Prusiner, S. B. (1990) Differential release of cellular and scrapie prion proteins from cellular membranes by phosphatidylinositol-specific phospholipase C. *Biochemistry* **29,** 5405–5412.
19. Safar, J., Ceroni, M., Gajdusek, D. C., and Gibbs, C. J., Jr. (1991) Differences in the membrane interaction of scrapie amyloid precursor proteins in normal and scrapie- or Creutzfeldt-Jakob disease-infected brains. *J. Infect. Dis.* **163,** 488–494.
20. Bolton, D. C., Meyer, R. K., and Prusiner, S. B. (1985) Scrapie PrP 27–30 is a sialoglycoprotein. *J. Virol.* **53,** 596–606.
21. Manuelidis, L., Valley, S., and Manuelidis, E. E. (1985) Specific proteins associated with Creutzfeldt-Jakob disease and scrapie share antigenic and carbohydrate determinants. *Proc. Natl. Acad. Sci. USA* **82,** 4263–4267.
22. Oesch, B., Westaway, D., Wälchli, M., McKinley, M. P., Kent, S. B. H., Aebersold, R., et al. (1985) A cellular gene encodes scrapie PrP 27–30 protein. *Cell* **40,** 735–746.
23. Locht, C., Chesebro, B., Race, R., and Keith, J. M. (1986) Molecular cloning and complete sequence of prion protein cDNA from mouse brain infected with the scrapie agent. *Proc. Natl. Acad. Sci. USA* **83,** 6372–6376.
24. Endo, T., Groth, D., Prusiner, S. B., and Kobata, A. (1989) Diversity of oligosaccharide structures linked to asparagines of the scrapie prion protein. *Biochemistry* **28,** 8380–8388.
25. Taraboulos, A., Rogers, M., Borchelt, D. R., McKinley, M. P., Scott, M., Serban, D., et al. (1990) Acquisition of protease resistance by prion proteins in scrapie-infected cells does not require asparagine-linked glycosylation. *Proc. Natl. Acad. Sci. USA* **87,** 8262–8266.
26. Thotakura, N. R. and Bahl, O. P. (1987) Enzymatic deglycosylation of glycoproteins. *Methods Enzymol.* **138,** 350–359.
27. Somerville, R. A. and Ritchie, L. A. (1990) Differential glycosylation of the protein (PrP) forming scrapie-associated fibrils. *J. Gen. Virol.* **71,** 833–839.
28. Elbein, A. D. (1987) Inhibitors of the biosynthesis and processing of N-linked oligo saccharide chains. *Annu. Rev. Biochem.* **56,** 497–534.
29. Shyng, S.-L., Huber, M. T., and Harris, D. A. (1993) A prion protein cycles between the cell surface and an endocytic compartment in cultured neuroblastoma cells. *J. Biol. Chem.* **268,** 15,922–15,928.
30. Jeffrey, M., Goodsir, C. M., Bruce, M. E., McBride, P. A., Scott, J. R., and Halliday, W. G. (1992) Infection specific prion protein (PrP) accumulates on neuronal plasmalemma in scrapie infected mice. *Neurosci. Lett.* **147,** 106–109.
31. Shyng, S.-L., Heuser, J. E., and Harris, D. A. (1994) A glycolipid-anchored prion protein is endocytosed via clathrin coated pits. *J. Cell Biol.* **125,** 1239–1250.
32. Jeffrey, M., Goodsir, C. M., Bruce, M. E., McBride, P. A., Fowler, N., and Scott, J. R. (1994) Murine scrapie-infected neurons *in vivo* release excess prion protein into the extracellular space. *Neurosci. Lett.* **174,** 39–42.
33. Jeffrey, M., Goodsir, C. M., Bruce, M., McBride, P. A., Scott, J. R., and Halliday, W. G. (1994) Correlative light and electron microscopy studies of PrP localisation in 87V scrapie. *Brain Res.* **656,** 329–343.

34. Borchelt, D. R., Taraboulos, A., and Prusiner, S. B. (1992) Evidence for synthesis of scrapie prion protein in the endocytic pathway. *J. Biol. Chem.* **267,** 16,188–16,199.

35. Taraboulos, A., Raeber, A. J., Borchelt, D. R., Serban, D., and Prusiner, S. B. (1992) Synthesis and trafficking of prion proteins in cultured cells. *Mol. Biol. Cell* **3,** 851–863.

36. McKinley, M. P., Taraboulos, A., Kenaga, L., Serban, D., Stieber, A., DeArmond, S. J., et al. (1991) Ultrastructural localization of scrapie prion proteins in cytoplasmic vesicles of infected cultured cells. *Lab. Invest.* **65,** 622–630.

37. Caughey, B., Ernst, D., and Race, R. E. (1993) Congo red inhibition of scrapie agent replication. *J. Virol.* **67,** 6270–6272.

38. Caughey, B., Brown, K., Raymond, G. J., Katzenstien, G. E., and Thresher, W. (1994) Binding of the protease-sensitive form of PrP (prion protein) to sulfated glycosaminoglycan and Congo red. *J. Virol.* **68,** 2135–2141.

39. Ehlers, B. and Diringer, H. (1984) Dextran sulphate 500 delays and prevents mouse scrapie by impairment of agent replication in spleen. *J. Gen. Virol.* **65,** 1325–1330.

40. Farquhar, C. F. and Dickinson, A. G. (1986) Prolongation of scrapie incubation period by an injection of dextran sulphate 500 within the month before or after infection. *J. Gen. Virol.* **67,** 463–473.

41. Kimberlin, R. H. and Walker, C. A. (1986) Suppression of scrapie infection in mice by heteropolyanion 23, dextran sulfate, and some other polyanions. *Antimicrob. Agents Chemother.* **30,** 409–413.

42. Ladogana, A., Casaccia, P., Ingrosso, L., Cibati, M., Salvatore, M., Xi, Y. G., et al. (1992) Sulphate polyanions prolong the incubation period of scrapie-infected hamsters. *J. Gen. Virol.* **73,** 661–665.

43. Ingrosso, L., Ladogana, A., and Pocchiari, M. (1995) Congo red prolongs the incubation period in scrapie-infected hamsters. *J. Virol.* **69,** 506–508.

44. Scott, M. R. D., Butler, D. A., Bredesen, D. E., Walchli, M., Hsiao, K. K., and Prusiner, S. B. (1988) Prion protein gene expression in cultured cells. *Prot. Eng.* **2,** 69–76.

45. Robertson, M. N., Spangrude, G. J., Hasenkrug, K., Perry, L., Nishio, J., Wehrly, K., et al. (1992) Role and specificity of T-cell subsets in spontaneous recovery from friend virus-induced leukemia in mice. *J. Virol.* **66,** 3271–3277.

46. Chesebro, B., Wehrly, K., Caughey, B., Nishio, J., Ernst, D., and Race, R. (1993) Foreign PrP expression and scrapie infection in tissue culture cell lines. *Dev. Biol. Stand.* **80,** 131–140.

47. Harris, D. A., Huber, M. T., van Dijken, P., Shyng, S.-L., Chait, B. T., and Wang, R. (1993) Processing of a cellular prion protein: identification of N- and C-terminal cleavage sites. *Biochemistry* **32,** 1009–1016.

48. Lynch, C. M. and Miller, A. D. (1991) Production of high-titer helper virus-free retroviral vectors by cocultivation of packaging cells with different host ranges. *J. Virol.* **65,** 3887–3890.

49. Priola, S. A., Caughey, B., Race, R. E., and Chesebro, B. (1994) Heterologous PrP molecules interfere with accumulation of protease-resistant PrP in scrapie-infected murine neuroblastoma cells. *J. Virol.* **68,** 4873–4878.

50. Priola, S. A., Caughey, B., Wehrly, K., and Chesebro, B. (1995) A 60-kDa prion protein (PrP) with properties of both the normal and scrapie-associated forms of PrP. *J. Biol. Chem.* **270,** 3299–3305.

51. Raeber, A. J., Borchelt, D. R., Scott, M., and Prusiner, S. B. (1992) Attempts to convert the cellular prion protein into the scrapie isoform in cell-free systems. *J. Virol.* **66,** 6155–6163.

52. Kocisko, D. A., Come, J. H., Priola, S. A., Chesebro, B., Raymond, G. J., Lansbury, P. T., et al. (1994) Cell-free formation of protease-resistant prion protein. *Nature* **370,** 471–474.

53. Kocisko, D. A., Priola, S. A., Raymond, G. J., Chesebro, B., Lansbury, P. T., Jr., and Caughey, B. (1995) Species specificity in the cell-free conversion of prion protein to protease-resistant forms: a model for the scrapie species barrier. *Proc. Natl. Acad. Sci. USA* **92,** 3923–3927.

54. Bessen, R. A., Kocisko, D. A., Raymond, G. J., Nandan, S., Lansbury, P. T., Jr., and Caughey, B. (1995) Nongenetic propagation of strain-specific phenotypes of scrapie prion protein. *Nature* **375,** 698–700.

55. Bolton, D. C., Bendheim, P. E., Marmostein, A. D., and Potempska, A. (1987) Isolation and structural studies of the intact scrapie agent protein. *Arch. Biochem. Biophys.* **258,** 579–590.

56. Caughey, B. W., Dong, A., Bhat, K. S., Ernst, D., Hayes, S. F., and Caughey, W. S. (1991) Secondary structure analysis of the scrapie-associated protein PrP 27–30 in water by infrared spectroscopy. *Biochemistry* **30,** 7672–7680.

18

Immunohistochemistry of Resinated Tissues for Light and Electron Microscopy

Martin Jeffrey and Caroline M. Goodsir

1. Introduction

1.1. Use of Ultrastructural Immunocytochemistry in Investigations of the CNS

In spite of the widespread use of immunocytochemical methods and the commensurate increase in the range of available techniques, a consistent means of electron microscopical antigen localization in the lipid-rich central nervous system (CNS) has yet to be established. However, some successes have been achieved, most notably in neuroanatomy studies. Combined morphological and histochemical techniques for the study of neuronal microcircuits have been described *(1)*. Many of the antigens investigated in the CNS, such as neurotransmitters, appear not to withstand conventional electron microscopical fixation and embedding methods and consequently pre-embedding methods are employed *(2)*. Pre-embedding immunocytochemistry unfortunately suffers from the disadvantages of unpredictable antibody penetration and suboptimal morphological structure.

1.2. Application of Ultrastructural Immunocytochemistry to the Transmissible Spongiform Encephalopathies

Relatively few ultrastructural immunocytochemical studies have been performed on the transmissible spongiform encephalopathies. Immunogold studies of PrP localization performed on gluteraldehyde-fixed, plastic-embedded tissue confirmed the presence of PrP within individual amyloid filaments located in the subependyma and subpia of scrapie-infected hamsters *(3)*. Similar studies have been performed on the tissues of human patients affected with

From: *Methods in Molecular Medicine: Prion Diseases*
Edited by: H. Baker and R. M. Ridley Humana Press Inc., Totowa, NJ

Gerstmann-Sträussler-Scheinker disease (GSS) or with Creutzfeldt-Jakob disease (CJD). Although these studies were again able to show that fibrils of amyloid plaques are immunoreactive to PrP, the rather poor tissue quality obtained from immersion-fixed autopsy material is inimical to more detailed investigations.

Ultrastructural pre-embedding immunolocalization studies have been performed on scrapie-infected cells maintained in vitro *(4,5)*. PrP accumulates within cytoplasmic vesicles of infected cells and is also released into the culture medium. Biochemical studies of the localization of the transformation and accumulation of PrP in scrapie-infected cells maintained in vitro also show that the transformation occurs at the plasma membrane or along an endocytotic pathway to lysosomes *(6,7)*. Immunogold electron microscopy has also been very effectively employed to show accumulation of ubiquitin–protein conjugates in lysosomes of scrapie-infected mouse brains *(8)*.

Our own studies have concentrated on postembedding immunocytochemical investigations of PrP localization in scrapie-infected mice. Good localization of PrP has been achieved on specimens prepared for routine electron microscopy and the comments outlined herein are largely confined to our detection of antigen in aldehyde-fixed tissues of murine scrapie.

Immunoelectron microscopy is likely·to prove helpful in other CNS disease investigations. The βA4 protein of Alzheimer's disease is similar to PrP in that both are cell membrane proteins that are released from neurons and form extracellular amyloid fibrils. Immunogold studies of βA4 protein localization and aggregation, particularly in the new mouse model of the disease, are likely to further elucidate the relationship of amyloid fibril formation and neural deficits in Alzheimer's disease. Similarly, antibodies raised to virus proteins, or to particular bacterial antigens, may help to clarify the pathogenesis of infectious CNS disorders.

2. Materials and Methods

2.1. Theory and Practice of Ultrastructural Immunocytochemical Staining for PrP

2.1.1. Tissues and Experimental Design: Nonspecific Binding and Background Deposits

One of the main prerequisites for light microscopic immunocytochemical studies, that some form of antiserum and tissue control is essential, also applies to the immunocytochemistry of resinated sections and for electron microscopy. This is particularly important in studies of PrP, because none of the available antibodies is able to distinguish between normal (cell-associated protease or phospholipase-sensitive) and disease-specific (protease-resistant) PrP, nor have antibodies yet been raised to hypothetical disease-specific conformational changes of PrP.

The immunogold method is notoriously capricious and a low frequency of nonspecific binding of randomly distributed gold particles as well as occasional clusters of gold particles are sometimes found throughout the section (and also on plastic adjacent to the tissue specimen) even in the "cleanest" preparations. To the unwary these may convey spurious specificity. Such clusters may be:

1. Caused by imperfections in the specimen;
2. Reactions between the etching agents and the fabric of the grid;
3. Nonspecific binding to aldehydes; and
4. Other reasons that currently are unknown.

Therefore, it is particularly important that ultrastructural studies of PrP localization are correlated with individual structures identified in immunocytochemically stained "thick" (1 μm) sections viewed at the light microscope level. Accumulations of PrP may then be clearly associated with the disease-specific patterns of PrP recognized in the particular disease under investigation. It is of particular value to deal with infections that give rise to characteristic or invariant lesion patterns, such as is found in several of the murine scrapie models *(9)* and in bovine spongiform encephalopathy (BSE).

Although disease-specific PrP accumulations are detected within the cytoplasm of infected neurons maintained in vitro, disease-specific PrP accumulations in vivo are generally found within the neuropil, although intracytoplasmic PrP is reported in several transmissible spongiform encephalopathies. Some intracytoplasmic PrP within neurons is unassociated with disease since it may be detected at specific neuroanatomic sites of both scrapie-inoculated mice and in controls *(9)*. In cattle, intracytoplasmic PrP crossreacts with ceroid lipofuscin of both BSE-affected animals and in cattle with other types of neurological disease and in controls *(10)*. However, intracytoplasmic PrP may also be found in human patients with CJD and in cats with feline spongiform encephalopathy (FSE). It seems probable that at least some of the PrP found within neurons in these latter diseases may represent disease-specific accumulation, but the subcellular sites of intraneuronal disease-specific PrP accumulation in vivo have not yet been determined.

2.1.2. Technical Comments

2.1.2.1. TISSUE SOURCE

1. The embedding and/or the etching processes used for immunocytochemistry cause some deterioration in specimen quality compared with methods developed for optimal preservation of tissue used in morphological studies. Therefore, immunolocalization studies should, where possible, be performed on tissues obtained from perfused animals.

 2. Ultrastructural immunolabeling of mature amyloid fibrils is usually possible on
    immersion-fixed tissues obtained at biopsy or at necropsy.

2.1.2.2. Choice of Fixative

 1. Many antigenic epitopes for PrP are preserved in formalin, periodate/lysine/
    paraformaldehyde and gluteraldehyde/paraformaldehyde fixatives. However,
    some antibodies that may recognize epitopes in cryostat or parrafin wax-embed-
    ded tissues may not recognize the same epitope in tissues treated with harsher
    fixation or embedding protocols *(see the following)*.
 2. Optimal tissue structure for electron microscopy is obtained from fixation in
    gluteraldehyde/paraformaldehyde fixatives. Although not appropriate for all sub-
    cellular studies, most of our investigations have been performed on tissues fixed
    in mixed aldehydes.

2.1.2.3. Effect of Resination on Tissue Structure
and Presentation of Antibody Epitopes

Araldite, Durcopan, and related hard epoxy plastics are the embedding
medium of choice for electron microscopy. However, as previously stated,
these resins are impervious to water and have to be etched prior to immunolabeling. Other water-soluble resins, such as LR White and Lowcryl, may
also be employed for immunolabeling. However, tissues embedded in water-
soluble resins give inferior morphology and, more important, subsequent treatments with antigen-enhancing reagents causes the section to lift from the slide
and float off. Where antigens are not exposed or are adulterated with aldehyde
fixatives or with plastic embedding procedures, it is preferable to employ a
pre-embedding regimen rather than to employ one of the water-soluble resins.

Several methods for etching epoxy plastics are available:

 1. Hydrogen peroxide (H_2O_2, an oxidizing agent): The effect of this reagent on resin
    is to oxidize the hydrophobic alkane side chains of alcohols, aldehydes, and acids.
    This has a net effect of increasing the hydrophilicity of the resin. At LM level,
    when used on its own, this reagent gave much poorer labeling than other etching
    reagents. At EM level, for incubation periods of 6 min, this reagent produced no
    significantly enhanced immunostaining of PrP. Only weak labeling of plaques
    was achieved. Myelin became bleached, although other tissue structures were
    well preserved.
 2. Sodium ethoxide (removes resin): particularly good for 1-μm thick resinated sec-
    tions (Fig. 1), but much too harsh for electron microscopy studies. It is very dif-
    ficult to keep sections on grids following sodium ethoxide etching.
 3. Potassium methoxide18 crown 6/Aq. DMSO (removes resin): produces etching
    of 1-μm thick sections, which is almost as good as that of sodium ethoxide. It
    also combines good etching with maintenance of good tissue structure for elec-
    tron microscopy. This is the only reagent that may be used in virtually identical
    protocols for both light and electron microscopy studies of PrP localization.

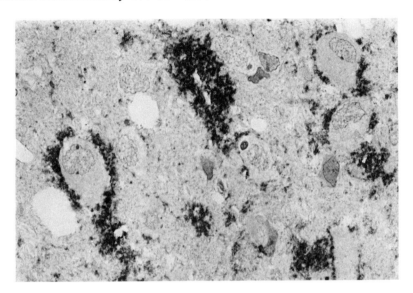

Fig. 1. Araldite-resinated 1-μm thick section, etched with sodium ethoxide, pretreated with formic acid, and immunostained by the peroxidase antiperoxidase method using the 1B3 antimurine SAF antibody. The section shows a PrP-releasing neuron from the lateral hypothalamus of a mouse infected with the 87V murine scrapie strain.

4. Sodium metaperiodate (a strong oxidizing agent): unmasks protein antigenic sites on gluteraldehyde-fixed, postosmicated tissue. This does not dissolve the araldite but punctures holes in it. Sodium metaperiodate etching retains the best tissue structure of all the etching procedures we have employed (Figs. 2 and 3), but was not as effective as potassium methoxide in revealing antigenic epitopes of PrP.

5. Triton/Tween (detergents): can be used for membrane permeabilization and therefore enhance penetration of the immunoreagents. In our experience, incorporating 0.2% Tween 20 in the wash buffer results in cleaner EM stained grids (fewer nonspecific gold particles deposited).

2.1.2.4. USE OF PRETREATMENTS TO ENHANCE OR REVEAL ANTIGENIC EPITOPES

The detection of PrP deposits in tissues prepared for light or electron microscopy and routinely immunostained without pretreatments is limited to the detection of plaques. We have employed the following pretreatments for detection of disease-specific PrP:

1. Formic acid: Good tissue structure was obtained using 98% formic acid incubated for 30 min for light microscopy and 10 min for electron microscopy. A wide range of patterns of disease-specific PrP have been detected, including preamyloid forms.

2. Hydrated autoclaving: Good enhancement of immunoreactivity has been achieved at light microscopy. Because this method is the most simple to perform

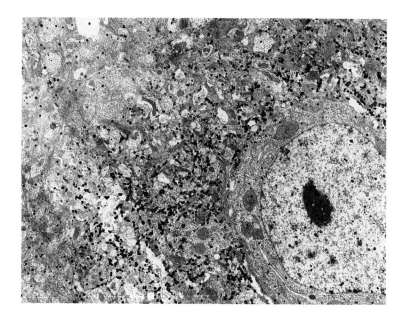

Fig. 2. An 80-nm thick section from the lateral hypothalamus of a mouse infected with the 87V strain of murine scrapie. This section has been etched with sodium metaperiodate and pretreated with formic acid, immunostained for 1B3 antiserum using 1 nm immunogold, and enhanced for 5 min with silver. Section shows the electron microscopical counterpart to Fig. 1 in which part of a morphologically normal neuron is surrounded by disease-specific PrP accumulation.

it has been used for routine screening of PrP distribution in 1-μm thick sections. However, at the EM level, tissue structure is considerably compromised.

2.1.2.5. EFFECT OF ANTIBODY

The effect of employing different antibodies for the detection of PrP at LM level is described in detail elsewhere in this volume. However, in our laboratory, several antibodies for the detection of PrP in the brains of BSE, FSE, ovine scrapie, and murine scrapie-infected tissues have been investigated. Brief details are presented in the following:

1. 1A8 and 1B3 antibodies *(11,12)* are high-titer polyclonal rabbit sera raised to murine scrapie-associated fibrils. They recognize several sites of the proteinase-resistant core of the PrP molecule *(13)* and recognize murine disease-specific PrP in all fixative and embedding protocols. These two antibodies also recognize PrP^{BSE}, PrP^{FSE}, and PrP^{SC} of natural sheep scrapie.
2. 3F4 *(14)* detects feline PrP in paraffin-wax-impregnated sections but is lost in gluteraldehyde-fixed tissues.

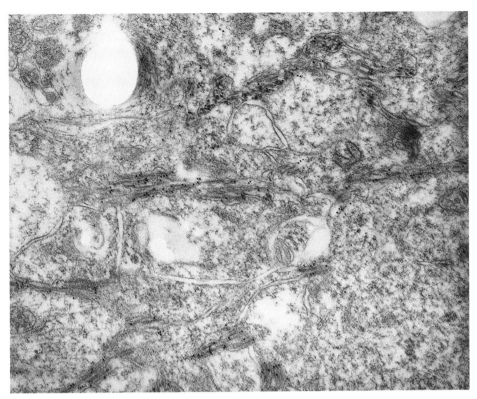

Fig. 3. An 80-nm thick section from the lateral hypothalamus of a mouse infected with the 87V strain of murine scrapie etched with sodium metaperiodate and pretreated with formic acid, immunostained for 1B3 antiserum using 10-nm colloidal gold. Section shows association of 10 nm colloidal gold particles with small bundles of amyloid fibrils forming in between neurites and glial cell processes.

3. N-terminal peptide antibodies are preserved in gluteraldehyde and paraformaldehyde fixatives, but much of the reactivity is lost on embedding in araldite. Peptide-specific antibodies to the protease-resistant core amino acids of PrP (PrP^{res}) are preserved under the same conditions.
4. A peptide antiserum raised to sheep PrP residues 141–153 (SP30) was effective in detecting PrP in FSE-affected brains. However, a second sheep peptide antiserum raised to PrP residues 219–233 (SP40) was not so effective against disease-specific PrP in FSE but stained disease-specific PrP in murine scrapie and BSE. Neither of these two antisera worked well for sheep scrapie.

In summary, prior to investigation of PrP localization in the electron microscope it is necessary to identify the antibody that best recognizes the disease-specific form of PrP to be detected. The effectiveness of that particular antiserum in recognizing specific PrP epitopes under various denaturing, etching, and fixation regimens then must be determined.

2.1.3. Immunohistochemical Methods and Visualization of Reaction Product: Light Microscopy and Electron Microscopy

1. To detect antigen in plastic-embedded thick sections it is possible to employ either the peroxidase-antiperoxidase (PAP) methods or to use avidin biotin complex (ABC).
2. In our own laboratory, the PAP method is preferred because it tends to give rise to fewer problems with background staining.
3. The ABC method is more sensitive and may have advantages where only low levels of PrP are present, such as in early preclinical disease, or where antibody is in short supply.
4. For light microscopy, reaction product may be visualized by 3-3, diaminobenzidine, or alkaline phosphatase.
5. Colloidal gold particles of different diameters may be used to immunostain tissues at the light level and also for electron microscopy.
6. For sensitive detection of gold, epi-polarization of stained sections may be employed.
7. One- and five-nanometers diameter particles may be enhanced with silver (immunogold silver method) for both light and electron microscopy.
8. Adjusting the silver incubation time alters the size of the silver–gold complexes to suit different purposes.
9. Silver incubation of approx 5 min produces large silver–gold complexes that can be seen at low magnifications allowing the distribution of the antigen to be determined over a large area of tissue (Fig. 2).
10. Silver incubations of approx 3 min gives much smaller metallic aggregates that permit more accurate localization of the antigen.
11. Gold particles of 10- to 20-nm diameter can be seen readily without silver enhancement. These permit accurate localization of antigens (Fig. 3).
12. The smaller the diameter of the gold probe, the better is the penetration of the colloidal gold particle into the tissue.

2.1.4. Tissue Thickness

Immunostaining on routine paraffin wax-embedded tissues is generally carried out on 5–7 μm thick sections. The antibodies generally penetrate the entire thickness of the section. Semithin, plastic-embedded sections usually are only 0.5–1 μm thick. This means that an antigen is only about one-fifth as abundant in these (1 μm thick) sections when compared with routine paraffin wax-stained sections. Stained sections for electron microscopy are only 1/40 of the thickness of a 1-μm section (60–70 nm). Clearly, therefore, some of the patterns of antigen distribution seen at light microscopy are not readily appreciated in semithin sections, and, in order to detect antigen in a 60-nm thick section, there must be significant amounts of the antigen available in tissues. Tissues that show only scant PrP accumulation at LM level generally are unsuitable for studies performed at the ultrastructural level.

2.2. Specimen Protocols

2.2.1. Protocol for Light Microscopy Immunostaining of PrP in Plastic Embedded Tissues

1. Fix tissues (preferably by perfusion) in freshly prepared 3% paraformaldehyde/ 1% gluteraldehyde in phosphate-buffered saline (PBS), pH 7.4 (0.1*M* sodium phosphate buffer, pH 7.4) *(15)*.
2. For perfused brains, remove brain and fix overnight in a fresh solution of the perfusate at 4°C.
3. After fixation, dissect 1-mm cubes of brain from appropriate areas and place in PBS, pH 7.4.
4. Wash in PBS and postfix in 2% osmium tetroxide and wash in distilled water.
5. Dehydrate tissues in a graded acetone series and infiltrate with resin contained in beem molds.
6. Polymerize resin at 60°C for 48 h.
7. Cut sections at 1 μm and mount on poly-L-lysine coated slides or on vectabond-treated slides (Vector Labs, Burlingame, CA).
8. Prepare solution of saturated sodium hydroxide in absolute ethanol, diluted 1:1 with absolute ethanol *(16)* and submerge slides in the solution for 60 min. Saturated NaOH solution is prepared by adding 15 g sodium hydroxide to 100 mL absolute ethanol, stirring for 1 h, and allowing the solution to age in a capped vessel at room temperature for 5 d before decanting the supernatant for use.
9. Block endogenous peroxidase for 10 min with 3% hydrogen peroxide in methanol.
10. Treat with 98% formic acid for 30 min.
11. Apply appropriate primary antibody (for example, 1A8 at 1/1600 dilution) in the peroxidase-antiperoxidase immunocytochemical staining procedure.
12. Develop antibody reaction product with 3-3' diaminobenzidine.

2.2.2. Protocol for Immunostaining Grids for Electron Microscopy Using an Avidin 10 nm Gold Method

1. Steps 1–6 as described.
2. For serial examinations of structures identified at light microscopy, take serial 1-μm and 80-nm sections from appropriate blocks and place latter sections on 400 mesh nickel grids.
3. On grid staining: All steps are performed at room temperature on drops on a sheet of dental wax. Do not permit grids to dry out throughout the entire procedure.
4. Etching step: Float grids face down on a droplet of sodium metaperiodate *(17)* for 60 min (or treat with potassium methoxide 18 crown-6/Aqueous DMSO *(18)* for 10 min).
5. Rinse grids in filtered distilled water for 5 × 3 min.
6. Block endogenous peroxidase and deosmicate with 3% H_2O_2 in methanol for 10 min.

7. Rinse grids in filtered distilled water for 5 × 3 min.
8. Treat with 98% formic acid for 10 min.
9. Rinse grids in filtered distilled water for 5 × 3 min.
10. Rinse grids in washing buffer before immunostaining for 2 × 10 min.
11. Quench residual aldehyde groups with $0.2M$ glycine in PBS, pH 7.4, for 3 min.
12. Rinse the grids on washing buffer for 5 min.
13. Incubate the grids on blocking solution for 30 min.
14. Rinse the grids on washing buffer for 5 min.
15. Incubate grids with primary antibody (e.g., α SAF 1B3 diluted 1:100 or 1A8 at 1:400) in incubation buffer for 1 h at room temperature.
16. Rinse the grids on washing buffer for 5 × 6 min.
17. Incubate with biotinylated antirabbit IgG diluted 1:100 in incubation buffer for 1 h.
18. Rinse the grids on washing buffer for 3 × 15 min and on PBS for 3 × 5 min.
19. Incubate with Extravidin 10 nm Colloidal Gold (Sigma, St. Louis, MO) diluted 1:10 in incubation buffer for 1 h.
20. Rinse the grids on washing buffer for 5 × 6 min.
21. Postfix grids with 2.5% gluteraldehyde in PBS for 10 min.
22. Rinse the grids on PBS for 3 × 5 min and on excess distilled water for 2 × 5 min.
23. Counterstain with uranyl acetate and lead citrate.

2.2.2.1. REAGENTS

1. Washing buffer: 0.8% bovine serum albumin (BSA) in distilled water; 0.1% IGSS quality gelatin in distilled water; 2 mM sodium azide (NaN_3) in distilled water; adjust the pH to 7.4 with 1N NaOH.
2. Incubation buffer: 0.8% BSA in distilled water; 0.1% IGSS quality gelatin in distilled water; 1% normal goat serum diluted in distilled water; 2 mM sodium azide (NaN_3) in distilled water; adjust the pH to 7.4 with 1N NaOH.
3. Blocking solution: As described, but substitute 5% normal goat serum for 1% normal goat serum.

3. Patterns of PrP Accumulation Revealed by Immunoelectron Microscopy

Our studies of the subcellular localization of disease-specific PrP in ME7 and 87V murine scrapie have shown that PrP accumulates initially at the plasma membrane of presumed scrapie-infected cells. The antibodies employed are not able to distinguish between conformationally modified PrP and normal PrP. However, we can infer that the PrP detected is disease specific because accumulations of PrP are not found in control mice brains at the same neuroanatomic sites. Disease-specific PrP then spreads throughout the extracellular matrix surrounding scrapie-infected cells and is present around many different types of glial and neuritic processes adjacent to the cell. Both the apparent PrP-releasing cell and the processes around which PrP accumulates are morphologically normal, at least initially. With the continued accumulation of PrP, individual amyloid filaments are found within the extracellular space. At about this time microglial cell processes become conspicuous and possess lysosomes that contain PrP *(10,19–21)*. Studies with N-terminal peptide antibodies to PrP show that N-terminal truncation of PrP is not a prerequisite of amyloid filament formation or in accumulation of preamyloid forms of PrP (Jeffrey and Goodsir, unpublished).

References

1. Bolam, J. P. and Ingham, C. A. (1990) Combined morphological and histochemical techniques for the study of neuronal microcircuits. *Handbook Chem. Neuroanat.* **8,** 125–198.
2. Priestly, J. V. (1984) Pre-embedding ultrastructural immunocytochemistry: immunoenzyme techniques. *Immunolab. Elect. Microscopy* **4,** 37–52.
3. Wiley, C. A., Burrola, P. G., Buchmeier, M. J., Wooddell, M. K., Barry, R. A., Prusiner, S. B., et al. (1987) Immunogold localization of prion filaments in scrapie-infected hamster brains. *Lab. Invest.* **57,** 646–656.
4. McKinley, M. P., Taraboulos, A., Kenaga, L., Serban, D., Stieber, A., DeArmond, S. J., et al. (1991) Ultrastructural localisation of scrapie prion proteins in cytoplasmic vesicles of infected cultured cells. *Lab. Invest.* **65,** 622–630.
5. Taraboulos, A., Serban, D., and Prusiner, S. B. (1990) Scrapie prion proteins accumulate in the cytoplasm of persistently infected cultured cells. *J. Cell Biol.* **110,** 2117–2132.
6. Caughey, B., Raymond, G. J., Ernst, D., and Race, R. E. (1991) N-terminal truncation of the scrapie-associated form of PrP by lysosomal protease(s): implications regarding the site of conversion of PrP to the protease-resistant state. *J. Virol.* **65,** 6597–6603.
7. Caughey, B. and Raymond, G. J. (1991) The scrapie-associated form of PrP is made from a cell surface precursor that is both protease- and phospholipase-sensitive. *J. Biol. Chem.* **266,** 18,217–18,223.
8. Laszlo, L., Lowe, J., Self, T., Landon, M., McBride, P. A., Farquhar, C., et al. (1992) Lysosomes as key organelles in the development of prion encephalopathies. *J. Pathol.* **166,** 333–341.
9. Bruce, M. E., McBride, P. A., and Farquhar, C. F. (1989) Precise targeting of the pathology of the sialoglycoprotein PrP, and vacuolar degeneration in mouse scrapie. *Neurosci. Lett.* **102,** 1–6.
10. Jeffrey, M., Halliday, W. G., and Goodsir, C. M. (1992) A morphometric and immunohistochemical study of the vestibular complex in bovine spongiform encephalopathy. *Acta Neuropathol.* **84,** 651–657.
11. Farquhar, C. F., Somerville, R. A., and Ritchie, L. A. (1989) Post-mortem immunodiagnosis of scrapie and bovine spongiform encephalopathy. *J. Virol. Methods* **24,** 215–222.
12. Farquhar, C. F., Somerville, R. A., Dorman, J., Armstrong, D., Birkett, C., and Hope, J. (1993) A review of the detection of PrP. *Transmissible Spongiform*

Encephalopathies: Proceedings of a Consultation on BSE with the Scientific Veterinary Committee of the CEC, pp. 301–303.

13. Langeveld, J. P. M., Farquhar, C. F., Pocchiari, M., Birkett, C., Bostock, C., and Meloen, R. H. (1993) Antigenic sites of bovine prion protein. *Transmissible Spongiform Encephalopathies: Proceedings of a Consultation on BSE with the Scientific Veterinary Committee of the CEC,* pp. 315–321.

14. Kascsak, R. J., Rubenstain, R., Merz, P. A., Tonna-DeMasi, M., Fersko, R., Carp, R. L., et al. (1987) Mouse polyclonal and monoclonal antibody to scrapie-associated fibril protein. *J. Virol.* **61,** 3688–3693.

15. Karlsson, U. and Schultz, R. L. (1965) Fixation of the central nervous system for electron microscopy by aldehyde perfusion. 1. Preservation with aldehyde perfusates versus direct perfusion with osmium tetroxide with special reference to membranes and the extracellular space. *J. Ultrstruct. Res.* **12,** 160–186.

16. Mar, H. and Wight, T. N. (1988) Colloidal gold immunostaining on deplasticised ultra-thin sections. *J. Histochem. Cytochem.* **36,** 1387–1395.

17. Bendayan, M. and Zollinger, M. (1983) Ultrastructural localisation of antigenic sites on osmium-fixed tissues applying the protein A-gold technique. *J. Histochem. Cytochem.* **31,** 101–109.

18. Iwadare, T., Harada, E., Yoshino, S., and Arai, T. (1990) A solution for removal of resin from epoxy sections. *Stain Technol.* **65,** 205.

19. Jeffrey, M., Goodsir, C. M., Bruce, M. E., McBride, P. A., Fowler, N., and Scott, J. R. (1994) Murine scrapie-infected neurons in vivo release excess PrP into the extracellular space. *Neurosci. Lett.* **174,** 39–42.

20. Jeffrey, M., Goodsir, C. M., Bruce, M. E., McBride, P. A., Scott, J. A., and Halliday, W. G. (1994) Correlative light and electron microscopy studies of PrP localisation in 87V scrapie. *Brain Res.* **656,** 329–343.

21. Jeffrey, M., Goodsir, C. M., Bruce, M. E., McBride, P. A., and Farquhar, C. F. (1994) Morphogenesis of amyloid plaques in 87V murine scrapie. *Neuropathol. Appl. Neurobiol.* **20,** 535–542.

Index

Methods in Molecular Biology™

Methods in Molecular Biology™ manuals are available at all medical bookstores. You may also order copies directly from Humana by filling in and mailing or faxing this form to: Humana Press, 999 Riverview Drive, Suite 208, Totowa, NJ 07512 USA, Phone: 201-256-1699/Fax: 201-256-8341.

☐ 60. **Protein NMR Protocols**, edited by *David G. Reid, 1996* • 0-89603-309-0 • Comb $69.50 (T)

☐ 59. **Protein Purification Protocols**, edited by *Shawn Doonan, 1996* • 0-89603-336-8 • Comb $64.50 (T)

☐ 58. **Basic DNA and RNA Protocols**, edited by *Adrian J. Harwood, 1996* • 0-89603-331-7 • Comb $69.50 • 0-89603-402-X • Hardcover $99.50

☐ 57. **In Vitro Mutagenesis Protocols**, edited by *Michael K. Trower, 1996* • 0-89603-332-5 • Comb $69.50

☐ 56. **Crystallographic Methods and Protocols**, edited by *Christopher Jones, Barbara Mulloy, and Mark Sanderson, 1996* • 0-89603-259-0 • Comb $69.50 (T)

☐ 55. **Plant Cell Electroporation and Electrofusion Protocols**, edited by *Jac A. Nickoloff, 1995* • 0-89603-328-7 • Comb $49.50

☐ 54. **YAC Protocols**, edited by *David Markie, 1995* • 0-89603-313-9 • Comb $69.50

☐ 53. **Yeast Protocols:** *Methods in Cell and Molecular Biology*, edited by *Ivor H. Evans, 1996* • 0-89603-319-8 • Comb $69.50 (T)

☐ 52. **Capillary Electrophoresis:** *Principles, Instrumentation, and Applications*, edited by *Kevin D. Altria, 1996* • 0-89603-315-5 • Comb $74.50

☐ 51. **Antibody Engineering Protocols**, edited by *Sudhir Paul, 1995* • 0-89603-275-2 • Comb $69.50

☐ 50. **Species Diagnostics Protocols:** *PCR and Other Nucleic Acid Methods*, edited by *Justin P. Clapp, 1996* • 0-89603-323-6 • Comb $69.50

☐ 49. **Plant Gene Transfer and Expression Protocols**, edited by *Heddwyn Jones, 1995* • 0-89603-321-X • Comb $69.50

☐ 48. **Animal Cell Electroporation and Electrofusion Protocols**, edited by *Jac A. Nickoloff, 1995* • 0-89603-304-X • Comb $64.50

☐ 47. **Electroporation Protocols for Microorganisms**, edited by *Jac A. Nickoloff, 1995* • 0-89603-310-4 • Comb $69.50

☐ 46. **Diagnostic Bacteriology Protocols**, edited by *Jenny Howard and David M. Whitcombe, 1995* • 0-89603-297-3 • Comb $69.50

☐ 45. **Monoclonal Antibody Protocols**, edited by *William C. Davis, 1995* • 0-89603-308-2 • Comb $64.50

☐ 44. ***Agrobacterium* Protocols**, edited by *Kevan M. A. Gartland and Michael R. Davey, 1995* • 0-89603-302-3 • Comb $69.50

☐ 43. **In Vitro Toxicity Testing Protocols**, edited by *Sheila O'Hare and Chris K. Atterwill, 1995* • 0-89603-282-5 • Comb $69.50

☐ 42. **ELISA:** *Theory and Practice*, by *John R. Crowther, 1995* • 0-89603-279-5 • Comb $59.50

☐ 41. **Signal Transduction Protocols**, edited by *David A. Kendall and Stephen J. Hill, 1995* • 0-89603-298-1 • Comb $64.50

☐ 40. **Protein Stability and Folding:** *Theory and Practice*, edited by *Bret A. Shirley, 1995* • 0-89603-301-5 • Comb $69.50

☐ 39. **Baculovirus Expression Protocols**, edited by *Christopher D. Richardson, 1995* • 0-89603-272-8 • Comb $64.50

☐ 38. **Cryopreservation and Freeze-Drying Protocols**, edited by *John G. Day and Mark R. McLellan, 1995* • 0-89603-296-5 • Comb $79.50

☐ 37. **In Vitro Transcription and Translation Protocols**, edited by *Martin J. Tymms, 1995* • 0-89603-288-4 • Comb $69.50

☐ 36. **Peptide Analysis Protocols**, edited by *Ben M. Dunn and Michael W. Pennington, 1994* • 0-89603-274-4 • Comb $64.50

☐ 35. **Peptide Synthesis Protocols**, edited by *Michael W. Pennington and Ben M. Dunn, 1994* • 0-89603-273-6 • Comb $64.50

☐ 34. **Immunocytochemical Methods and Protocols**, edited by *Lorette C. Javois, 1994* • 0-89603-285-X • Comb $64.50

☐ 33. **In Situ Hybridization Protocols**, edited by *K. H. Andy Choo, 1994* • 0-89603-280-9 • Comb $69.50

☐ 32. **Basic Protein and Peptide Protocols**, edited by *John M. Walker, 1994* • 0-89603-269-8 • Comb $59.50 • 0-89603-268-X • Hardcover $89.50

☐ 31. **Protocols for Gene Analysis**, edited by *Adrian J. Harwood, 1994* • 0-89603-258-2 • Comb $69.50

☐ 30. **DNA–Protein Interactions**, edited by *G. Geoff Kneale, 1994* • 0-89603-256-6 • Paper $64.50

☐ 29. **Chromosome Analysis Protocols**, edited by *John R. Gosden, 1994* • 0-89603-243-4 • Comb $69.50 • 0-89603-289-2 • Hardcover $94.50

☐ 28. **Protocols for Nucleic Acid Analysis by Nonradioactive Probes**, edited by *Peter G. Isaac, 1994* • 0-89603-254-X • Comb $59.50

☐ 27. **Biomembrane Protocols:** *II. Architecture and Function*, edited by *John M. Graham and Joan A. Higgins, 1994* • 0-89603-250-7 • Comb $64.50

☐ 26. **Protocols for Oligonucleotide Conjugates:** *Synthesis and Analytical Techniques*, edited by *Sudhir Agrawal, 1994* • 0-89603-252-3 • Comb $64.50

☐ 25. **Computer Analysis of Sequence Data:** *Part II*, edited by *Annette M. Griffin and Hugh G. Griffin, 1994* • 0-89603-276-0 • Comb $59.50

☐ 24. **Computer Analysis of Sequence Data:** *Part I*, edited by *Annette M. Griffin and Hugh G. Griffin, 1994* • 0-89603-246-9 • Comb $59.50

☐ 23. **DNA Sequencing Protocols**, edited by *Hugh G. Griffin and Annette M. Griffin, 1993* • 0-89603-248-5 • Comb $59.50

☐ 22. **Microscopy, Optical Spectroscopy, and Macroscopic Techniques**, edited by *Christopher Jones, Barbara Mulloy, and Adrian H. Thomas, 1993* • 0-89603-232-9 • Comb $69.50

☐ 21. **Protocols in Molecular Parasitology**, edited by *John E. Hyde, 1993* • 0-89603-239-6 • Comb $69.50

☐ 20. **Protocols for Oligonucleotides and Analogs:** *Synthesis and Properties*, edited by *Sudhir Agrawal, 1993* • 0-89603-247-7 • Comb $69.50 • 0-89603-281-7 • Hardcover $89.50

☐ 19. **Biomembrane Protocols:** *I. Isolation and Analysis*, edited by *John M. Graham and Joan A. Higgins, 1993* • 0-89603-236-1 • Comb $69.50

☐ 18. **Transgenesis Techniques:** *Principles and Protocols*, edited by *David Murphy and David A. Carter, 1993* • 0-89603-245-0 • Comb $69.50

☐ 17. **Spectroscopic Methods and Analyses:** *NMR, Mass Spectrometry, and Metalloprotein Techniques*, edited by *Christopher Jones, Barbara Mulloy, and Adrian H. Thomas, 1993* • 0-89603-215-9 • Comb $69.50

☐ 16. **Enzymes of Molecular Biology**, edited by *Michael M. Burrell, 1993* • 0-89603-322-8 • Paper $59.50

☐ 15. **PCR Protocols:** *Current Methods and Applications*, edited by *Bruce A. White, 1993* • 0-89603-244-2 • Paper $54.50

☐ 14. **Glycoprotein Analysis in Biomedicine**, edited by *Elizabeth F. Hounsell, 1993* • 0-89603-226-4 • Comb $64.50

☐ 13. **Protocols in Molecular Neurobiology**, edited by *Alan Longstaff and Patricia Revest, 1992* • 0-89603-199-3 • Comb $59.50

☐ 12. **Pulsed-Field Gel Electrophoresis:** *Protocols, Methods, and Theories*, edited by *Margit Burmeister and Levy Ulanovsky, 1992* • 0-89603-229-9 • Hardcover $69.50

☐ 11. **Practical Protein Chromatography**, edited by *Andrew Kenney and Susan Fowell, 1992* • 0-89603-213-2 • Hardcover $59.50

☐ 10. **Immunochemical Protocols**, edited by *Margaret M. Manson, 1992* • 0-89603-270-1 • Comb $69.50

☐ 9. **Protocols in Human Molecular Genetics**, edited by *Christopher G. Mathew, 1991* • 0-89603-205-1 • Hardcover $69.50

☐ 8. **Practical Molecular Virology:** *Viral Vectors for Gene Expression*, edited by *Mary K. L. Collins, 1991* • 0-89603-191-8 • Paper $54.50

☐ 7. **Gene Transfer and Expression Protocols**, edited by *Edward J. Murray, 1991* • 0-89603-178-0 • Hardcover $79.50

☐ 6. **Plant Cell and Tissue Culture**, edited by *Jeffrey W. Pollard and John M. Walker, 1990* • 0-89603-161-6 • Comb $69.50

☐ 5. **Animal Cell Culture**, edited by *Jeffrey W. Pollard and John M. Walker, 1990* • 0-89603-150-0 • Comb $69.50

Name _____

Department _____

Institution _____

Address _____

City/State/Zip _____

Country _____

Phone #_____ Fax #_____

"T" denotes a tentative price. Prices listed are Humana Press prices, current as of October 1995, and do not reflect the prices at which books will be sold to you by suppliers other than Humana Press. All prices subject to change without notice.
UK, Europe, Middle East, and Africa: Order directly from Chapman & Hall by faxing to: +44-171-522-9623.

Postage & Handling: *USA Prepaid (UPS):* Add $4.00 for the first book and $1.00 for each additional book. *Outside USA* (Surface): Add $5.00 for the first book and $1.50 for each additional book.

☐ **My check for $_____ is enclosed** *(Drawn on US funds from a US bank).*

☐ Visa   ☐ MasterCard   ☐ American Express

Card # _____

Exp. date _____

Signature _____